ADVANCES IN
ANALYTICAL GEOCHEMISTRY

Volume 1 • 1993

ADVANCES IN ANALYTICAL GEOCHEMISTRY

Editors: MARIAN HYMAN
MARVIN W. ROWE
Department of Chemistry
Texas A&M University
College Station, Texas

VOLUME 1 • 1993

Greenwich, Connecticut *London, England*

55 Old Post Road, No. 2
Greenwich, Connecticut 06836

JAI PRESS LTD.
The Courtyard
28 High Street
Hampton Hill, Middlesex TW12 1PD
England

ISBN: 1-55938-332-1

Manufactured in the United States of America

CONTENTS

LIST OF CONTRIBUTORS	vii
INTRODUCTION TO THE SERIES: AN EDITOR'S FOREWORD *Albert Padwa*	ix
PREFACE *Marian Hyman and Marvin W. Rowe*	xi
AMS IN THE EARTH SCIENCES: TECHNIQUE AND APPLICATIONS *R. C. Finkel and M. Suter*	1
RESONANCE IONIZATION MASS SPECTROMETRY FOR GEOCHEMICAL APPLICATIONS *B. L. Fearey, C. M. Miller, and N. S. Nogar*	115
X-RAY FLUORESCENCE SPECTROMETRY *Gerald R. Lachance*	155
NEUTRON ACTIVATION ANALYSIS *Gregory W. Kallemeyn*	193
THERMAL ANALYSIS *Patrick K. Gallagher*	211
INDEX	259

LIST OF CONTRIBUTORS

B. L. Feary — Isotope and Nuclear Chemistry Division
Los Alamos National Laboratory
Los Alamos, New Mexico

R. C. Finkel — Nuclear Chemistry Division
Lawrence Livermore National Laboratory
Livermore, California

Patrick K. Gallagher — Departments of Chemistry and Materials Science and Engineering
The Ohio State University
Columbus, Ohio

Gregory W. Kallemeyn — Institute of Geophysics and Planetary Physics
University of California
Los Angeles, California

Gerald R. Lachance — Geochemical Consultant
Nepean, Ontario, Canada

C. M. Miller — Isotope and Nuclear Chemistry Division
Los Alamos National Laboratory
Los Alamos, New Mexico

N. S. Nogar — Chemical and Laser Sciences Division
Los Alamos National Laboratory
Los Alamos, New Mexico

M. Suter — Paul Scherrer Institute
Zurich, Switzerland

INTRODUCTION TO THE SERIES: AN EDITOR'S FOREWORD

The JAI series in chemistry has come of age over the past several years. Each of the volumes already published contains timely chapters by leading exponents in the field who have placed their own contributions in a perspective that provides insight to their long-term research goals. Each contribution focuses on the individual author's own work as well as the studies of others that address related problems. The series is intended to provide the reader with in-depth accounts of important principles as well as insight into the nuances and subtleties of a given area of chemistry. The wide coverage of material should be of interest to graduate students, postdoctoral fellows, industrial chemists and those teaching specialized topics to graduate students. We hope that we will continue to provide you with a sense of stimulation and enjoyment of the various sub-disciplines of chemistry.

Department of Chemistry
Emory University
Atlanta, Georgia

Albert Padwa
Consulting Editor

PREFACE

This is the first in a series entitled *Advances in Analytical Geochemistry*. The aim is to provide an exchange of information among the diverse fields involved in geochemical studies, including geology, chemistry, physics, oceanography, biology, and others. Analytical geochemistry makes use of an extensive breadth of experimental techniques in which substantial developments have been particularly notable in recent years. *Advances in Analytical Geochemistry* will serve this expanding subject by presenting timely and authoritative reviews on the recent developments and new techniques, as well as imaginative application of more established techniques. Future volumes will broaden and update this interdisciplinary field.

The five articles in the present inaugural volume illustrate the range of contemporary techniques used in geochemical analysis. The first two articles deal with isotopic measurements. The lead review article by Finkel and Suter discusses, in considerable detail, accelerator mass spectrometry which is used today in vastly different applications in geochemistry, as well as in archaeology. Instrumentation is described and each major isotope of geochemical importance (^{10}B, ^{14}C, ^{26}Al, ^{36}Cl, ^{41}Ca, ^{129}I, etc.) and its application is detailed. Resonance ionization mass spectrometry, the second article in this issue, like accelerator mass spectrometry, is rapidly finding important applications in geochemistry. Fearey, Miller and Nogar explain the technique, covering both the uses of pulsed and continuous wave lasers, as well as specific applications of the technique to geochemical studies.

There is a common need for qualitative identification and quantitative determination of the elements, whether present in major, minor or trace levels in geochemical samples. The third and fourth articles in this volume describe elemental techniques which are useful for this purpose. Both are capable of multi-element analysis and both are element specific. The third review, by LaChance, gives a detailed analysis of the theoretical and practical application of x-ray fluorescence for elemental analysis. Kallemeyn then follows in the fourth article with a discussion of neutron activation analysis, both instrumental and radiochemical, and briefly mentions applications in meteorite, lunar and oceanographic studies. In the fifth and final article in this book, Gallagher treats the utilization of a wide variety of techniques embodied in the term "thermal analysis" as applied to geochemical problems. Thermogravimetry, thermomagnetometry, evolved gas detection and analysis, differential thermal analysis and differential scanning calorimetry, as well as other miscellaneous thermal methods, are covered. As Gallagher states, refering to thermal analysis techniques, "Rarely does a single method provide the complete and unequivocal answer." That holds equally true for any analysis technique used in an area as complex as geochemistry.

It is hoped that these five articles, chosen from a wide range of available techniques, give an indication of the breadth, scope and variety of contemporary analytical geochemistry and also an indication of what is planned in future volumes of the series.

Marian Hyman
Marvin W. Rowe
Editors

AMS IN THE EARTH SCIENCES:
TECHNIQUE AND APPLICATIONS

R. C. Finkel and M. Suter

I.	Introduction: General	2
II.	The Technique of Accelerator Mass Spectrometry	3
	A. Instrumentation	3
	B. Performance	28
	C. Comparison with Counting	36
III.	Sources of Radionuclides in Nature	38
	A. Cosmogenic Radionuclides	38
	B. Primordial and Radiogenic Radionuclides	46
	C. Anthropogenic Radionuclides	48
IV.	Distribution of Radionuclides in Nature	49
	A. Extraterrestrial Origin	49
	B. Atmospheric Origin	49
	B. In Situ Origin	51
V.	Concepts and Problems Underlying the Application of Long-Lived Radioisotopes to Geophysics	52
	A. Production Variations	52
	B. Transport and Deposition	53
	C. Dating	53
	D. Caveats	57

Advances in Analytical Geochemistry
Volume 1, pages 1–114

ISBN: 1-55938-332-1

VI. Isotope Encyclopedia ... 57
A. ^{10}Be and ^{7}Be ... 59
B. ^{14}C ... 70
C. ^{26}Al ... 80
D. ^{36}Cl ... 85
E. ^{41}Ca ... 91
F. ^{129}I ... 94
G. Isotopes Less Studied by AMS ... 97
VII. Conclusions ... 99
Acknowledgments ... 100
Notes ... 100
References ... 100

I. INTRODUCTION

Accelerator mass spectrometry (AMS) of long-lived radioisotopes has become an important tool in environmental sciences. Worldwide, several thousand measurements are performed every year. Radiocarbon studies make up the majority of applications, for which reduction of sample size is the main advantage compared to conventional decay counting techniques. In addition, the efficient detection of radioisotopes with longer half-lives, such as ^{10}Be, ^{26}Al, ^{36}Cl, ^{41}Ca, and ^{129}I, has opened up a large variety of new applications which could not have been undertaken with decay counting. In the geosciences, radioisotopes are used both to date geophysical and geochemical events and as tracers of geophysical processes. Applications cover the spectrum of geoscience, including studies of climate history, investigations of magma generation processes in subduction zones, determination of terrestrial ages of meteorites embedded in Antarctic ice, unravelling the history of solar activity cycles, and many other areas.

We intend this article for scientists and students who want to be informed about the potential and the limitations of AMS. It gives an introduction both to the AMS technique and to the wide range of applications in which AMS has been used.

The article is organized into two sections. In the first section instrumental techniques are described based on the present status of operating systems. In the second section applications are discussed first in general terms and then on a nuclide-by-nuclide basis. The two sections are almost independent of each other, so that the reader can easily read one section or the other depending on his or her interest.

In the technical section the basic principles of the AMS method are discussed in sufficient detail to allow the reader to understand and interpret AMS data. The level of general performance which has currently been attained is discussed and the factors which limit sensitivity, efficiency, and accuracy are given. This discussion should enable the reader to estimate the potential of AMS for a specific application.

In the applications section the mechanisms of isotope production are discussed and guidelines given which enable isotope concentrations to be estimated in different reservoirs. A brief description of the points to be considered when preparing a sample for AMS measurement is then followed by selected examples of geophysical applications. We hope to give some idea of the scope of the technique and also allow an assessment of the feasibility of using AMS to investigate a particular problem.

A special word regarding our selection of references: Our goal in selecting our bibliography is to provide access to the literature for the interested reader rather than to give a picture of the development of a research area. We have therefore emphasized recent literature and review articles and placed less emphasis on pioneering early work. We hope our colleagues will forgive us this slight, which can be remedied by referral to the review articles cited. Some attempt has been made to include recent developments by citing review articles that have occurred since the preparation of this chapter.

II. THE TECHNIQUE OF ACCELERATOR MASS SPECTROMETRY

A. Instrumentation

Principles

One principle of mass spectrometry is based on the fact that ions with the same energy are deflected in a magnetic field according to their masses. This effect allows ions of different mass to be separated and thus permits quantitative information about the elemental and isotopic composition of the sample to be obtained. The sensitivity reached by conventional mass spectrometry is not, unfortunately, high enough to allow the detection of naturally produced, long-lived radioisotopes, which occur at abundance ratios (radioactive isotope/stable element) of 10^{-10}–10^{-16}. The background which limits the sensitivity of conventional mass spectrometry arises from several sources. First, molecules and isobars can occur with masses which are extremely close to that of the nuclide being determined. For example, $M/\Delta M = 1134$ for $^{12}CH_2$–^{14}C and $M/\Delta M =$ 84,000 for ^{14}N–^{14}C. Such relative mass differences can in principle be resolved by spectrometers operating with very high mass resolution. However, this high mass resolution is achieved by expedients, such as narrowing slits, which lead to a severe reduction in efficiency. Second, long-lived radioisotope concentrations are very low, which means that a typical sample will have only a small number of atoms. Therefore spectrometers must be operated at as high an efficiency as possible. Another limitation to the sensitivity of conventional mass spectrometry arises from background counts caused by scattering on walls, apertures, and residual gas in the mass spectrometer.

Accelerator mass spectrometry overcomes most of these problems and therefore extends the sensitivity of conventional mass spectrometry by many orders of magnitude. In AMS, particle accelerators, originally developed for nuclear physics research, are used to accelerate the sought-after nuclide to a high energy, at which molecules can be easily destroyed and isobars and neighboring masses can be identified by nuclear detection techniques. In addition, scattering cross sections are much smaller at high energies leading to a further background reduction. Because interferences are removed by techniques other than deflection in magnetic and electric fields, relatively low resolution mass spectrometers can be used as components of AMS systems. The transmission and thus the efficiency for detecting rare nuclides can therefore be kept very high.

The first experimental use of an accelerator for isotope identification was in 1939 by Alvarez and Cornog, who used a cyclotron for the detection of ^{3}He (Alvarez and Cornog, 1939). In 1977, Muller reintroduced AMS and proposed that ^{14}C dating might be possible with a cyclotron if the ^{14}C and ^{14}N isobars could be separated by using range techniques (Muller, 1977). At the same time experiments were begun to investigate the use of tandem accelerators for the detection of long-lived radioisotopes (Purser et al., 1977). First results appeared for ^{14}C (Bennet et al., 1977; Nelson et al., 1977). Today most AMS work is done with tandem accelerators. Several dedicated facilities have been built and accelerators available from nuclear research have been adapted for AMS. About two dozen accelerators around the world are routinely used for the detection of rare isotopes with AMS, and another dozen have been used occasionally for such experiments.

We will concentrate our further discussion on the standard tandem accelerator technique as used at most facilities and only briefly mention other special approaches.

The principle of tandem accelerator mass spectrometry is shown in Figure 1. The purified and pretreated material under investigation is loaded into the ion source and bombarded with Cs ions to produce a negative ion beam. Because in many cases interfering elements do not form stable negative ions, the source can act as a first filter. Next the mass(es) of interest are usually selected with a first magnetic mass spectrometer (second filter). The ions which pass the mass filter are accelerated to the high voltage terminal of the tandem van de Graaff accelerator. There they pass through a thin carbon foil or a narrow canal filled with gas at low density. This causes electrons to be stripped away and most ions to end up in a positive, multiply-charged state. This stripping process, which causes molecules to be destroyed to a large extent because multiply-charged molecules ($q \geq 3^{+}$) are not stable, constitutes the third filter. The positive ions are then accelerated back to ground potential and analyzed in a high energy mass spectrometer (fourth filter). Finally the ions are identified in a detector system (fifth filter). With this multiple filtering system, the high sensitivity required can be obtained, while at the same time the efficiency is also kept high. In the

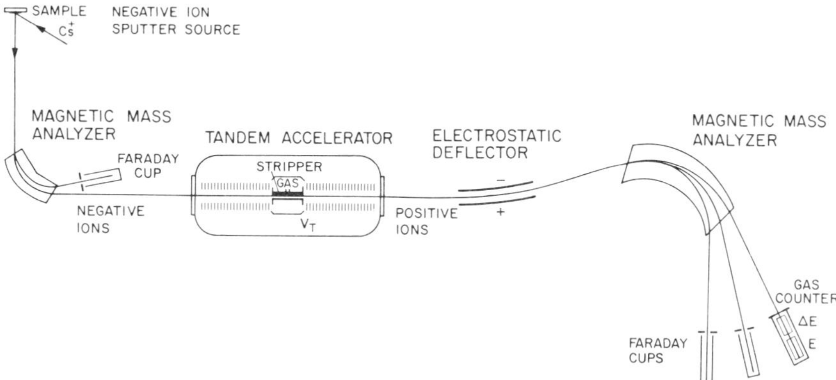

Figure 1. Schematic diagram of an AMS-facility. Negative ions are produced from the material under investigation in a Cs-sputter source. These ions are mass analyzed in a first magnetic spectrometer. In the first section of the tandem accelerator the negative ions are accelerated to high energy. In the stripper canal of the high voltage terminal the fast moving ions lose electrons and are converted to a positive charge state. In the second section the ions are further accelerated. With electrostatic deflector and magnetic spectrometer stages they are mass analyzed and finally detected and identified by nuclear counter techniques.

following sections the individual components and techniques will be discussed in more detail. Additional technical information and references can be found in the review articles presented at the AMS conferences which have been held periodically since 1977 (Litherland, 1984; Wölfli, 1987; Suter, 1990).

Ion Source

In the ion source a negative ion beam is generated from the target material under investigation. Currently, Cs-sputter sources are most often used for this stage. In a Cs-sputter source the target is bombarded with positive Cs ions. In the sputtering process a fraction of the emitted atoms and molecules becomes negatively charged and is extracted and focused by an appropriate electric field configuration. Cs deposited on the sample enhances the formation of negative ions. For AMS it is important that the source be simultaneously optimized for high negative ion yield, high current, and small beam emittance (small spot size and angular divergence). Stability is also essential, especially for high precision measurements. For those elements which do not form negative ions easily, molecular ions can be used; for example, BeO^-, CaH_3^-.

Several source designs have been used in AMS. In the original sources, many of which are still in use, Cs ions are formed by diffusing Cs vapor through a hot porous tungsten frit. Various geometries are in use for focusing the Cs ions on

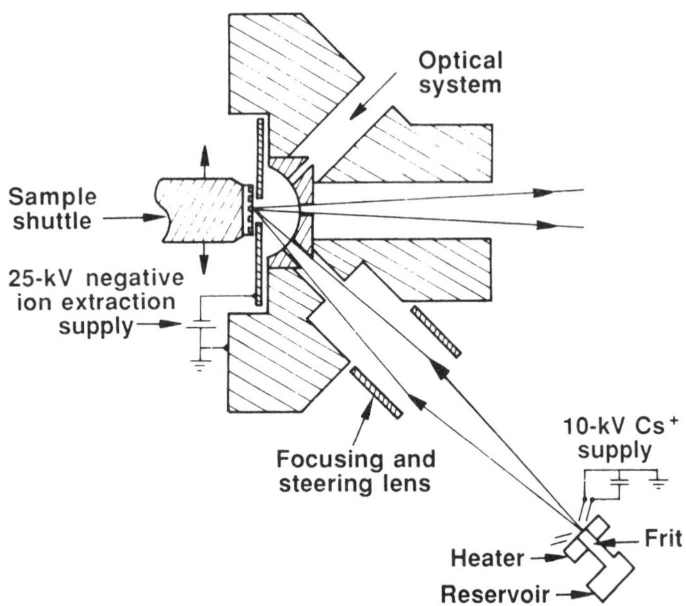

Figure 2. Geometry of modified-cone frit-type Cs-sputter source used for AMS. (Elmore and Phillips, 1987)

the target: symmetric injection from the front (Chapman geometry), and asymmetric injection from the side or from the back by reflecting the Cs beam (Figure 2). The Cs energies are typically 10–40 kV and the Cs currents are in the 100 μA range for these designs.

Newer designs (so-called high intensity sources) contain larger surface ionizers, rather than frits, onto which the Cs vapor is spread directly (Figure 3). The surface ionizers operate in general with higher Cs^- currents (mA range) but smaller energies (1–8 kV) than the frit type sources. These designs give currents which, for some species, are an order of magnitude larger than currents from the older types (Middleton, 1983, 1984, 1989; Alton, 1990). With some modification this type of source can also be used with gas feeding. This is of special interest for ^{14}C measurements because gas feeding allows the direct use of CO_2 rather than requiring costly and time-consuming conversion to graphite as required by the standard sources. The main problem with high intensity sources is cross contamination, especially when operating with gases or materials having a high vapor pressure. New, open designs and good pumping have allowed these effects to be greatly reduced. Gas sources are now successfully used for routine operation (Middleton et al., 1989; Bronk and Hedges, 1990).

To obtain the high sample throughput required by many Earth science prob-

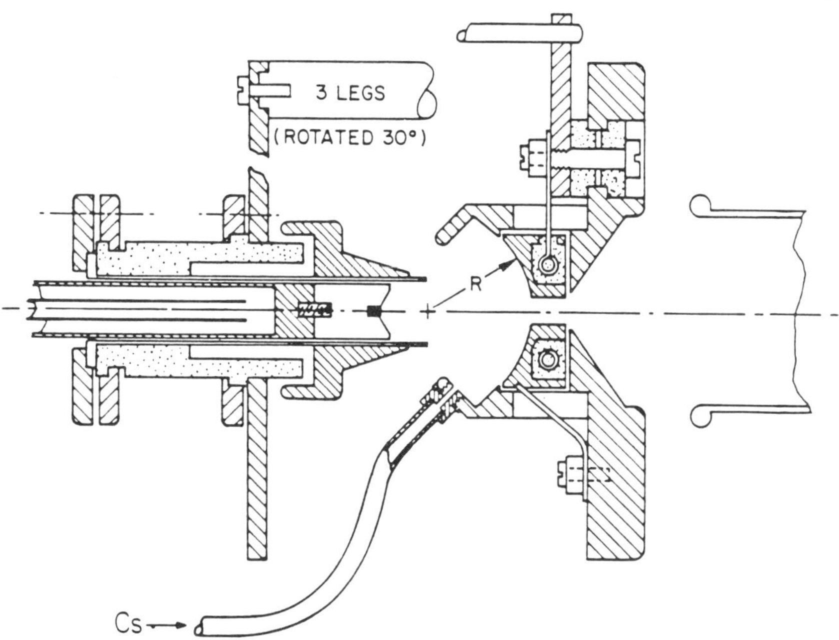

Figure 3. Schematic diagram of high intensity Cs-sputter source. Cs-vapor is sprayed onto a hot spherical surface-ionizer. With this arrangement very intense Cs beams can be focused on a small spot of the target. (Middleton, 1989)

lems, an automated sample loading and changing system is needed. The best are multitarget stages, which can be loaded through a lock so that high vacuum is not affected during the loading procedure, and so that the high voltages required for source operation do not have to be turned off. However, there are difficult technical problems involved in constructing such automated sample changers. The sample has to be at high potential during sputtering and, therefore, the changer has to operate in a high potential environment. The targets have to be cooled, and a good vacuum must be maintained. Cross-talk between samples must be avoided. Cs vapor and sputtered sample material, which can cover movable parts or insulators and lead to failures, must be taken account of. For automatic sample changing, multiposition sample holders of various types are used. Target wheels or X-Y-stages are one possibility. Cassette systems allow the exchange of whole batches of samples without interrupting the measurements (Balzer et al., 1984). High intensity sources impose special requirements on multisample systems. The first such systems, proposed in 1988 (Purser et al., 1988), have recently become available (Proctor et al., 1990), but further improvements in the sample handling systems would be desirable.

Injector

After extraction from the ion source, the ions are normally analyzed with a first magnetic spectrometer. The intensity of neighboring masses and also of molecules, which can be a significant fraction of the sputtered material, can thus be greatly reduced. Although all the ions are accelerated in the ion source by the same electric field, the beam formed is not quite monoenergetic. In some cases energy is transferred from the Cs ions to the negative target ions leading to a tail toward higher energies (sputtering tail) (Litherland, 1984). Sputtered molecules can be in excited states which break up during or after the acceleration in the ion source. The fragments will then have to share the acceleration energy and will thus have a lower energy. Tails toward lower and higher energy can cause interferences to appear from neighboring masses (Figure 4). By adding an electrostatic filter, which allows selecting only ions of the correct energy, (E/q), these tails can be significantly reduced. This standard technique in conventional mass spectrometry has been introduced to AMS at the Isotrace Laboratory in Toronto (Litherland, 1984) using a 45° spherical sector analyzer, and recently also at the ETH/PSI facility using a 90° analyzer built for AMS (Synal et al., 1991).

The Tandem Accelerator

Tandem accelerators belong to the class of electrostatic accelerators in which the particles are accelerated in a static electric field. In order to reach ion energies in the MeV range, megavolt (MV) potentials are needed. To obtain these high voltages without arcing, the accelerator is mounted in a tank filled with insulating gas. Older tanks were designed for tank pressures of 16 to 20 bars using mixtures of CO_2 (~22%) and N_2 (~78%). Today most tandem accelerators use SF_6 gas at pressures of 2 to 4 bars. In tandem accelerator systems, negative ions are injected and are then accelerated to the positive high voltage terminal (Figure 5). There, several electrons are removed in a stripper (gas or foil) and the positively charged ions further accelerated to ground potential. The advantage of this type of accelerator is that the ion source can be outside the tank and is therefore easy to maintain. The beam intensity is constant in time and high optical transmission can be maintained. Due to the polarity change of the ions in the terminal the high potential can be used twice for acceleration (thus the name tandem for this type of accelerator). A disadvantage is that in the stripping process various charge states are populated but only one can be selected for final analyses. This lowers the efficiency. Also, elements which do not form negative ions easily, such as noble gases, cannot be accelerated with this type of accelerator.

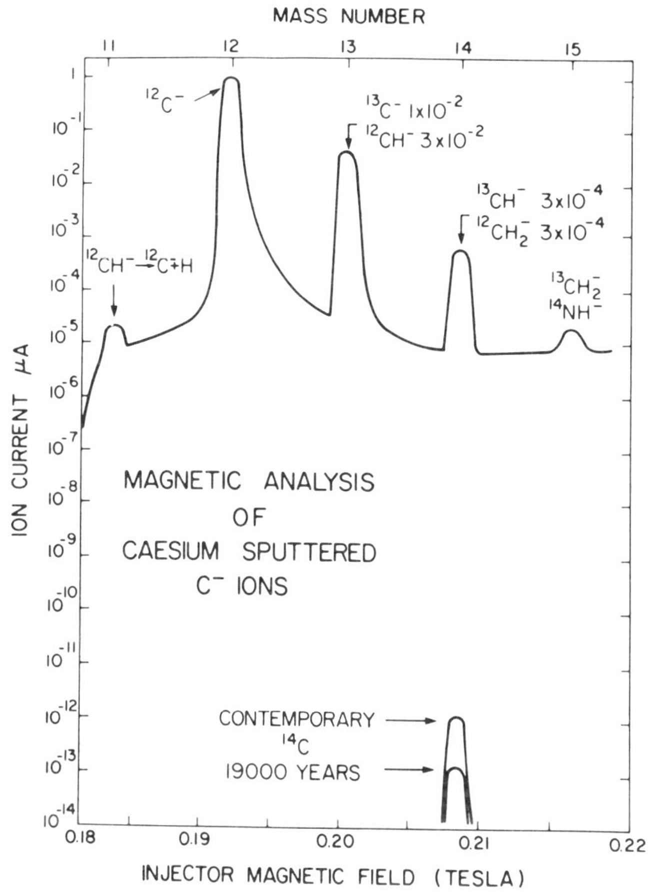

Figure 4. Momentum scan at the injector magnet, showing tails due to sputtering and breakup of molecules. (Litherland, 1984)

The total energy, E, of the accelerated ion with final charge state, q, is related to the terminal voltage, V, in the following way:

$$E = (1 + q)eV$$

where the 1 comes from the initial acceleration of an ion with charge -1 and the q from the second acceleration of the charge state of the ion after stripping. If negatively charged molecules with mass $M = (M1 + M2)$ are used, the energy of the fragment M1 is:

$$E = \left(\frac{M1}{M} + q\right)eV$$

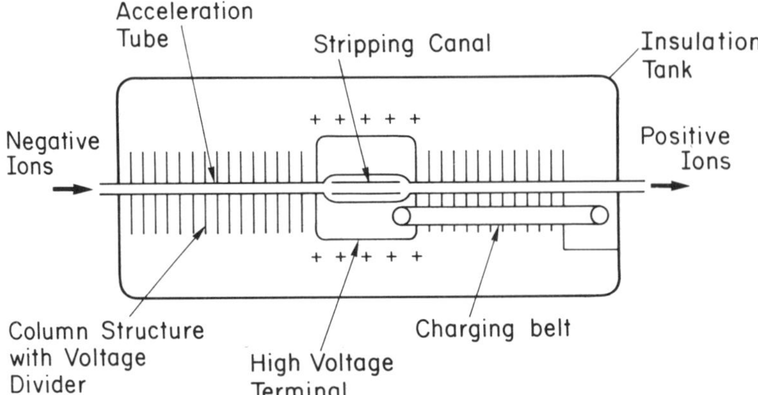

Figure 5. Schematic diagram of a tandem accelerator. The negatively charged ions are accelerated to the positive high voltage terminal. In the stripper canal the ions lose electrons and the positive ions are further accelerated in the second part of the accelerator. With a resistor chain along the acceleration tube, a constant field gradient can be maintained. The high voltage is generated by a fast moving rubber belt or chain.

Two different principles are used to generate the high voltage on the terminal. The original van de Graaff tandem accelerators use a fast-moving belt to transport electric charge to the high voltage terminal (current source) (Figure 5). In newer versions the rubber belts have been replaced with chain systems, called pelletrons. Several older facilities have been upgraded with chain systems. The disadvantage of the belt system is that charge transport depends on the properties of the belt and is not always constant. In addition, the fast-moving belt or chain produces dust, which has to be removed from time to time. Due to the slow response time of the primary mechanical charge transport system, fast voltage fluctuations are regulated by a well-controlled corona discharge from the terminal to the tank wall, allowing stabilization in the 1 kV range.

The second method used to produce the high voltage for electrostatic accelerators is the cascade-generator, in which an RF (radio frequency) signal is capacitatively coupled to a series of stacked diode circuits which rectify the RF and sum the signal to a high DC voltage. This idea was first used by Cockroft and Walton in 1930 for a 300-keV accelerator, and later in "dynamitrons" to produce MV potentials. In recent devices this type of power supply has been mounted into a separate leg of a tandem accelerator leading to a T-shaped vessel which gives the name Tandetron to this design (Figure 6). Several newly designed AMS facilities are equipped with Tandetrons operating at around 2 MV (Purser et al., 1988; Kieser, 1989). These accelerators have the advantage of simplified operation and

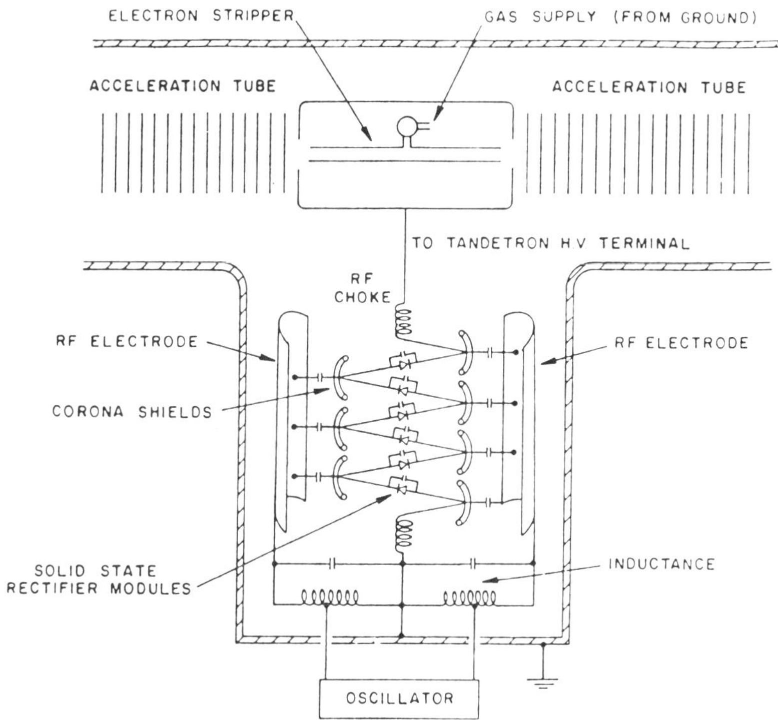

Figure 6. Schematic diagram of Tandetron: In these accelerators the high voltage is generated by a cascade-generator. (Turner et al., 1987)

maintenance because no movable parts are used. They use voltage sources which can be well regulated and also support much higher currents than the mechanical charge transport systems. On the other hand, these devices can release much more energy in a spark, which can cause severe damage. Therefore, careful conditioning before operating at high voltages is required.

MV voltages cannot be measured accurately with standard measuring devices, therefore generating voltmeters are used whose operation is based on a electric field measurement at the tank wall. These devices may have small drifts which lead to corresponding drifts in energy. Energy drifts can also result from variation of the thickness of the stripper foil, which causes variation in the energy loss of the penetrating ion. By measuring the beam position after the high energy analyzing magnet, this type of drift can be monitored and corrected for. In nuclear physics experiments, slit systems are used to measure the beam tails on the left and right hand sides in order to derive information on the beam position. In AMS the slits are normally wide open to minimize transmission losses. Therefore, Faraday cups have been developed which allow precise current and position measurements (White et al., 1981; Suter et al., 1984c).

The gas stripper normally consists of a narrow gas canal which is open on both sides. Originally all gas flowing out of the canal was pumped through the acceleration tube, leading to significant beam losses due to charge-changing processes with gas in the tubes. Nowadays pumping systems are installed in the terminal. With a turbo molecular pump, the gas streaming out of the stripper can be recirculated to a large extent (Figure 7). Thin carbon foils (2–20) $\mu g/cm^2$) can also be used for stripping, but radiation damage destroys them after a few hours of operation, requiring a remote-controlled foil-changing system.

The stripping efficiency can be optimized by the appropriate choice of energy, charge state, stripper medium, and thickness (Betz, 1972). Extensive studies on charge-state distributions have been performed at the ETH for some elements which are relevant for AMS (Stoller et al., 1983; Hofmann et al., 1984; Hofmann, 1987; Hofmann et al., 1987b). In Figure 8 the stripping efficiency of ^{14}C ions of charge state 3^+ and 4^+ are shown for O_2 and Ar gas and for C-foils. The curves are evaluated from ^{12}C data under the well established assumption that the charge state distributions as a function of velocity are to first order independent of the mass. The maximum for 3^+ is around 3 MeV and 4^+ peaks around 6–7 MeV. It is interesting to note that the maximum for 4^+ is significantly higher than that of 3^+. This behavior can be explained by the electron shell structure of carbon. To reach 5^+ a K-shell electron would have to be removed. The cross

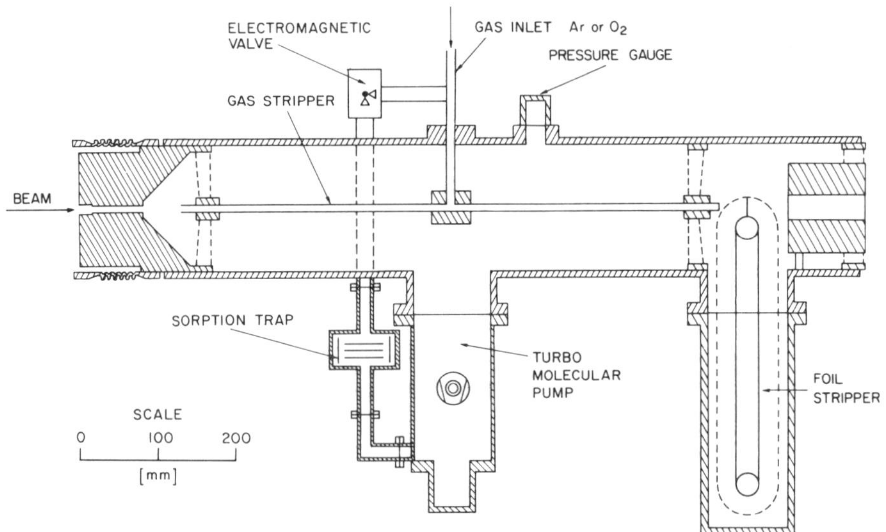

Figure 7. Stripper arrangement, with terminal pumping and foil stripping capability. (Bonani et al., 1990)

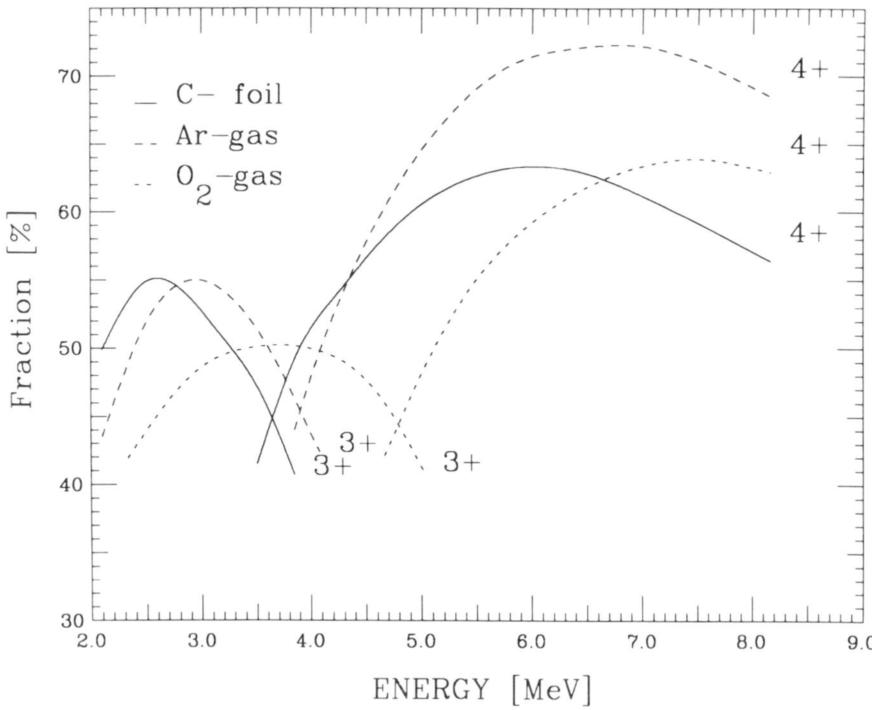

Figure 8. Stripping yield for ^{14}C as a function of energy for equilibrium conditions. (Suter, 1990)

section for this process is almost an order of magnitude lower than that for L-shell electrons. This decrease in cross section prevents the population of the 5^+ state and thereby enhances the population of the 4^+ state. The large difference between O_2 and Ar gas was unexpected. In general the dependence on target gas is weak and it is omitted in most semiempirical formulas (Betz, 1972; Betz, 1983). In the case of carbon, however, the 4^+ yield reaches 70% between 6 and 8 MeV with an Ar stripper, which is much higher than the maximum for O_2 gas. It also exceeds the yield obtained with C-foils. This higher yield makes the large tandem accelerators which can operate above 6 MV very attractive for ^{14}C measurement when compared to Tandetrons operating normally around 2 MV, where the stripping yield is about 40%. The yields shown are only valid when charge-state equilibrium conditions are reached. This requires thicknesses of 1.2 $\mu g/cm^2$ at about 2 MeV (Figure 9) and 1.6 $\mu g/cm^2$ at 6 MeV.

When gas stripping is used the thicknesses required for equilibrium conditions would lead to unacceptably high pressures in the acceleration tubes without appropriate pumping in the terminal (Bonani et al., 1990). The reduction of the vacuum in the acceleration tube leads to losses due to charge-exchange processes

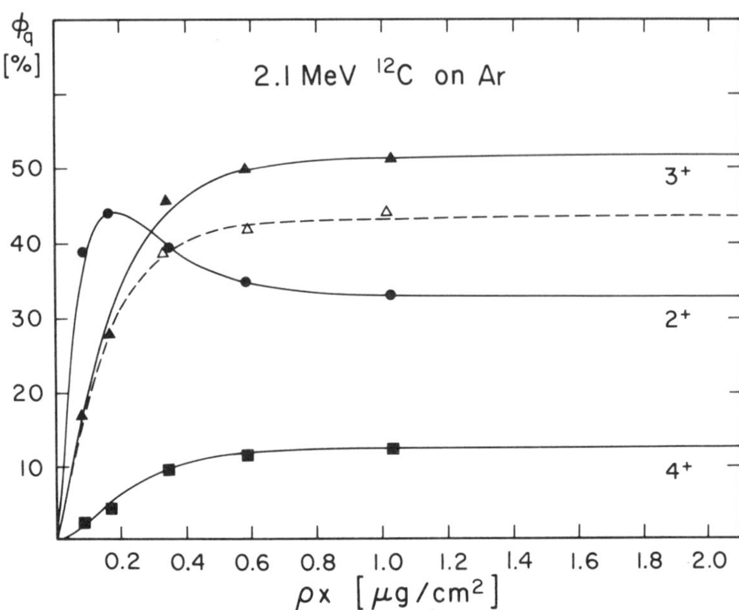

Figure 9. Stripping yields for ^{12}C at 2.1 MeV (full symbols, solid lines) and $^{12}C^{+3}$ at 1.8 Mev (open symbols, dashed line) as a function of gas thickness. (Hofmann et al., 1987b)

in the accelerator. These losses can be up to 30% without terminal pumping, and even with such a pump they can be of the order of 10 to 15% (Bonani et al., 1990).

For ^{10}Be measurements, BeO^- molecules are injected because Be does not form stable negative ions (Kvale et al., 1985). Hydrides, which give only small currents, are difficult to prepare. Due to the larger mass of BeO molecules compared to Be, significantly higher terminal voltages are required to reach the maximum yield for charge state 3^+. When foils are used, the fast dissociation of the molecules (Coulomb explosion) leads to additional energy and angular straggling. With a combination of a low density gas followed by a thin foil, the effect of the Coulomb explosion can be reduced and higher transmissions are obtained (Klein et al., 1982). The maximum at charge state 3^+ can be reached around 8–10 MV with efficiencies around 60% (Figure 10) for O_2 gas and foils. To date, Ar gas has not been investigated in detail for BeO. With Tandetrons operating around 2 MV, the yield for 3+ is very low. However, by using 2+ and a post stripper to destroy molecules, good results can be obtained (Raisbeck et al., 1984).

For heavier isotopes the stripping yield can be predicted quite well with semiempirical formulas (Betz, 1972; Betz, 1983). The maximum of the most

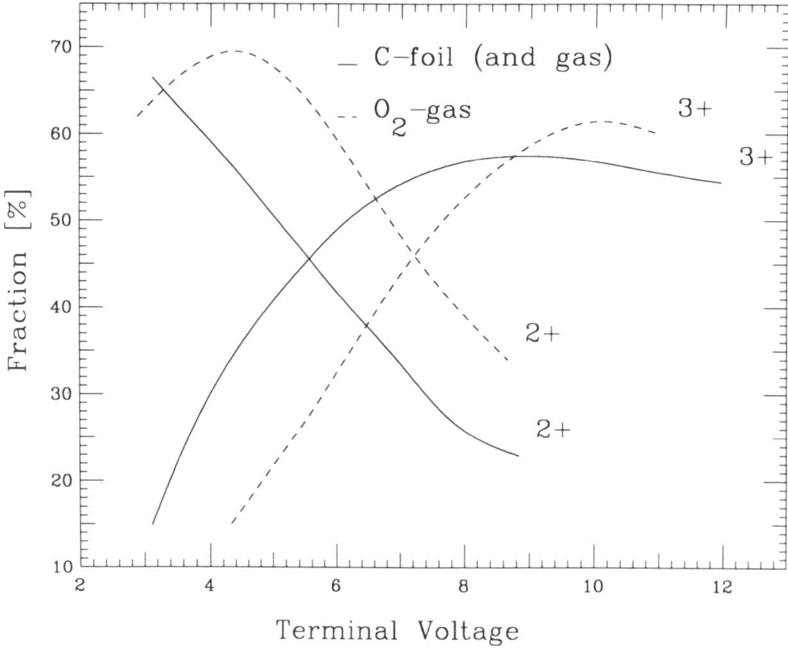

Figure 10. Stripping yield for BeO as a function of terminal voltage. (Suter, 1990)

intense charge state reaches about 35–40% for Al, Si, and Cl, and 25–30% for I. Foils can produce higher charge states than gases at the same energy.

High Energy Mass Analysis

After the accelerator, the ion beam consists of various components having different charge states and different energies. Beside the main components, molecular fragments are also found and, due to charge changing processes in the acceleration tubes, a small continuous background of species with various energy and charge state combinations is present. For this reason a magnetic spectrometer alone is generally not sufficient to perform the final mass separation. In most cases an electrostatic deflector is also included, selecting ions with the same E/q. In some configurations these devices are placed in front of the magnet, allowing the same beam path for rare and stable isotopes up to the final spectrometer (Figure 1). This design permits monitoring the transmission and stability of the electrostatic deflector by repetitively measuring the abundant isotopes. Instead of electrostatic deflectors, Wien filters (crossed electric and magnetic field) have also been used at several laboratories (White et al., 1981; Klein et al., 1982).

The advantage of these devices is that they do not deflect the beam and can therefore be simpler to construct. However, Wien filters require switching when they are expected to pass the abundant isotope also.

Detectors

The high energy of the ions which have passed through the accelerator allows further analysis to be performed in a final detector system. The detector, either a gas counter or a solid state detector, gives a signal which is proportional to the total energy which is deposited. This energy determination can provide additional background suppression because it allows the removal of (M/q) ambiguities; for example, between $^{14}C^{4+}$ and $^{7}Li^{2+}$. These backgrounds must not be too high, however, or the detector and associated electronics will be overloaded, degrading the energy resolution. With a ΔE detector in addition to the total E detector additional information can be obtained from the specific energy loss, dE/dx, of the ion. This additional information makes possible a determination of the atomic number. A time-of-flight detector allows measurement of the ion velocity. Together with the total energy measurement this velocity measurement allows an additional mass determination to be performed which is independent of the previous analyzing steps.

ΔE-E Gas Counter. Gas counters have been very successfully used in AMS for energy and specific energy loss measurements (Figure 11). They work according to the following principles: When fast moving ions pass through a gas, atoms

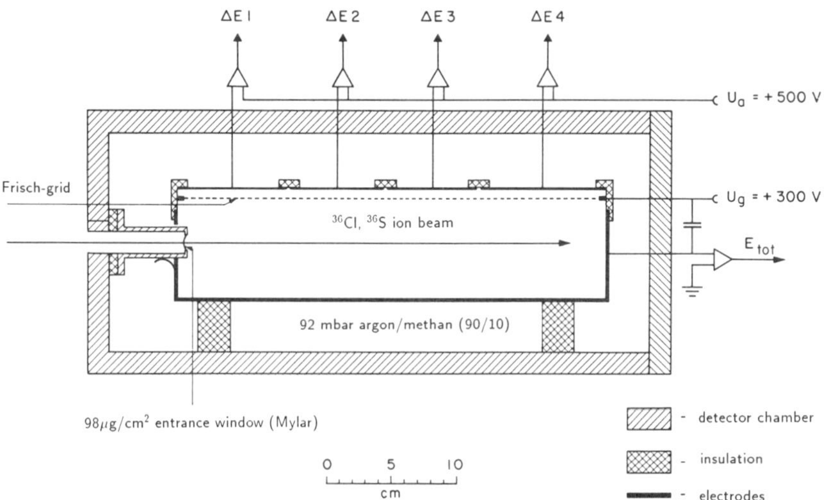

Figure 11. Gas counter with 4 ΔE electrodes. (Synal, 1989)

present along the track of the ion are ionized. The number of electron-ion pairs produced is proportional to the energy loss. The average energy loss needed for producing one charge pair depends on the gas type and lies between 20 to 30 eV. By applying a homogeneous electric field perpendicular to the incoming ion path, the charge produced can be collected at a set of electrodes. Appropriate operating conditions (gas mixture, pressure, and electric field) can be found which avoid recombination or uncontrolled charge multiplication and therefore allow measurement of an integrated current pulse which is proportional to the energy loss of the ion in the associated detector section. The electrons drift with high velocities, in the cm/μs range, whereas the ions produced move much slowly. In order to separate the fast electron signal from the slow ion signal, and to make the electron signal independent of the site of ionization, a fine-meshed grid with high transmission (called a Frisch grid) is placed in front of the anode. The anode is thereby well shielded from ion currents, and the current pulse induced by the electrons has a well-defined shape which is independent of the location at which the ionization occurred. This design provides good energy resolution. By dividing the anode into sections, ΔE signals can be derived. The detector signals are analyzed with standard nuclear electronics. The current pulses are integrated in a preamplifier and shaped and filtered in the following main amplifier so that they can be digitized in a multichannel analyzer or the ADC of a computer system.

The resolution of the total energy signal depends on the energy loss and straggling in the entrance window, which normally consists of a thin mylar or polypropylene foil. Heavier ions (such as Al, Cl) loose a few MeV even in the thinnest usable foils (0.7 μm). Thickness variation which can easily be of the order of 10% gives rise to an energy spread of a few 100 keV. The stochastic nature of the interaction with the target electrons, and of energy loss in nuclear collisions, leads to an additional spread in the energy signal. Electronic noise leads to a further reduction of the resolution. All these factors limit the resolution to typically 1%.

Isobar separation. The rate of energy loss, which depends on the atomic number of the ion, can be calculated from stopping power data (Northcliffe and Schilling, 1970; Ziegler et al., 1985). This allows the identification of ions according to element. By splitting the anode in two or more parts both energy loss signals (ΔE) and the total energy (E) can be derived. The difference of the ΔE signals between two neighboring elements becomes smaller for heavier elements but increases with increasing energy. By the appropriate choice of electrode length and gas pressure the difference in ΔE signals between isobars can be optimized.

Isobar suppression depends not only on the peak separation but also on the peak width in the ΔE spectrum. The main contribution to this width is energy loss fluctuations related to variations in the charge state of the incoming ion, and to the random nature of the interaction with the target electrons (Bohr straggling). Based on models and experimental data, semiempirical formulas have been derived describing these processes (Schmidt-Böcking, 1978; Schmidt-Böcking and H. Hornung, 1978). With this information the suppression of various isobars has been calculated as a function of energy (Figure 12) (Suter, 1990). Isobar pairs of light ions ($Z < 14$) can be easily separated with small accelerators (2–3 MV). ^{36}Cl and ^{41}Ca require energies of 40–80 MeV for reasonable separation (6–10 MV tandem accelerators). Heavier isobars cannot be sufficiently separated with this technique at energies attainable with normal tandem arrangements.

Other factors have to be considered for more precise predictions. At high count rates, pulse pileup, base line shifts, and incomplete charge collection also reduce resolution. Nuclear collisions, in which a significant amount of energy is transferred to the atoms of the target gas, lead to significant scattering which causes tails that would have to be included in more reliable models.

For radioisotope measurements in which no severe isobar interferences are present in the final detector, it is sufficient to record one ΔE signal and the

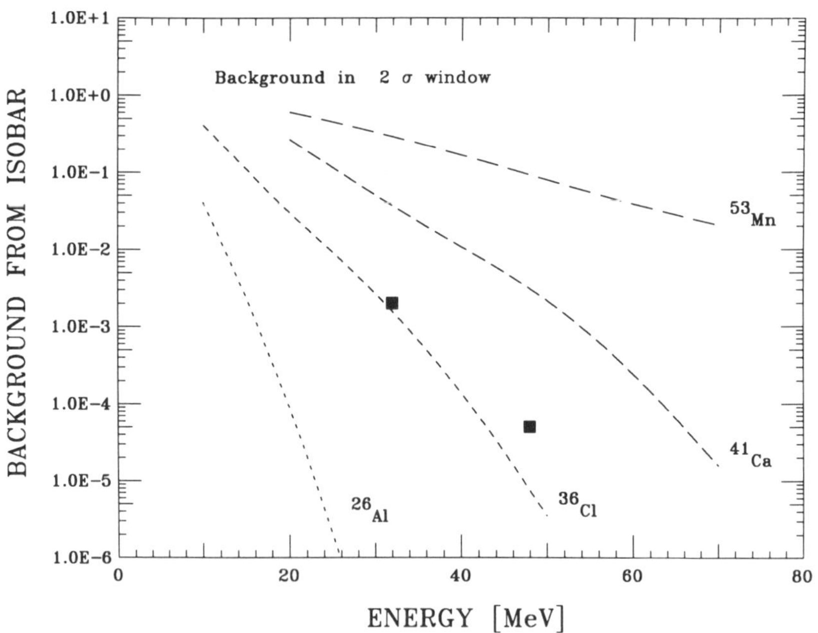

Figure 12. Isobar suppression for various radioisotopes as a function of the beam energy for a split-anode detector, based on calculations with a model which includes energy-straggling in the entrance window and the counter gas. For comparison, two experimentally determined suppression factors for ^{36}Cl are also shown (■). (Suter, 1990)

residual energy. These data are normally stored as two-dimensional spectra (histogram) (Figure 13). The peaks of interest are integrated over rectangular areas. The spectrum, given in Figure 13, shows ^{14}C, ^{13}C, and ^{12}C peaks which all have different total energies corresponding to equal momenta, indicating that the energy resolution of the electrostatic deflector was not sufficient but the momentum analysis in the magnet was satisfactory. The background peaks of the stable

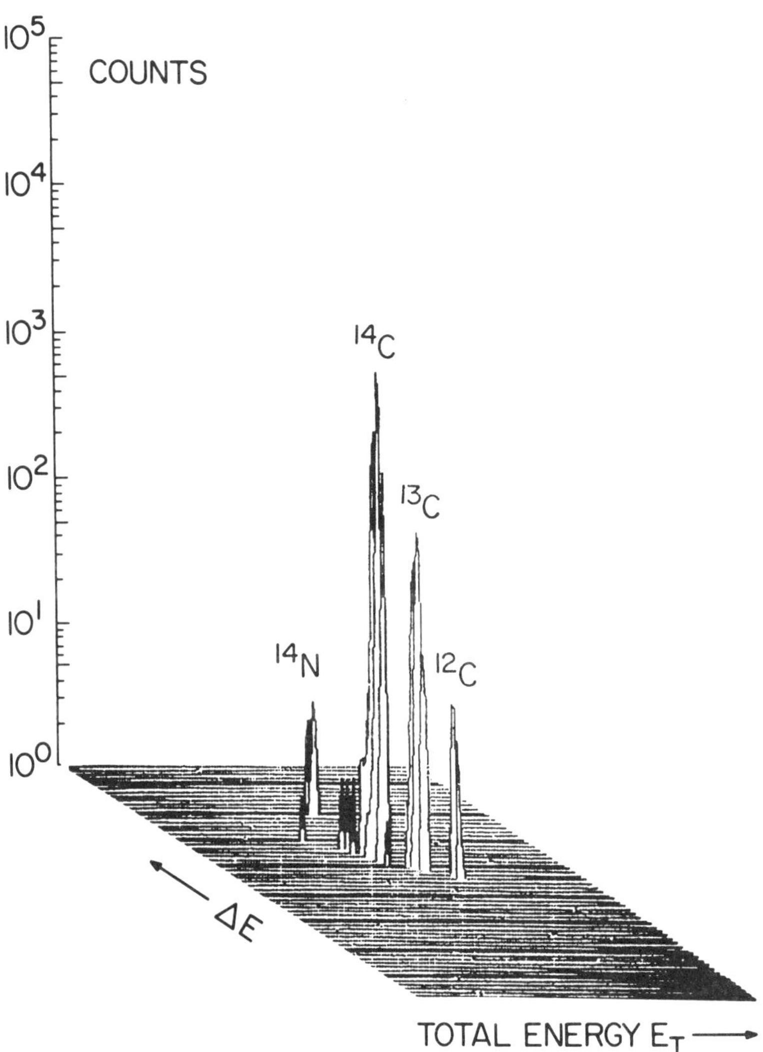

Figure 13. Two-dimensional spectrum from a modern C sample, obtained with a ΔE-E counter.

isotopes come from molecular components injected simultaneously with ^{14}C (^{13}CH and $^{12}CH_2$). The N peak originates from $^{14}NH^-$ ions. They have almost the same total energy as ^{14}C, but they can be easily separated from ^{14}C by dE/dx-techniques.

For ^{36}Cl and ^{41}Ca isobaric and isotopic interferences are a more severe problem. Therefore in most cases multi-ΔE systems are used (Elmore and Phillips, 1987; Synal et al., 1987; Fink et al., 1990a). By setting energy windows for each measured parameter, background can be further reduced. Total energy can be measured by a separate signal derived from the cathode and the Frisch grid which are capacitively coupled together (Figure 11). This signal is more precise than adding up all ΔE signals electronically.

Bragg Curve Spectrometer. This is a gas counter in which the charge produced is collected in the direction of the incoming ions. Due to the different drift times to the Frisch grid, the time evolution of the current pulse at the anode corresponds to the dE/dx curve of the stopped ion. With appropriate electronics it is possible to derive ΔE information and perform isobar separation with this type of counter.

Semiconductor Detectors. Si-detectors (mainly surface barrier detectors) generate current pulses which are significantly larger than the corresponding signal of a gas counter. This is especially helpful at low energies, when electronic noise is a dominant contribution to the resolution. In several systems a Si-counter is used as a final stop detector in combination with a gas ΔE counter.

For heavy ion detection several disadvantages of Si detectors have to be considered. Nuclear collisions cause structural damage of the semiconductor which lead to a reduction in performance after a dose 10^6–10^7 ions/cm^2. Energy loss and associated straggling in the dead-layer at the surface of the detector reduce the energy resolution and introduce some nonlinearities in the energy scale (pulse height defects). Nuclear stopping, which does not lead to charge-pair production, reduces the pulse height (pulse height defects) and affects the resolution. All these effects cause the resolution at higher energies to be no better than with gas counters.

Si-detectors can in principle also be used for ΔE determination. But there are technical problems in constructing thin devices (μm) with the required homogeneity.

Passive Absorber. The dE/dx technique can also be applied using passive absorbers by stopping the ions without actually measuring the energy loss. This technique can be used when the interfering isobar has a higher Z and can be completely stopped in the absorber while the ions under investigation are transmitted. This technique has been introduced in cyclotron-AMS to separate ^{14}N from ^{14}C, and is today a standard method for the suppression of boron in ^{10}Be

measurements (Klein et al., 1982). Gas absorbers have the advantage of being easily adjustable in thickness. However, hydrogen present in the absorber can cause background due to the nuclear reaction $^{10}B(p,\alpha)^{7}Be$ (Klein et al., 1982). Passive dE/dx has also been applied without complete stopping. For example, ^{10}Be can be detected with small accelerators very successfully by using a 200-μg/cm^2 carbon foil, followed by a momentum analysis (with a magnet) and a ΔE-E counter. In this case the difference in charge state distribution between ^{10}Be and ^{10}B enhances the suppression. On the other hand, one has to accept a reduction in efficiency (Raisbeck et al., 1984). Passive foil absorbers have also been used in combination with magnetic spectrographs for ^{32}Si (Kutschera et al., 1980a), ^{60}Fe (Kutschera et al., 1984), and ^{205}Pb (Ernst et al., 1984; Henning and Schuell, 1988) detection.

Time-of-Flight Detectors. Time-of-flight detectors are very effective for heavy ion identification in AMS. For typical particle energies (0.1–2 MeV/AMU) used in AMS, the ion velocities are in the range of cm/ns. By measuring the flight path over 1–3 m with a time resolution of a few 100 ps, precise velocity information can be obtained. The start detector consists of a thin carbon foil. When fast ions pass through this foil, low energy secondary electrons are emitted. These can be detected with electron multiplier detectors (Figure 14). Today microchannel plates are used in most cases (Leskovar, 1977; Wiza, 1979). For the stop signal, a similar detector can be used. Alternatively, the final

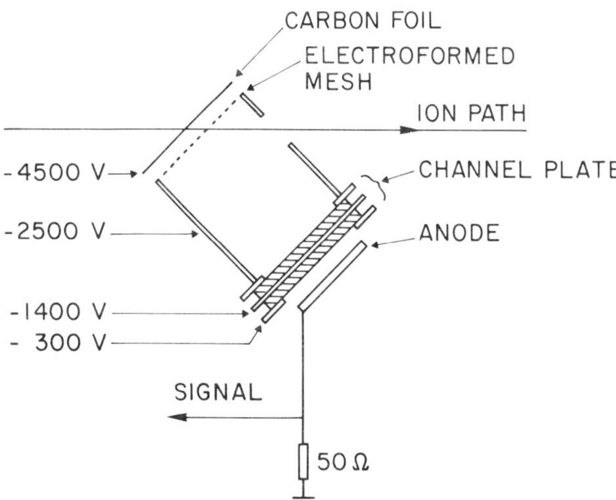

Figure 14. Example of a time-of-flight detector. When ions pass through the thin carbon foil secondary electrons are produced which can be collected with microchannel detector.

E or ΔE detector can provide the fast timing signal. Various configurations have been developed to collect and detect the secondary electrons efficiently and with the best possible time resolution.

The disadvantage of time-of-flight detectors is that the ions have to pass through a thin foil in the start detector. This leads to angular straggling which can cause counting losses in the final detector. This effect can be reduced by adding a lens, but due to the nonsingular charge-state distribution, perfect focusing is not possible. To reduce the effect of energy straggling, an isochronous beam transport system between start and stop detector has been proposed (Kilius and Litherland, 1987).

Normalization Procedure

In most cases radioisotope concentrations are calibrated with standards. Because sputter rates and thus the currents may vary from sample to sample and do not necessarily remain constant in time, an adequate normalization to the currents of the abundant isotope is needed. Various procedures have been developed to accomplish this:

1. In the simplest arrangement, the current of the stable isotope(s) is measured at the low energy side of the accelerator after the first magnet. The stable isotope current can be monitored periodically by a Faraday cup which moves into the beam for measurement and out of the beam to allow the rare isotope to pass. Alternatively, the magnet can be designed to allow the stable isotope Faraday cup to remain in place when the rare isotope is injected into the accelerator. In this case both isotopes can be measured simultaneously. The disadvantage of measuring the stable isotope current at the low energy side of the accelerator is that there is no direct way to estimate possible variations in transmission through the accelerator to the final detector. In addition, molecular beam components may lead to errors.

2. This disadvantage can be eliminated by monitoring the most abundant isotope(s) on the high energy side of the accelerator. The effect of systematic errors can thus be reduced and potential measurement problems can be recognized more easily. It is possible to monitor stable isotopes on the high energy side either sequentially or simultaneously with the rare isotope measurement. At the beginning of AMS development, sequential measurement procedures with slow cycling were introduced and are still used at many laboratories. This method has the problem that the high beam currents of the abundant isotopes can overload the accelerator, especially the older tandems. It is possible to reduce the intensity of abundant isotope beams by passing them through a fine-meshed grid before injection into the accelerator (Gove et al., 1980). Another problem which remains however, is that short-term fluctuations of the beam, which can result

from discharges in accelerator or ion source, cannot be continuously monitored. In addition, the time lost for the stable isotope measurement can be significant. Stable isotope measurements take several seconds and are repeated periodically at intervals of 20 seconds to several minutes.

When magnetic focusing and steering elements are used, all the elements have to be switched to pass isotopes with different momentum to charge ratios. An alternative is to cycle the terminal voltage so that all isotopes of interest have the same momentum. For dedicated AMS arrangements, electrostatic lenses and steerers, which do not distinguish between isotopes of the same charge, have been preferred to avoid the need for switching the power supplies of other beam elements.

At the time of the first AMS conference in 1977, several authors had already proposed using a fast-switching procedure to reduce the effect of loading on the accelerator by the intense stable isotope beams, and to enable an almost continuous measurement of the isotope ratio with little dead-time in the counting of the rare isotope (Gove, 1978). In designing a fast-switched system, one has to make the pulses short enough so that the stability of the terminal voltage is not affected (Suter et al., 1984b; Suter et al., 1989). For negative ion beams in the range of 10 to 100 μA, the total loading currents (positive and negative) are on the order of 100 μA to 1 mA. During a short pulse, this current, I, is drained from the terminal capacitance C, which is in the order of 100 pF, leading to voltage drop (ΔV) which grows linearly with pulse duration Δt:

$$\Delta V = I\Delta t/C$$

This relation shows that for voltage drops smaller than 1 kV, the pulses have to be in the 100 μs to 1 ms range. Such a fast-switching can be achieved at the source magnet by adjusting the energy of ions before they enter the magnetic field. This adjustment is performed by a high voltage pulse U applied on the magnet chamber box and on symmetrically placed apertures (Figure 15). In order to obtain the same momenta for two different masses, M1 and M2, with energies E1 and E2, the acceleration voltages, V1 and V2, have to fulfill the following relation:

$$V1/V2 = E1/E2 = M2/M1$$

For example, when the magnet is set for ^{14}C ions with an energy of 40 kV, the energy of ^{12}C ions has to be 46.67 keV if they are to be deflected the same as ^{14}C. This means that a voltage pulse of 6.67 kV has to be applied. Appropriate fast-switching high voltage power supplies have been designed by several groups (White et al., 1981; Balzer, 1984; de Haas et al., 1987). Special current measuring devices are needed for integrating these short pulses (Suter et al., 1984b; de Haas et al., 1987). The pulse repetition rate is usually in the range of 5

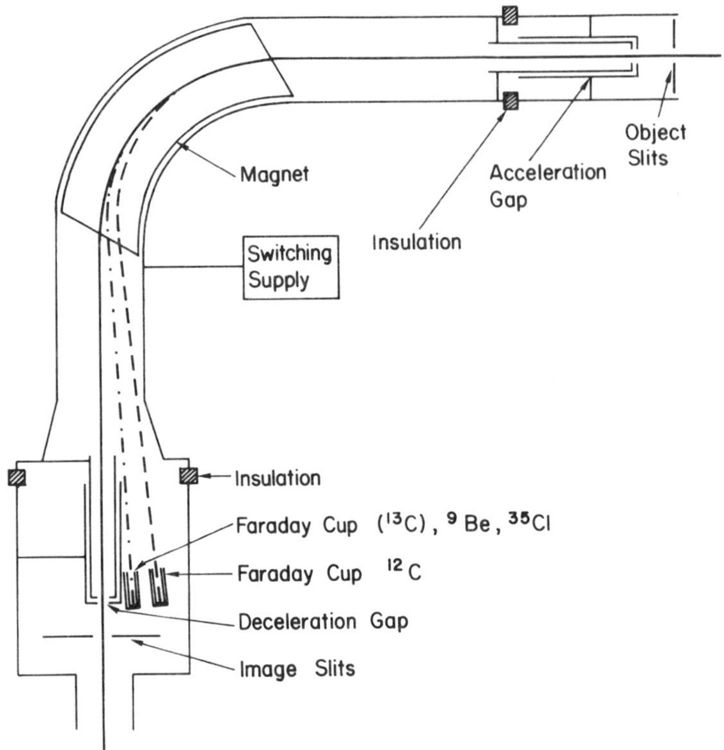

Figure 15. Injector for sequential isotope injection. By applying high voltage pulses to the magnet chamber and the associated apertures the injected beam can easily be switched from the rare isotope to the most abundant one. (Suter et al., 1989)

to 20 Hz, allowing the use of more than 99% of the time for the counting of the rare isotope (Suter et al., 1984b).

Simultaneous injection of more than one isotope into the accelerator has also been used (Lobb et al., 1981; Southon et al., 1986). In this technique a magnet system first separates all the masses and then recombines only the isotopes of interest. This method has the advantage of being really simultaneous, but it requires a significantly better filtering system at the high energy end of the accelerator and has to be able to handle the large intense stable isotope beam in a continuous mode. Such a simultaneous injection system is used in a new dedicated facility at Woods Hole in order to provide high precision measurements (Purser et al., 1988; Litherland and Kilius, 1990; Purser and Smick, 1990).

3. When using molecular beams at the injection side, there exists the possibility to monitor continuously a specific fraction of the abundant isotope when the accompanying element has more than one stable isotope. This is the case for ^{10}Be measurements; $^{10}Be^{16}O$ and $^{9}Be^{17}O$ are injected simultaneously, allowing a continuous monitoring of the ^{9}Be fraction at the high energy side (Imamura et al., 1984; Klein and Middleton, 1984).

Special Techniques and Accelerators

In some special cases, other accelerator types have been used or special techniques applied for background suppression.

Gas-Filled Magnet. When fast-moving ions of a well-defined energy pass through a gas which is in a magnetic field, the trajectory followed depends on the atomic number of the ion, because the average charge is related to the charge of the nucleus. This property has been applied to isobar separation in AMS for the detection of ^{41}Ca, ^{59}Ni, and ^{126}Sn at Argonne National Laboratory using an Enge-split-pole-spectrograph (Henning et al., 1987; Kutschera et al., 1989; Paul et al., 1989; Paul, 1990) (Figure 16). Tests with a similar magnet have been performed at Rochester for separating ^{36}Cl from ^{36}S (Kubik et al., 1989). At the AMS facility in Zurich, ^{32}S has been suppressed by four orders of magnitude in ^{32}Si measurements with 44-MeV ions using a 90° magnet (Suter et al., 1989). Monte Carlo simulation programs allow modeling of the general behavior (peak position and half-width) quite well (Paul et al., 1989), but the detailed shape of the tails, which is important for determining the extent of suppression, is not given by these calculations. Experimental studies which have been performed indicate that initial expectations of attaining good isobar separations at low energy can probably not be realized. Angular straggling, which scales roughly with 1/E, dominates the peak width at low energies. In addition, the relatively large energy loss in the magnet gas and associated windows makes identification in the final detector difficult. In general, one can say that a gas-filled magnet is advantageous only when sufficient energy is available, and when high count rates would degrade the resolution of the ΔE-E counter.

Complete Stripping Technique. A further possibility for isobar identification is available when the ions can be completely stripped ($q = Z$). Isobars with Z lower than the nuclide to be determined will then have a lower charge state. Therefore a simple charge analysis after a stripper foil with a magnetic field allows a separation with very low background. In order to reach a significant probability to remove all the inner-shell electrons, the projectile velocity has to be on the order of the classical Bohr velocity of these electrons. This requires energies in the 100 MeV range for ^{26}Al (Raisbeck and F. Yiou, 1979), ^{36}Cl (Kubik et al., 1984), ^{41}Ca (Steinhof et al., 1987), ^{53}Mn (Korschinek et al., 1987), and ^{59}Ni (Henning et al., 1981; Faestermann et al., 1990), which can only

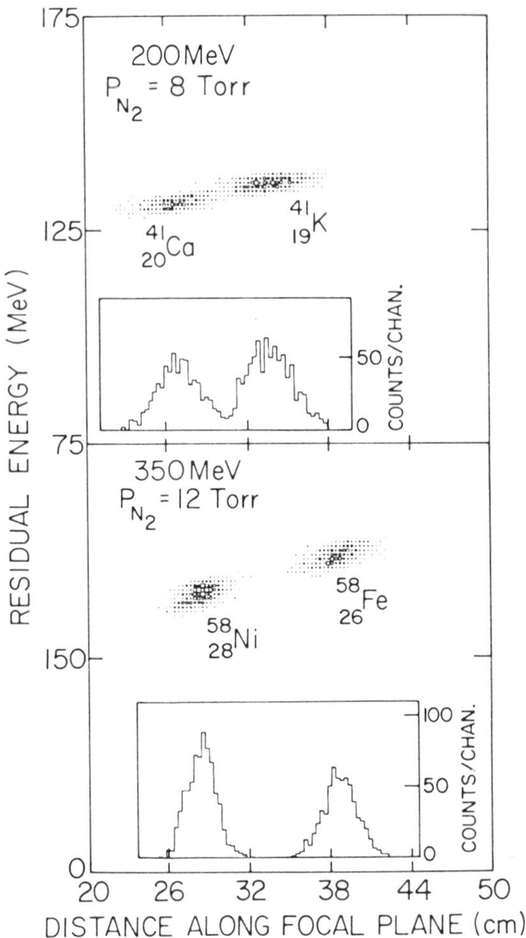

Figure 16. Position and energy spectra obtained with a gas-filled magnet. (Paul et al., 1989)

be obtained at a few large facilities. Only for light ions such as Be is full stripping possible with small accelerators (Figure 17). Yiou and Raisbeck have demonstrated that ^{7}Be ($t_{1/2}$ = 53.3 days) can be detected nearly background free with AMS using this technique and with a detection efficiency comparable to the decay counting method. This technique is especially suited for relative measurements of ^{7}Be and ^{10}Be, as needed for studies of atmospheric residence times (Raisbeck and Yiou, 1988b).

Lasers. Several laser techniques have been developed for the detection of low concentrations and/or small numbers of specific atoms. Combining these tech-

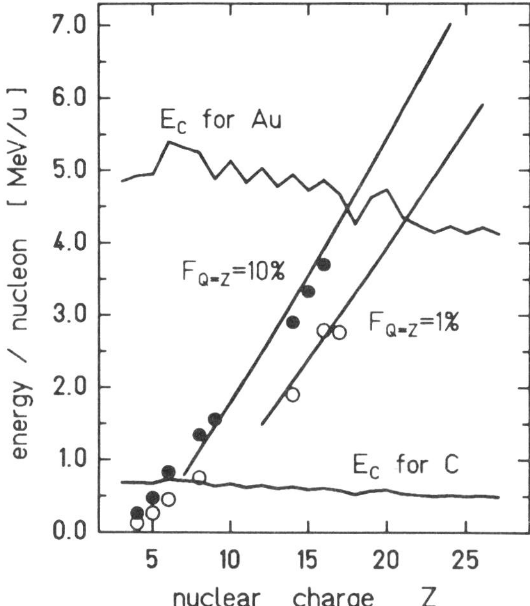

Figure 17. Energy dependence of full stripping probabilities $F_{Q=Z}$ of 1% and 10% in carbon targets for different projectile ions. The symbols show experimental values (Shima et al., 1986), the lines are calculated according to (Steinhof et al., 1989). For comparison, the Coulomb barrier E_C for gold and carbon targets is also displayed. These values were calculated for the isotopes with the largest abundance. (Steinhof, 1990)

niques with AMS has been proposed to get a further improvement in sensitivity. The main problem is that most laser techniques operate in a pulsed mode, leading to a low-duty cycle. Recent initial experiments have been reported whose goal is to reduce the isobaric background in ^{36}Cl measurements by electron detachment of the interfering S^- in a laser field. The technique works in principle but, due to the low duty cycle of the laser, the efficiency is presently too low by orders of magnitude for practical applications (Berkovits et al., 1989, 1990).

RF-Fields for Acceleration and Background Suppression. In a LINAC or a cyclotron the ions are accelerated in an electric field which varies in time (RF-fields). The requirement of strict phase matching of the accelerated beam bunches makes these types of accelerators excellent mass filters. With multiply-charged ions *M*/*q* ambiguities lead to interferences which may reduce the sensitivity. On the other hand, these species can be used for tuning the accelerator (as pilot beams) and for normalization (Kutschera et al., 1989; Steinhof, 1990). In the beginning of AMS, pioneering experiments were made with cyclotrons. In recent years cyclotrons have lost importance. New attempts to use mini-cyclotrons for AMS are being made (Bertsche et al., 1987; Chen et al., 1989a, b; Bertsche et al., 1990; Subotic et al., 1990). Due to their small size, mini-cyclotrons would be very attractive, but first the sensitivity and normalization problems have to be satisfactorily resolved.

At several larger tandem facilities, linear arrangements of RF-cavities have been added to enhance the accessible energy range. Two of these facilities have been used for AMS (Kutschera et al., 1989; Faestermann et al., 1990). They allow the elimination of isotopic interferences due to the high mass filtering power. The high energies make possible a better isobar separation. These facilities have primarily been used for studies of heavier isotopes such as ^{36}Cl, ^{4} ^{1}Ca, ^{59}Ni, ^{60}Fe, and ^{126}Sn.

Accelerators consisting of LINACS have been used at GSI for ^{41}Ca (Steinhof et al., 1987) and ^{205}Pb detection (Ernst et al., 1984). Tests with noble gases are in progress at Argonne National Laboratory (^{39}Ar,^{81}Kr, ^{85}Kv) (Kutschera et al., 1989).

Single-Ended van de Graaff. There are also electrostatic accelerators in use which have a source for positive ions in the high voltage terminal (single-ended van de Graaff). In these facilities the electric field is used only once for acceleration and, therefore, the final energy of the ion beam is smaller. A further disadvantage is that sample loading and changing is more difficult. Only a few exploratory experiments with methane, to estimate deuterium contents (Fowler et al., 1987), and with tritium detection have been performed with this type of accelerators (Jiang et al., 1984).

Stable Isotope Detection. AMS is not restricted to the detection of long-lived radioisotopes. It can also be used for stable isotopes at low concentrations which are not accessible with conventional mass spectrometry due to molecular or isobaric interferences. Trace element information is important especially in mineralogy. In solid-state physics material properties also depend in some cases on the trace element contents. In many of these applications lateral and depth distribution of the trace elements are of interest. Therefore special ion sources with Cs-microbeam stages with scanning capability are needed, similar to those developed for secondary ion mass spectrometry (SIMS). In order to avoid background produced by sputtering of electrodes or ionization of residual gas, special design and ultrahigh vacuum systems are needed.

B. Performance

The performance of an AMS system depends very strongly on the specific instrumentation in use and varies from facility to facility. Increased understanding of the factors which are important in determining performance and advances in technology have enabled upgrading programs to be instituted at a number of facilities which will lead to a general improvement in performance in the next few years. For these reasons, we have chosen to discuss performance in terms of the technical factors and the fundamental physics which limit the sensitivity, efficiency, and precision which are attainable. These theoretical limits can be compared with the current state of the art to see what improvements might be expected in the near future. The important data are summarized in Table 1.

Background and Sensitivity

The background measured in the final detector limits the lowest isotopic ratios which can be determined. It also limits the minimum number of atoms detectable. The background can be divided into two categories:

1. Contamination: Ions of the isotope of interest, which do not come from the original sample material.
2. Instrumental background: Other particles which reach the final detector to produce signals which mimic the real ones, and are therefore wrongly identified.

Background levels are determined by measuring blank samples which have been prepared in the same way as actual samples, but using material thought to contain none of the rare isotope under investigation.

Table 1. Performance Limits of AMS

Isotope	*Sample material*	*Extracted Ions*	*Negative Currents* μA^1	*Negative Ion Yield*[2]	*Stripping Yield*[3]	*Overall Efficiency*	*Background Level*
^{10}Be	BeO	BeO^-	0.5–8	1.0–1.8%	64%	10^{-5}–1%	10^{-15}
^{14}C	C	C^-	20–100	5–10%	72%	0.5%–5%	10^{-15}
^{26}Al	Al_2O_3	Al^-	0.2–6	0.1–0.25%	40%	10^{-4}–10^{-3}	10^{-15}
^{32}Si	Si	Si (SiH_3^-)	10–100		40%	10^{-3}–1%	10^{-12}
^{36}Cl	AgCl	Cl^-	10–200	10–16%	35%	10^{-3}–2%	10^{-15}
^{41}Ca	CaH_2	CaH_3^-	0.2–10	0.2%	35%	10^{-4}–10^{-3}	10^{-15}
^{53}Mn	MnO_2	MnO^-	1–8		32%	10^{-5}–10^{-6}	10^{-12}
^{60}Fe	Fe	Fe^-	1–8	1.2%	32%		10^{-12}
^{59}Ni	Ni	Ni^-	5–50		32%		10^{-10}
^{129}I	AgI	I^-	5–100		30%	10^{-3}–1%	10^{-14}

[1] First value: reported currents obtained with frit type sources. Second value: currents obtained with high intensity sources (Middleton, 1989).

[2] First value: reported yields obtained with frit type sources (Suter et al., 1985). Second value: Reported yields obtained with high intensity sources (Middleton, 1989).

[3] Highest possible stripping yield for charge state 3+ or higher under equilibrium conditions (Hofmann et al., 1984, Hofmann et al., 1987b, Suter 1990). For heavier isotopes the yield is estimated.

Contamination. Isotopic contamination can be introduced into a sample during sample collection, sample pretreatment, or final target preparation. Preparation contamination can be introduced from labware or from the general environment. This type of background presently limits ^{14}C detection to levels which correspond to ages of 40- to 55-thousand years.

Furthermore, cross contamination can occur in the ion source, especially when the element to be studied has a high vapor pressure and when the sputter region is not well pumped (^{36}Cl). Special care to minimize the cross talk problem has to be taken when designing gas sources (Middleton, 1984; Bronk and Hedges, 1987; Middleton et al., 1989; Bronk and Hedges, 1990).

The rare isotope can also be introduced when a carrier has to be added to samples to provide sufficient material for processing. This has been reported for ^{10}Be (Middleton et al., 1984). Commercially available Be contains ^{10}Be at the 10^{-14} level. By using specially selected carrier material, a reduction to the 10^{-15} level is possible. It seems that the background found for ^{129}I is also determined primarily by the radioisotope content of the carrier. This contamination at present restricts the sensitivity of $^{129}I/I$ measurements to a few times 10^{-14} (Boaretto et al., 1990).

Instrumental Background: Mass Separation. Mass separation is normally performed with magnetic spectrometers placed at the low and high energy side of the accelerator. In addition, the final detector system allows a further analysis and is able to resolve ambiguities among different ions with the same (M/q). The separation therefore depends very strongly on the design of the relevant components: (1) the electric and magnetic filters; (2) the appropriate apertures; (3) the vacuum conditions in these devices; and (4) the gas pressure in the acceleration tubes. With proper layout and a sufficient number of filtering stages, mass interferences are normally not the limiting factor for lighter ions (^{10}Be, ^{14}C), but for heavier ions these types of interferences are more severe (^{36}Cl, ^{41}Ca, etc.).

Molecules having the same mass as the isotope under investigation break up in the terminal and the fragments may lead to (M/q) ambiguities at the high energy analyzing system which cannot be resolved by deflection in static electric and magnetic fields. These interferences can be greatly reduced by the appropriate choice of the final charge state. For illustration the following examples are given:

1. $^{7}Li_2^-$ molecules can interfere with $^{14}C^-$ at the low energy side. When even charge states for ^{14}C are selected, they will coincide at the high energy side with Li ions having half the charge. But when selecting C^{4+} above 5 MV this type of background does not cause any serious problems because Li is stripped predominantly into charge state $3+$ rather than $2+$.
2. For ^{36}Cl there are potential severe interferences from C_3 molecules, which can cause problems when a charge state which is a multiple of 3 is selected. More complex and less obvious interferences have been found for

$^{41}CaH_3^-$ having the same mass as $^{32}S^{12}C^-$ (Fink et al., 1990c).
3. For $^{129}I^-$ problems arise with $^{97}Mo^{16}O_2^-$. When $^{129}I^{4+}$ is selected after the accelerator, $^{97}Mo^{3+}$ can interfere (Kilius et al., 1990).

Other problems are caused by hydrides of heavy elements. The final energy difference between the isotope of interest and the hydride of neighboring isotopes with smaller mass is minor. Therefore discrimination requires a high resolution in the final analyzing system (Kilius et al., 1990).

When energy filtering with electrostatic deflectors alone is not sufficient, a measurement of the total energy in the final detector can provide additional background suppression and also allows the removal of (E/q) ambiguities. However, if these backgrounds are too high, the detector and associated electronics can be overloaded, degrading the energy resolution. Time-of-flight detectors combined with the energy analysis allow an additional mass identification.

Instrumental Background: Isobar Separation. In several cases isobaric interferences limit the sensitivity of AMS. There exist several techniques to suppress this type of background. Each method has its range of application and its limits. In general a combination of several techniques is applied.

A first rejection already occurs during the sample preparation and purification, when chemical differences can be used to separate isobars. Chemical segregation by a factor of 10^6 can be attained. Isobars can also be removed in the ion source, because in some cases they do not form negative ions. In these cases isobaric background is in general not a problem (^{14}C-^{14}N, ^{26}Al-^{26}Mg, ^{129}I-^{129}Xe). In some cases molecules can be used to suppress the formation of beams of the interfering isobar ($^{41}CaH_3^-$, $^{32}SiH_3^-$). For ^{10}Be detection, the interfering ^{10}B can be stopped in a hydrogen-free absorber material. Any hydrogen present leads to background caused by the nuclear reaction $^{10}B(H,\alpha)^7Be$. Full stripping techniques can be used to remove interfering ^7Li in ^7Be measurements. Suppression of isobaric interferences in heavier ions (^{36}Cl-^{36}S, ^{32}Si-^{32}S) is mostly attained using the dE/dx-technique, for which the suppression power depends strongly on atomic number and the beam energy (cf. Section II-A). For ^{36}Cl a sulfur suppression of about 10^{-4} has been achieved at 48 MeV and about 10^{-6} at 80 MeV, leading to background levels of 10^{-14} or 10^{-15} at 80 MeV. Using a complete stripping technique, similar backgrounds have been reached at 150 MeV. The separation of heavier isobars, such as ^{53}Cr in ^{53}Mn measurements, is significantly more difficult, and only the largest facilities, which can attain very high energies, are capable of reaching a useful suppression.

Efficiency and Throughput

A high detection efficiency is important when working with small samples and generally advantageous for obtaining high throughput. High counting rates are needed to reach low statistical errors and high accuracy. Neglecting losses during sample preparation, the overall detection efficiency can be divided in two parts: the yield of the ion source, and the transmission to the final detector (including the stripping efficiency).

Source Efficiency. The source efficiency is usually expressed as the ratio of the number of negative ions of interest to the total number of sputtered sample atoms. In reality one has to consider that not all the target material can be used for analysis; the current usually drops after the target has been sputtered for a certain time and also the stability of the beam worsens with time due to cratering effects. The background caused by sputtering of the sample holder material can also limit how much of the sample can be used. This is especially true for ^{36}Cl. In some cases sample conditioning (burn-in) is required to avoid cross talk or, especially with gas sources, to reach stable operating conditions. All these factors can reduce the overall efficiency of the source by a factor of 2 to 4 compared to the sputter yield determined on test benches.

R. Middleton has measured yields for the negative ion formation in high intensity sources which he has developed (Middleton, 1984; Middleton, 1989; Middleton et al., 1989). Some values relevant for AMS are given in Table 1. With other sources of the frit-type somewhat smaller efficiencies have been measured at ETH (Suter et al., 1985) (see also Table 1). They show the same trends, which have also been confirmed by other authors (Bronk and Hedges, 1987). For elements having a low negative ion yield with sputter sources (e.g., Al), the use of positive ion sources combined with a charge changing cell might lead to a better efficiency (Litherland, 1980).

Transport Losses and Stripping Efficiency. The stripping efficiency can be optimized by the appropriate choice of energy, charge state, stripper medium and thickness as shown in Section II-A. The maximum efficiencies are also summarized in Table 1.

In principle, the optical transmission should be close to 100%, and is largely limited by engineering restrictions. The emittance of well designed sources can be kept sufficiently small to match the acceptance of the accelerator. Older tandems were primarily designed for light ions and in general do not have very high transmission for heavier ions. Thus they require significant upgrading to reach good optical properties. Beam losses of 50% or higher are not uncommon for these older facilities. With newer acceleration tube designs, computer-optimized injection optics, and well-matched stripper arrangements, transport losses smaller than 10% seem to be realistic.

Overall Efficiency. Based on current experience, one can characterize the overall detection efficiency for existing systems and, by including theoretical limits on optics and stripping yields, one can estimate maximum efficiencies which might be reached under optimal conditions. For ^{14}C typical efficiencies are around 1% for well-designed systems and might reach a few percent under favorable conditions. For Be, efficiencies range from several times 10^{-4} (Tandetrons) up to a few times 10^{-3} with larger accelerators. The possible maximum would be about 1%, which could be reached with optimized systems operating at a terminal voltage of 8 to 10 MV. Due to the low sputter yields of Al- and CaH_3^- ions the overall efficiencies are low, in the 10^{-3} range. High efficiency, at the percent level, is in principle possible for Cl, but in practice, much lower levels are achieved since the count rate of the interfering S increases enormously when the Cs beam hits the target holder.

High efficiency is advantageous, but in many applications efficiency is not the limiting factor. Sample preparation and handling of milligram quantities is in general much easier, more reliable, and more efficient to perform than dealing with smaller quantities. In many cases the background limits the smallest number of atoms detectable to about 10^5 to 10^6.

Throughput. When discussing the throughput, besides the transmission efficiency, one has to take into account beam currents, isotopic ratios, and desired precision. Typical beam currents obtainable with sputter sources are listed in Table 1. In addition, one also has to consider set-up time, sample loading and changing time, conditioning of the accelerator, and maintenance time of the source and accelerator in order to estimate annual throughput.

One of they key elements in determining throughput is certainly the ion source. To reduce downtime, several facilities are installing more than one source, allowing maintenance to be performed on one while the other is in operation. Fast and easy sample loading and changing is also essential.

Uncertainty, Precision, Accuracy

The uncertainty in the final result depends on the experimental arrangement and the measuring procedure applied. In most cases isotopic ratios are measured relative to that of standards. Absolute measurements are only needed when no standards are available, and/or when the half-lives are not precisely known. The specific problems related to absolute measurements are not included in this discussion. Papers dealing with half-life measurements contain more detail on this subject (Elmore et al., 1980; Kutschera et al., 1980a; Kutschera et al., 1984; Hofmann et al., 1990).

Due to the random distribution of the rare isotopes in the sample and the statistical nature of the sputtering and stripping process, in determining the total uncertainty we must consider statistical uncertainties in the measurement of the unknown samples, in the calibration (standards), and in the background subtraction (blank samples). These statistical uncertainties (σ) are related to the number of counts N, $\sigma = \sqrt{N}$.

When measurements are made relative to standards, many systematic errors are cancelled and it is variations in the systematic errors which primarily contribute to the final uncertainty. We can distinguish between time-dependent variation caused by voltage drifts, changes in the vacuum conditions or ion source parameters (temperature, Cs current etc.), and sample-related variation. The time-dependent effects can be recognized by monitoring short- and long-term stability. Sample-dependent effects are more difficult to recognize.

Sources of systematic error may already arise during sample preparation. The degree of contamination might not be fully controlled and may vary from sample to sample. For example, the effect of cross contamination may depend on the isotopic ratio of the previous samples. Chemical mass fractionation might vary from sample to sample due to parameters which are not sufficiently well controlled. The chemical composition of the final target and the structure of the material (grain size and surface) can affect the sputter process (Nadeau et al., 1984). The importance of this effect can be deduced from observations that current can vary from sample to sample, even for samples thought to have been prepared in the same manner. These variations can also effect the ion optical parameters, leading indirectly to changes in the mass fractionation. When a carrier is added, uncertainties due to different chemical behavior of carrier and sample material must also be considered.

Source conditions such as temperature and Cs current also affect mass fractionation. They not only influence the sputter yield but also the phase space of the beam and therefore the optics. Cratering effects at the sample surface can be reduced by spreading the sample material over an area of several mm in diameter and then wobbling the sample continuously (Suter et al., 1984a; Bonani et al., 1987; Suter et al., 1989) or stepwise (Beukens et al., 1986). In this way surface effects can be averaged and reduced.

If all measured isotopes had the same beam path from the source to the final spectrometer, no mass fractionation in the beam transport system could occur, even if the phase space of the beam changed or power supplies drifted. But due to residual magnetic fields present along the beam path (Earth's magnetic field, magnetic suppression in acceleration tubes), the beams of the different isotopes can become separated by a small amount. Additionally, the switching system at the first magnet or at the parallel injection arrangement can lead to mass-dependent optical transport due to design errors, misalignments, or deviations between the real field configurations and the desired ones. The effect of these ion optical differences can be greatly reduced when the transmission is high and does not vary due to minor parameter variations (so-called "flat topping") (Purser et al., 1981). It has been proposed that limiting the phase space close to the source, before any significant separation can occur, would also be a way of obtaining flat-top transmission through the following optics.

Significant mass fractionation is possible in the stripping process (Hofmann et al., 1984; Hofmann et al., 1987b), but variations are in general small when gas stripping is used and equilibrium conditions are reached (Bonani et al., 1990). With foils, however, fluctuations from homogeneity and time dependent thickness variations can lead to significant errors. Foils are therefore not well suited for high precision measurements.

Systematic errors in the final detector system have to be considered at high count rates. Pulse pileup and dead-time effects can lead to counting losses, and high background peaks can lead to tails under the peak of interest which generate additional uncertainties due to the subtraction procedure. Errors can also arise in the beam current measurements. Especially when using the fast-switching procedure, the current integration can be nonlinear due to the specific characteristics of the operational amplifiers.

Current-dependent effects have been found by several groups. This behavior can be caused by various mechanisms. Loading effects can cause changes in the field gradients in the accelerator and voltage drops in beam guiding elements can affect the optics. Enlargement of the beam due to space charge could be another reason for such effects.

Measuring Procedure

By choosing these appropriate measuring procedures, systematic errors are easier to recognize.

1. Stable isotope ratio measurements such as $^{13}C/^{12}C$ are not limited by statistical uncertainties. Therefore they can be used as very sensitive indicators to recognize drifts and fluctuations in the mass fractionation. For precise measurements it is advantageous to measure the $^{13}C/^{12}C$ ratio in addition with a conventional mass spectrometer (Suter et al., 1984a; Bonani et al., 1987; Kromer et al., 1987). Consistent results with the two techniques indicate that mass fractionation in all processes has been kept constant for the unknown samples and standards. $^{13}C/^{12}C$ ratios measurements are also needed for making the required mass fractionation corrections to obtain the proper ^{14}C-ages (Stuiver and Polach, 1977). These corrections are appropriate for mass fractionation occurring in nature, in sample preparation and in the sputtering process, but are not necessarily applicable for stripper mass fractionation and ion optical effects, which may not be linear in mass.
2. By measuring one or more isotopes at both the low energy side and high energy side the transmission can be monitored.
3. Dividing the measurements into intervals (cycles) allows detection of short term fluctuations originating from sparks or discharges, or detection of drifts. Current-dependent effects are also easily visible.
4. Long-term stability can be checked by repeating the measurement of the standards and unknown samples at longer time intervals.
5. The reproducibility of the target preparation can be controlled by making more than one target.
6. Current-related effects can be discovered and corrected for by representing the isotopic ratios and the transmission as a function of the beam current.

Precision of a few percent can be reached relatively easily when an appropriate normalization system is used and when stable power supplies are available. This precision is satisfactory for many applications. Higher precision, primarily required for ^{14}C studies, demands a more careful optimization of the equipment with consideration of the systematic error sources discussed above. To check the performance, internal consistency tests have to be made. Systematic comparison of the measured $\delta^{13}C$ with AMS and conventional mass spectrometry show that the mass fractionation can be kept constant at the 1-per-mil level (Suter et al., 1984a; Bonani et al., 1987). Inter-laboratory comparisons give information on accuracy. Several tests of this type have shown that 0.2–0.3% accuracy is possible by AMS (Beukens et al., 1986; Bonani et al., 1987; Kromer et al., 1987; Damon et al., 1989; Beukens, 1990). Measurements at this level of accuracy are still very time consuming due to the statistical uncertainties and the limited beam current available. The quoted high precision measurements have been performed without using high intensity sources. Sources delivering ^{12}C currents in the 100 μA range could reduce the measuring time and the statistical uncertainty significantly, but possible loading and space-charge effects as well as other systematic errors have not yet been fully explored for this type of source.

For radiocarbon dating, standard material is available from NIST (formerly NBS), and well-established procedures are used for reporting results and errors (Stuiver and Polach, 1977). For other isotopes, one has not yet reached this stage. Many laboratories are using their own AMS standards obtained by dilution of NIST standards at much higher concentration, or produced by nuclear reactions and calibrated by decay counting or based on production cross sections. When converting activity data to isotope concentration, one has to use the half-life, which is not precisely known for many isotopes. Therefore, one has to be aware, when comparing absolute values measured at various laboratories, that uncertainties in the standard are not usually included in the quoted errors. It would be desirable if publications containing AMS data would give the method of normalization.

C. Comparison with Counting

The long-lived radioisotopes now accessible with AMS were previously detected only by decay counting, which has in general much lower efficiencies but is, on the other hand, technically much simpler and easier to perform. We will briefly discuss the fundamental difference between the two techniques and their advantages and disadvantages.

Efficiency

In the counting technique, the decay rate dN/dt is proportional to the number of radioactive atoms in the sample N_{sample} and inversely proportional to the half-life, $t_{1/2}$.

$$dN_{sample}/dt = -\lambda N_{sample} \text{ where } \lambda = \ln(2)/t_{1/2}$$

The number of detected atoms is n_{det}, assuming every decay is detected;

$$n_{det} = \epsilon_{count} N_{sample}$$

and

$$\epsilon_{count} = \ln(2)t/t_{1/2} \qquad \text{for } t \ll t_{1/2}$$

where t is the measuring time. The measuring time divided by the mean life, λ^{-1}, can be viewed as a maximum detection efficiency, ϵ_{count}, for a given counting time. Nuclides with longer half-lives are therefore more difficult to measure. In addition one has to consider that detection efficiency is not always 100% (not all decays are detected). The "maximum detection efficiency" for ^{14}C with a half-life of 5730 years, when counting 1 week is about 2.5×10^{-6}.

In AMS one can define the counting efficiency as the fraction of ions detected in the final detector from a sample put into the ion source.

$$N_{det} = \epsilon_{AMS} N_{sample}$$

For carbon ϵ_{AMS} can be more than 1%. That is almost 4 orders of magnitude higher than for counting, allowing a reduction from gram-sized samples to milligram-sized samples. For isotopes with longer half-lives (^{10}Be, ^{26}Al, ^{36}Cl, and ^{41}Ca etc), the efficiency improvement can be even higher, opening the possibility of detecting these isotopes in natural concentrations in many reservoirs which were not accessible before.

Background

Both decay counting and AMS are limited by background. In decay counting, background radiation (cosmic rays and environmental radioactivity) leads to undesired pulses in the detector. With good shielding the environmental background can be reduced. By using counters surrounding the real detector and operating in anti-coincidence, highly energetic cosmic events can be recognized and suppressed. Additional background suppression can be obtained by using underground laboratories. The background in such situations can be measured very accurately, because it can be assumed to be nearly constant over time. Therefore a reliable subtraction can be made, and it is possible to measure activities which are significantly below the background count rate.

In AMS, isotopic ratios are measured. The background measured with blank samples is also expressed as an isotopic ratio. But the background sources are more complex and more difficult to hold constant than with decay counting. For the small samples used in AMS, contamination can be important. Background in the final detector can also originate from neighboring masses or isobars. All these backgrounds can be strongly tied to the sample preparation, which is not well enough under control to allow measurement of ratios below the background level.

Errors

In both techniques statistical errors have to be accounted for a similar way. In general the data are normalized to standards prepared and measured in the same way as the samples, so that most systematic errors are cancelled. The primary error source is the variation of systematic errors. The reliability of the data therefore depends strongly on how well the systematic errors can be kept constant. An AMS facility is more complex than decay counting systems, and therefore more difficult to control.

Equipment shared for decay counting is much smaller, simpler, and cheaper to buy and to operate. AMS facilities are large, with higher installation, operating, and maintenance costs.

III. SOURCES OF RADIONUCLIDES IN NATURE

Radioisotopes found on the Earth can be classified into four categories, depending on their origin: cosmogenic, primordial, radiogenic, and anthropogenic.

A. Cosmogenic Radionuclides

Introduction: Fluxes (GCR and SCR)

Cosmogenic isotopes are formed by nuclear reactions induced by cosmic rays. The processes involved have been discussed by Lal and Peters (1967). Much of the following discussion derives from this reference, which remains one of the best general introductions to the topic. The nucleonic component of the cosmic rays consists of energetic charged particles, primarily protons (~85%) and α-particles (~14%), but also including heavier nuclei (1%). Cosmic ray particles originate either in the sun, in which case they are called solar cosmic rays (SCRs), or outside the solar system; for example in supernova explosions in which case they are called galactic cosmic rays (GCRs). SCRs, which are produced in sporadic solar flare events, have characteristic energies of 10–100 MeV and mean fluxes of about 70 particles/cm^2sec (Nishiizumi et al., 1987). The spectrum of SCR particles can be represented fairly well by a power law in the kinetic energy per nucleon, $E^{-\gamma}$, where γ varies between 2 and 4 for proton energies between 20 and 80 MeV (Reedy et al., 1983). Galactic cosmic rays have characteristic energies of 1 to 10 GeV, and mean fluxes of about 3 particles/cm^2sec. At $E > 5$–10 GeV/nucleon, the spectrum of GCR particles can also be approximated roughly by a power law in energy, $E^{-2.5}$ (Reedy et al., 1983). As can be seen in Figure 18, SCRs form a significant part of the cosmic ray flux in the vicinity of the Earth at energies below 300 MeV (Reedy et al., 1983).

Depth Dependence of the Cosmic Ray Flux

Secondary Production. When charged cosmic ray particles penetrate matter (e.g., a meteorite, the surface of the moon, or the Earth's atmosphere), they lose energy by two competing processes: nuclear interaction and ionization (more

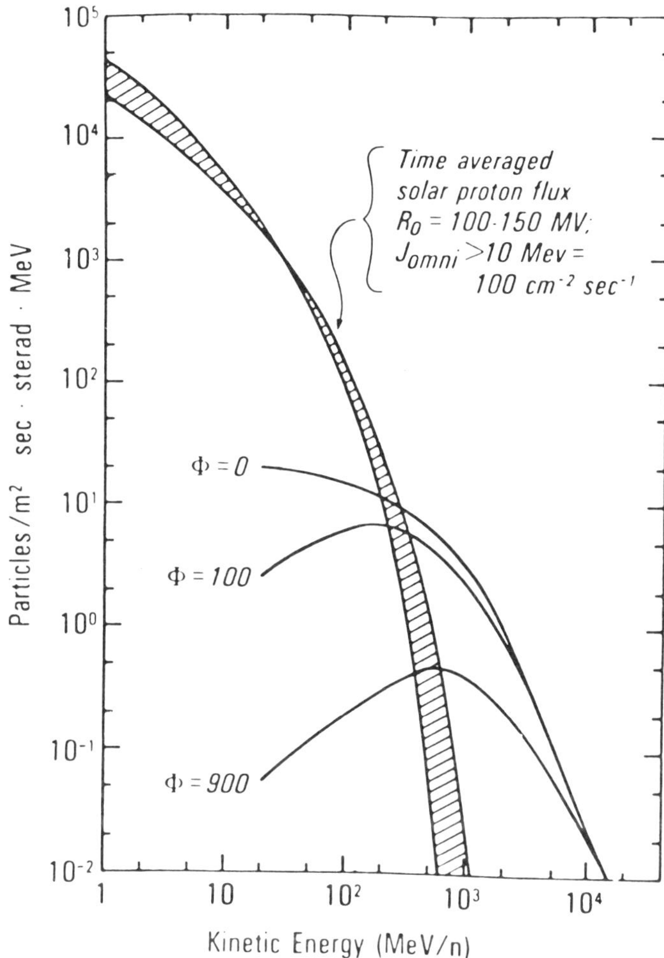

Figure 18. Galactic cosmic ray (GCR) flux and solar cosmic ray flux (SCR) outside the Earth's atmosphere. The long-term average fluxes of solar protons are much larger at low energies than the GCR-proton fluxes. Fluxes are given for different solar modulation levels ($\phi = 0$ refers to no modulation; $\phi = 900$ is typical of the GCR fluxes during solar maximum). R_o is a parameter which indicates the energy spectrum of the solar proton flux. Larger R_o, indicates more energetic protons. See reference for details. (Reedy et al., 1983)

precisely, electromagnetic interaction with the target electrons). The relative importance of these two processes is a function of the energy of the cosmic ray particle. At high energy, nuclear interaction is most important. The interaction of high energy GCR particles with light target nuclei can best be described in terms of collisions between the cosmic ray particle and free nucleons in the target nucleus. The incident particle retains a large fraction of its incident energy and secondary particles are produced, the majority of which have energies of less than 500 MeV. The cosmic ray flux at depth in a body (e.g., the Earth's atmosphere or the surface of the moon), then consists of a combination of primary cosmic ray particles and secondary particles which have been produced by nuclear reactions in the overlying matter.

Absorption Mean Free Path. GCRs, with a mean energy of 3 GeV, travel in the mean through about 100 g/cm^2 in undergoing a nuclear interaction. This distance is much shorter than the range calculated when only energy loss by ionization is considered, which is about 2000 g/cm^2. Thus a GCR proton has a high probability of interacting with an atomic nucleus of the body it is penetrating before its energy is reduced to the point that it no longer can cause a nuclear reaction. Production of nuclides occurs up to a depth of several meters in the moon and in meteorites and on the Earth through the atmosphere all the way down to sea level. The low energy SCRs lose energy more rapidly by ionization than by nuclear interaction and produce isotopes only in the outer layers of solid bodies or in the upper part of the atmosphere.

The secondary nucleons and mesons can initiate further nuclear reactions and a cascade of particles is produced. This means that as cosmic rays penetrate the Earth's atmosphere, nuclide production at first increases as the cosmic ray flux (primaries + secondaries) grows. Low energy secondary particles travel only a short distance before interacting with nuclei of the target or being stopped by ionization. Since most of the secondary protons lose their energy primarily by ionization, neutron-induced reactions dominate in importance in the production of cosmogenic isotopes.

The high energy flux continues to produce secondaries over a long range. After the cosmic rays penetrate about 100 g/cm^2, an equilibrium condition is reached between the high energy component, which is determined by particles propagating over long distances, and the low energy component which propagates over only short distances. The result is that the shape of the energy spectrum of particles remains almost constant, but the flux decreases with depth (Lal and Peters, 1967).

The scale length over which this absorption occurs is determined by the interaction mean free path, λ, of the incident particles. Because an incident high energy nucleon ($E > 1$ GeV) loses on the average only 45% of its energy in each collision (Pal and Peters, 1964), it may still be capable of initiating nuclear reactions. The actual absorption mean free path, Λ, is, therefore, longer than the interaction mean-free path. The absorption mean free path is just the distance required to diminish the flux of particles to 1/e of its incident value. The absorption mean free path is important because it allows one to estimate the flux, $\phi(x)$, of high energy particles at any depth in a body in a relatively simple fashion;

$$\phi(x) = \phi(d = 0)e^{\frac{-d}{\Lambda}}$$

where d is the depth in g/cm^2. Since, as we have discussed above, the energy dependence of the cosmic ray flux is independent of depth below about 100 g/ cm^2, it is possible to use this expression, which strictly applies only at high energy, to normalize the spectrum over the entire energy range. In the Earth's atmosphere, for example, $\lambda = 80$ g/cm^2 and $\Lambda \approx 125$ g/cm^2. Isotope production will decrease a factor of ten for every 280 g/cm^2 traversed. Sea level is 1030 g/cm^2 atmospheric depth and at 2600 m is about 750 g/cm^2. Therefore production rates are 10 times higher at 2600 m than at sea level. Figure 19 shows the cosmic ray flux as a function of depth in the atmosphere at different latitudes and as a function of depth below the Earth's surface. Isotope production rates may, in general, be taken as being proportional to this flux. This process by which isotope production changes with depth has been discussed by Lal and Peters (1967) and by Reedy and Arnold (1972).

Muons. The above discussion does not consider muons. Because cosmic ray muons have a longer absorption mean free path than cosmic ray neutrons and protons, their relative flux increases with depth and they become important near sea level (Lal, 1991). Muons produce isotopes not only by direct μ-capture reactions [e.g., $^{39}K(\mu^-,p2n)^{36}Cl$], but also through the secondary neutron flux produced by the capture of slow negative muons in (μ^-,xn) reactions.

Deeper than about 3000 g/cm^2 in the Earth's crust, cosmic ray particles are attenuated to such an extent that other particle fluxes, derived from the decay of primordial nuclei, become dominant. These fluxes are discussed below.

Cosmogenic Radionuclide Production Mechanisms

Cosmic rays can produce isotopes by two general mechanisms according to the energy of the cosmic ray particle: high energy spallation (nucleon-nucleon) reactions, and low energy (nucleon-nucleus) reactions. Spallation reactions occur at incident particle energies above about 50 MeV. At high energies the incident cosmic ray particle interacts directly with the individual nucleons of the target nucleus. So much energy is given to the target that several particles or groups of particles can be directly emitted from the nucleus. The residual nucleus can then further decay by fragmentation or particle emission. The products of spallation reactions can consist of protons, neutrons, and pions, but also of complex nuclei, which can be stable or can have half-lives ranging from minutes or less to millions of years. Spallation reactions always produce products which are lower in mass than the target. At lower incident energies, of the same order of magnitude as the binding energy of an individual nucleon, the incident cosmic ray particle interacts with the nucleus as a whole leaving the nucleus in an excited

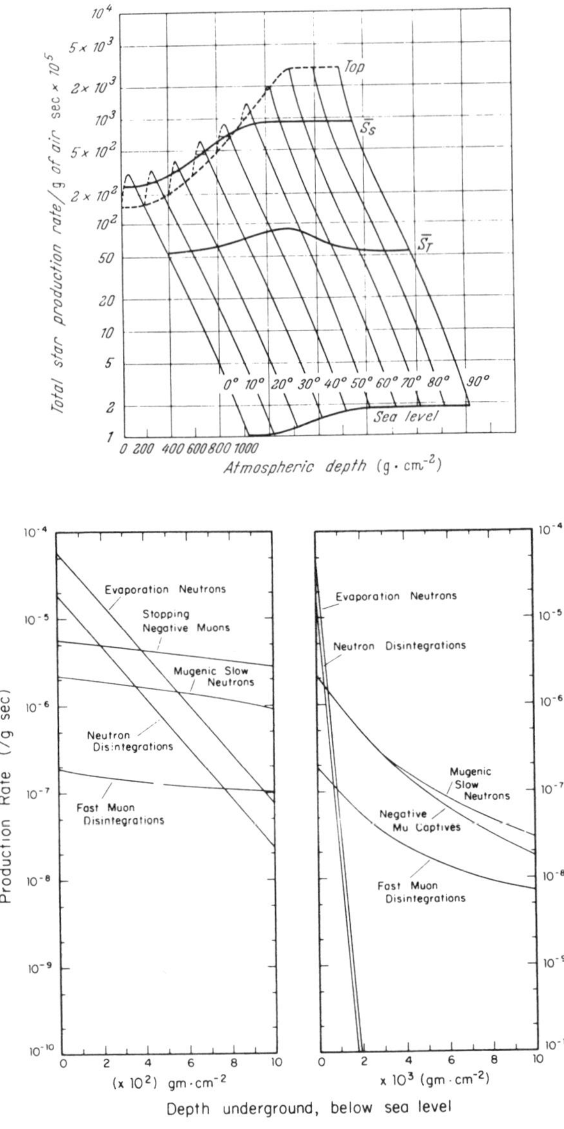

Figure 19. **a.** The rate of occurrence of nuclear disintegrations (equivalent to star production rate) in the atmosphere as a function of atmospheric depth and of latitude. Curves for different latitudes have been offset for clarity. Note that the production rate first increases with depth as secondaries are produced and then decreases with depth. **b.** The rates of occurrence of nuclear disintegrations due to nucleons and muons, and production rates of evaporation neutrons in fast-neutron-produced disintegrations in standard rock (density = 3.09 g/cm^{-3}, $Z^2/A = 6.3$) as a function of depth underground (below sea level) for geomagnetic latitudes ≥45°. The left side shows on an enlarged scale the production rates of 0-1 kg cm^{-2} depth (0–3.3 m rock equivalent). (Lal and Peters, 1967; Lal, 1988a)

state. The nucleus returns to its ground state by boiling off particles, mostly protons, neutrons, and α particles. The most important reaction of this type is the production of ^{14}C and ^{14}N in the atmosphere: $^{14}N(n,p)^{14}C$. Furthermore, pions produced in energetic spallation reactions decay to muons which can slow down and be captured by stable nuclei in the atmosphere or in the crust leading to additional transformations.

The products of the nuclear interactions which occur when cosmic rays pass through matter are determined primarily by the mass of the target nucleus and the energy of the incident cosmic ray particle. In geophysics we are primarily concerned with isotopes produced in the Earth's atmosphere, in rocks which make up the Earth's crust, and in extraterrestrial bodies such as meteorites and the moon. In the Earth's atmosphere the main target elements are N (78%), O (21%), and a small amount of Ar (0.9%). In terrestrial and extraterrestrial rocks, O, Fe, Si, and Mg make up 90% or more of the total mass, with lesser amounts of Al, Ca, Na, and K. Therefore nuclides of atomic weight below Fe are produced with the highest abundance and are the most important for geophysical studies.

Temporal Variations of the Cosmic Ray Flux

Cosmic Ray Flux Variations. The flux of SCRs and GCRs may vary with time. There is some evidence that the average SCR flux may have been a factor of four greater over the past 10^4 years than during the previous 10^6 years (Reedy et al., 1990) (cf Section VI-E). Recent work with ^{41}Ca has confirmed that there is no evidence for changes in the mean SCR flux between 10^5 years and 5×10^6 years (Klein et al., 1990; Reedy et al., 1990). Comparison of exposure ages of meteorites derived from ^{40}K–^{41}K with those derived from nuclides with shorter half-lives, ^{36}Cl, ^{39}Ar, ^{26}Al, ^{10}Be, and ^{41}Ca, indicate that the GCR flux may have increased by 35% 0.5 My or more ago (Fink et al., 1990b).

Geomagnetic Field. The cosmic ray flux in the Earth also depends on the extent of magnetic shielding. The magnetic field of the Earth deflects the charged particles. This deflection is a function of the momentum of the particle. High momentum particles are less deflected than those of low momentum. The geomagnetic field can be well-approximated by a dipole currently tilted at 11.5° to the axis of the Earth's rotation (Creer, 1988). The result is that the deflection of cosmic ray particles is smaller at the magnetic poles and a maximum at the equator, varying as $\cos^4\lambda$, where λ is the geomagnetic latitude. For detailed calculations it must be remembered that the geomagnetic dipole does not exactly align with the axis of rotation, so that the geomagnetic latitude is not exactly equal to the geographic latitude. The orientation of the geomagnetic dipole moves with time. In 1970, the north geomagnetic pole lay at about 78°N, 69°W (Pomerantz, 1971).

It is known that the geomagnetic field has varied over time. Although the course of that variation is a topic of active investigation, it seems fairly certain that the dipole component, which explains about 90% of the Earth's field, was higher than the 10,000-year average by a factor of about 1.3 at 2000 BP. The variation in the global mean dipole moment over the last 12,000 years as determined from archaeomagnetic studies is given in Figure 20. Today the dipole moment is slightly lower than the long-term average. Lal (Lal, 1988b) and Blinov (Blinov, 1988) have discussed the dependence of cosmic ray flux in the atmosphere on the geomagnetic field strength. If the geomagnetic dipole moment were to decrease to zero, global mean production rates would approximately double over their current values (O'Brien, 1979). On the other hand, if the geomagnetic dipole moment were to double, production rates would decrease to about 2/3 of current levels. The relationship between production rate, Q, and geomagnetic dipole moment, **M**, is approximately

$$\frac{Q_1}{Q_0} = \sqrt{\frac{\mathbf{M_0}}{\mathbf{M_1}}},$$

where the subscripts refer to two different times. This relation applies to changes in the dipole moment of not much more than a factor of four. More detailed calculations have been given by Lal (1988b). The change in nuclide production rates could vary somewhat from these values because the production cross sections, $\sigma(E)$, have different energy dependences for different isotopes. Figure 21 shows the nuclear disintegration rate, which is proportional to the nuclide

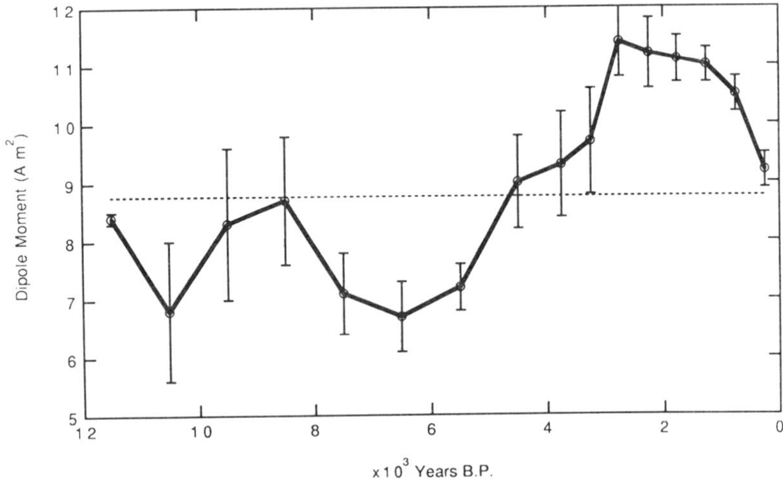

Figure 20. Global mean dipole moments with 95% confidence limits for 500 year intervals (to 200 BC) or 1000 year intervals (prior to 200 BC). (From archaeomagnetic data in McElhinny and Senanyake, 1982)

production rate, at different geomagnetic latitudes for different values of the geomagnetic dipole moment. (See also Section VI-B).

Solar Modulation. Production rates are also influenced by changes in the magnetic activity of the sun. The low energy galactic cosmic rays scatter on magnetic irregularities carried by the solar wind plasma. The energy spectrum of GCRs near the Earth is therefore significantly modified by these solar fields. Since these fields vary with time, the GCR flux at the Earth also varies with time. One obvious indicator of solar magnetic activity is the sunspot number, which is known to vary with an 11-year cycle, the Schwabe cycle. It must be kept in mind

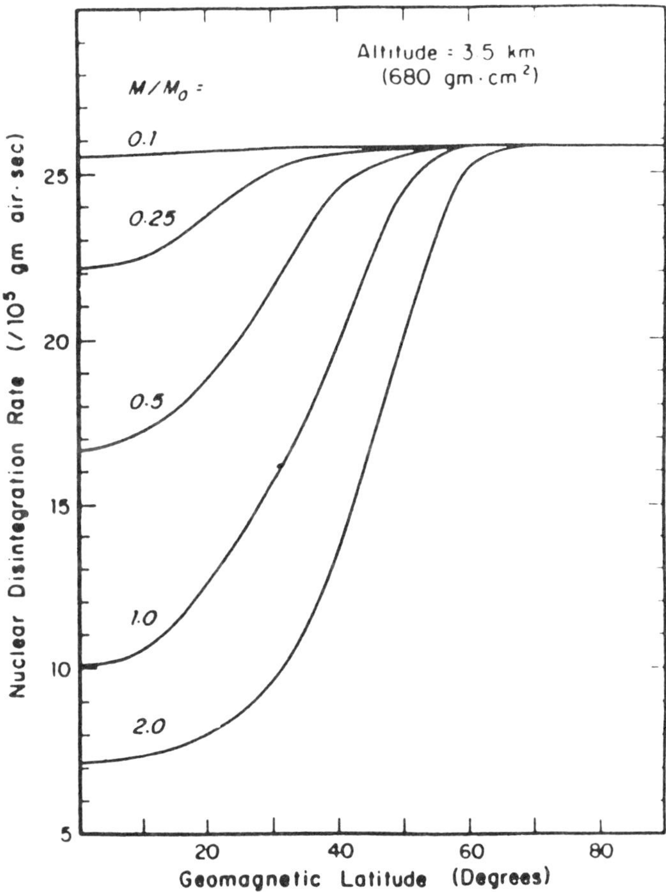

Figure 21. Total nuclear disintegration rate in the atmosphere at 3.5 km (680 g/cm^2) as a function of geomagnetic latitude for different intensities of the Earth's dipole field, M. The present field is designated by M_o. (Lal, 1991)

that both sunspot numbers and solar modulation of the galactic cosmic ray flux are secondary effects of an underlying mechanism. It is not necessarily the case that sunspot numbers can be used to predict precisely variations in solar modulation of the GCR flux. During periods of high sunspot activity, the strengthened solar magnetic field leads to a reduction of the GCR flux at the Earth, while during periods of low sunspot number the weaker solar field allows more GCR particles to reach the Earth. Calculations reveal that production variations of $\pm\sim 30\%$ are to be expected between typical solar minima and maxima (Lal and Peters, 1967; O'Brien, 1979; Castagnoli and Lal, 1980; Lal, 1987; Blinov, 1988; Lal, 1988b). There are recurring periods, lasting on the order of 100 years, when sunspot activity is almost completely absent. The best studied of these is the Maunder Minimum, which lasted from about 1645 to 1715, and which was a period of higher cosmic ray flux at the Earth, Figure 22 (Eddy, 1976). The influence of transport and deposition can strongly influence the production pattern, so that measurements in a given archive may show larger or smaller variations than the global average. (See also Section VI-A.)

B. Primordial and Radiogenic Radionuclides

Primordial isotopes are isotopes which were formed during the process of element formation in the early universe or in later episodes of stellar burning. They are found on Earth because their half-lives, on the order of 10^9 years, are roughly the same as the age of the Earth. The primordial isotopes include ^{40}K, ^{87}Rb, ^{238}U, ^{235}U, and ^{232}Th among others. These isotopes are found in the crust and mantle of the Earth. Their decay is capable of generating radioactive isotopes via a number of mechanisms, the most important of which are spontaneous fission, thermal-neutron induced fission of ^{235}U, and neutron- and α-induced reactions with light nuclei. Spontaneous decay of U and Th series nuclides is the source of the α particles and, via (α,n) reactions on light nuclei, can also be the source of neutrons which contribute to radionuclide production. These processes have been discussed by Andrews et al. (1989).

The magnitude of the various fluxes and the resultant isotope production rates depends strongly on the composition of the rock matrix. A specific example is given in Figure 23, which shows the ^{36}Cl concentration at equilibrium in a high-Ca granite as a function of depth. Shallower than about 3000 g/cm^2, cosmic ray neutrons and muons predominate as is clear from the depth dependence of the concentrations. Below 3000 g/cm^2, neutrons and α reactions caused by fluxes of particles derived from primordial radionuclide decay dominate. Isotope production due to primordial isotopes is, in general, several orders of magnitude smaller than surface production from cosmic rays. However, in some applications these primordial sources must be considered.

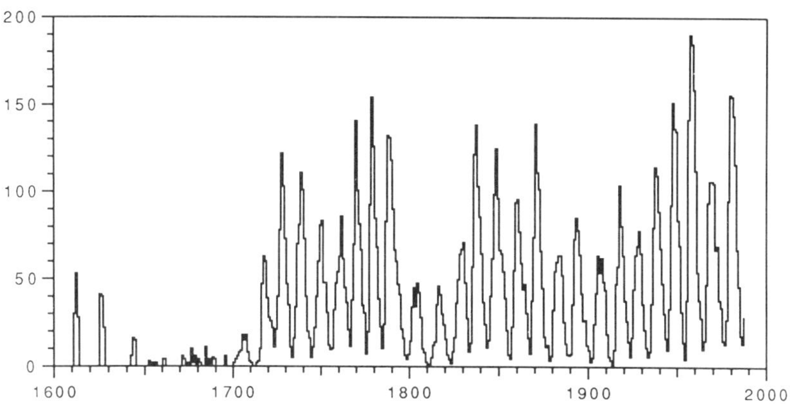

Figure 22. Estimated mean annual sunspot numbers from 1610–1988. Data between 1610–1715 from (Eddy, 1976), between 1715–1987 from National Geophysical Data Center (Anonymous, 1989).

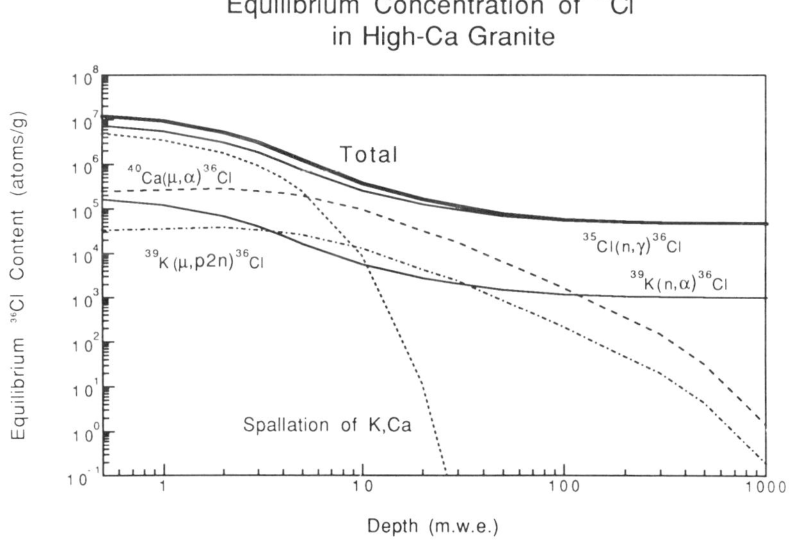

Figure 23. Equilibrium concentrations of ^{36}Cl in high-Ca granite as a function of depth. Calculated for depths relative to a rock surface exposed at sea level at a geomagnetic latitude of >60°. Assuming 200 ppm Cl, 3.4 × 10^6 atoms ^{36}Cl/g corresponds to a ^{36}Cl/Cl ratio of 1 × 10^6. (Data from Fabryka-Martin, 1988)

C. Anthropogenic Radionuclides

Anthropogenic isotopes are formed as the result of man's activities. This source includes nuclear weapons testing, use of radioisotopes as tracers in industry and research, reactor emissions, leakage of nuclear waste storage facilities, and other human activities (e.g., release of U-Th decay series nuclides from coal burning). Anthropogenic sources can sometimes be many orders of magnitude above natural levels and are of interest either because it is necessary to understand the potential detrimental effect of the radiation from these isotopes or, more often, because they can be used as inadvertent tracers to study geophysical problems. For example, the bomb pulse of ^{36}Cl, which led to deposition rates of 3 orders of magnitude above natural levels (Figure 24), has become a very useful hydrological tracer (cf Section VI-E). Other anthropogenic radionuclides which have found use in geophysical studies include ^{3}H, ^{14}C, ^{90}Sr, ^{137}Cs, and ^{129}I.

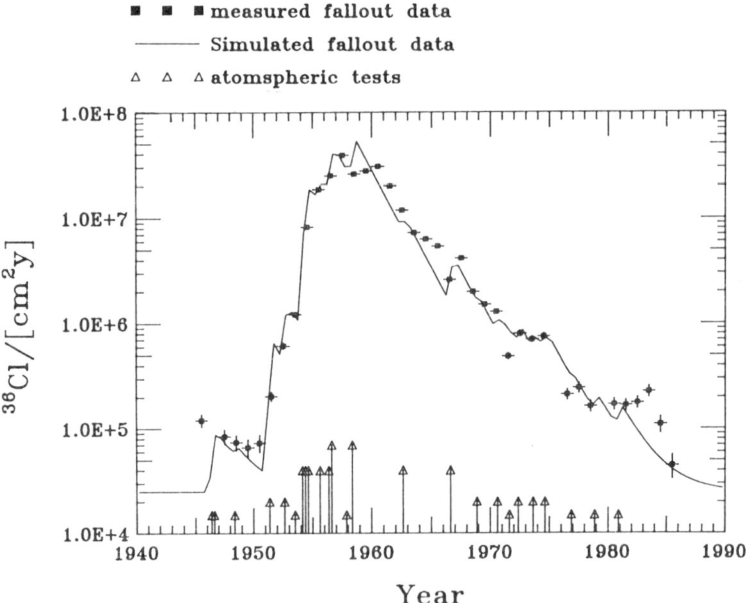

Figure 24. Comparison of ^{36}Cl fluxes (data points) with model fluxes calculated using a 4-box reservoir model for Dye 3, Greenland. Arrows indicate nuclear tests which released ^{36}Cl to the atmosphere. The bomb-test peak is approximately 3 orders of magnitude above natural levels. (Synal et al., 1990)

IV. DISTRIBUTION OF RADIONUCLIDES IN NATURE

The distribution of radioisotopes in nature is controlled not only by their production mechanisms but also by any subsequent migration of the nuclides from their sites of formation. Because the pathways depend strongly on the site of origin—in meteorites, in the atmosphere, in rocks—it is useful to consider separately nuclides of extraterrestrial, atmospheric, and crustal origin.

A. Extraterrestrial Origin

Radionuclides produced by activation of extraterrestrial materials (primarily the moon or meteorites) must first reach the Earth before they can be brought into our laboratories to be used as geophysical tracers. Aside from infrequent sampling expeditions such as the *Apollo* lunar landings, we must rely on natural events to bring extraterrestrial material to the Earth. These events include entry of interplanetary dust into the Earth's atmosphere ($\sim 10^4$ tons/year (Maurette et al., 1986)) and meteorite impacts. Three classes of extraterrestrial material have been identified on Earth: (1) meteorites in the size ranging $>$ 10 g; (2) stratospheric micrometeorites well below 1 μg; and (3) spherules and fragments in the μg range (Brownlee et al., 1977; Nishiizumi et al., 1991a). The interplanetary dust has been best studied in archives where its dilution with terrestrial material is a minimum: in stratospheric air filters (Brownlee et al., 1977), on glacial surfaces (Maurette et al., 1986), and in deep sea sediments (Millard and Findelman, 1970). Once at the Earth's surface the components of meteorites and interplanetary dust particles behave in a similar fashion to those of crustal rocks. Therefore the migration of extraterrestrial isotopes can be considered with the discussion of crustal rocks given below.

B. Atmospheric Origin

The distribution of radionuclides produced directly in the atmosphere is influenced by both atmospheric mixing and deposition processes and by the chemical nature of the nuclide (Figure 25). We can think of the atmosphere as consisting of two boxes: the stratosphere (where about 2/3 of the production occurs), and the troposphere (where about 1/3 of the production occurs). The lower boundary of the stratosphere lies at about 17 km at the equator and falls to 9 km at high latitudes. Because the shielding of the geomagnetic field is a minimum at the geomagnetic poles and a maximum at the equator, production rates in the stratosphere are highest at the poles. In the troposphere, production is almost uniform as a function of latitude because the particles which have sufficient energy to penetrate to tropospheric depths will not be influenced by the

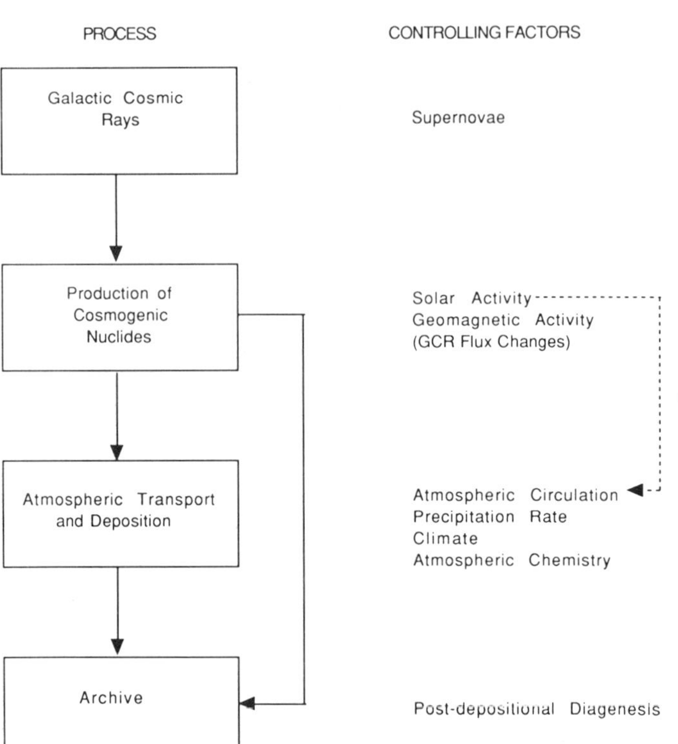

Figure 25. Processes controlling the terrestrial distribution of cosmogenic radionuclides.

geomagnetic field. The distribution of production rates in the atmosphere is illustrated in Figure 19.

Removal of aerosol-bound species from the stratosphere is primarily by advection and is very nonuniform in both time and space. Estimated residence times, which are primarily determined by the time scale for exchange of air between the stratosphere and troposphere, vary between 1 and 3 years (Reiter, 1975). The advective transport is primarily through Hadley cell circulation (ascending air in the warm tropics; descending air in the cooler temperate region) and tends to occur primarily along descending limbs of the Hadley cells in midlatitudes. The maximum transfer occurs in the winter-spring half-year. The validity of this mechanism of stratosphere-troposphere exchange is supported by the record of ^{90}Sr atmospheric concentrations (Sarmiento and Gwinn, 1986). These authors show that annual average ^{90}Sr concentrations peak at 40° latitude. Stratosphere-troposphere exchange may not be hemispherically symmetric. Measurements of ^{7}Be concentrations (Husain et al., 1979) suggest that mid-latitude injection is more active in the Northern Hemisphere than in the Southern.

Removal from the troposphere, the region in which weather occurs, is primarily by precipitation scavenging and dry deposition. Residence times are much smaller than in the stratosphere, being about 2 weeks to 1 month.

Atmospheric concentrations do not necessarily reflect production rate distributions because transport moves nuclides from their sites of production in the atmosphere. In the same way deposition processes can result in fallout concentrations which do not accurately reflect atmospheric concentrations. The concentration of isotopes in precipitation depends not only on the precipitation rate, but also on the seasonal distribution of precipitation and the chemistry of the nuclide. Those nuclides of atmospheric origin which hydrolyze readily; for example, Be and Al are removed from the atmosphere quickly and tend to remain where they are deposited, be it in sediments, on the surfaces of soil particles, or in glacial ice. On the other hand, isotopes of soluble elements, such as ^{36}Cl, tend to end up in the hydrosphere after being deposited on the Earth's surface. Nuclides which have volatile character have long residence times in the atmosphere and reach equilibrium among their various reservoirs. For example, ^{14}C becomes converted to $^{14}CO_2$ and then distributes itself among several reservoirs including the atmosphere, the oceans, and the biosphere.

Volatile radionuclides can remain in the atmosphere long enough to become homogeneously distributed and to reach isotopic equilibrium with their stable isotopes. One of the powers of the use of ^{14}C is that, within limits set by changes in production rates, the $^{14}C/C$ ratio is globally and temporally constant. Isotopes of elements such as chlorine or beryllium do not remain in the atmosphere long enough to become isotopically homogeneous. This makes interpretation of their concentrations much more difficult because there is no simple way to understand how transport and deposition affect their concentrations in various reservoirs.

C. In Situ Origin

Nuclides produced directly in rocks or brought to the Earth's surface with meteoritic material may either remain in place or migrate under the influence of weathering. Hydrolyzable nuclides such as ^{10}Be and ^{26}Al have a strong tendency to remain where they are produced unless groundwaters have unusually low pH, or unless physical weathering moves the particles to which they are attached. Nuclides such as ^{36}Cl and ^{129}I, on the other hand, readily move into groundwater and special precautions must be taken, such as choosing mineral phases which are impervious to weathering, if in situ concentrations are to be studied.

V. CONCEPTS AND PROBLEMS UNDERLYING THE APPLICATION OF LONG-LIVED RADIOISOTOPES TO GEOPHYSICS

Long-lived radioisotopes have been applied to tracing and dating geophysical systems in a wide range of settings. Several central concepts lie at the core of most of the current applications: production rate variations, transport and deposition processes, and dating. There are also several sources of uncertainty which constrain the use of radioisotope in Earth sciences. These concepts and uncertainties are discussed below.

A. Production Variations

We have already shown how variations in solar activity can cause variations in the production rate of isotopes in the Earth's atmosphere and on its surface. Measurement of concentration profiles in tree rings (^{14}C) (Stuiver et al., 1991) (cf Section VI-B) and ice cores (^{10}Be) (Beer et al., 1991b) (cf Section VI-A) has allowed us to extend our knowledge of solar variations to time periods before instrumental measurements became available. We now know, for example, that the Maunder minimum period (~1645–1715) was both a time during which almost no sunspots were observed, and a time during which the cosmic ray flux in the Earth's atmosphere was enhanced by about 20% (Stuiver and Quay, 1980). It was also a period of cold climate in Europe, which suggests that small changes in solar activity may also have climatic consequences. The recently demonstrated connection between sunspots and the solar constant (the overall solar energy emission) supports this deduction (Willson and Hudson, 1988). The isotope record in tree rings and ice cores is currently being studied to see how strong this connection might be and what the past pattern of variability might have been. In addition to the 100-year Maunder minimum-type variations, there are also shorter variations in solar activity, including the well-known 11-year Schwabe sunspot cycle. The origin of this cycle, which can also be detected in ^{10}Be concentrations in ice cores, is not fully understood. Work is currently underway to understand whether the 11-year variability is a permanent property of the sun, and whether, for example, modulation of isotope production continued even during Maunder minimum periods when sunspots were almost totally absent.

A second source of isotope variation is changes in the geomagnetic field. Both slow secular variations and more rapid changes which occur at times of field reversal are important. The magnetic reversals which have punctuated Earth history for at least the last 200 million years are important stratigraphic markers for dating marine sediments. The most recent of these reversal boundaries, the Brunhes-Matuyama, which occurred about 730,000 yr BP, is young enough to be studied by long-lived cosmogenic isotopes including ^{10}Be, whose production rate may increase by as much as a factor of two during zero field periods.

Measurement of ^{10}Be in sediments can possibly help define the mechanism of reversal. A few studies have been published (Raisbeck et al., 1979; Henken-Mellies et al., 1990) but much more work is needed.

Secular variations, slow changes in the geomagnetic field can be observed in the isotope record in tree rings (Stuiver et al., 1991), ice cores (Beer et al., 1988), and corals (Bard et al., 1990). Interpretation of paleomagnetic data must be undertaken with caution because it is difficult to infer the strength of the global dipole from a single paleomagnetic measurement and because postdepositional changes in the strength of the trapped field can occur (Tarling, 1988). Because interpretation of direct paleomagnetic measurements is so difficult, the isotope record can potentially be of great use in understanding geomagnetic history.

B. Transport and Deposition

The distribution of radioisotopes depends qualitatively on the manner in which the atmosphere, ocean, and hydrosphere mix, and how these systems are coupled together, and quantitatively on how rapidly the mixing and interchange occur. $^{10}Be/^{7}Be$ ratios in the atmosphere can tell us how and where the stratosphere and troposphere mix. Concentrations and seasonal variations of ^{10}Be and ^{36}Cl in ice cores can potentially give similar information over periods perhaps extending well into the last glacial period. Ice core measurements can also give information about paleoprecipitation rates (Raisbeck et al., 1981). Comparison of ^{14}C concentrations in planktonic (surface-living) and benthic (bottom-living) foraminifera can tell us how the oceans mixed in the past (Broecker et al., 1988). Did changes in the mode of deep water formation, which currently occurs primarily in the North Atlantic, influence climate changes associated with the beginning and end of the ice ages? Accumulation rates of Mn nodules can be studied by ^{10}Be and may be related to changes in climate via the redox potential of the oceans (Segl et al., 1984). ^{10}Be can be used to trace soil movement and to study the history of high stands of lakes and rivers (Brown, 1987). Because of the increasing influence of man on atmospheric composition and energy balance, the understanding which can be gleaned through these studies is of ever increasing importance. Paleoenvironmental studies using ^{10}Be and ^{14}C and perhaps other isotopes can tell us how the terrestrial system responded to past changes and offer some guidance as to what can be expected in the future.

C. Dating

Conventional Dating

Dating is in one sense the most straightforward use of radioisotopes. The radioactive decay rate of a nuclide is not influenced by any external variables, such as climate, chemical environment or physical location, and thus can in principle be used to derive an absolute age. The systematics of radioisotope production and decay are illustrated for the general case in Figure 26. The most commonly used technique is based on the exponential decrease of the concentration, A,

$$A = A_o \exp(-\lambda t)$$

where λ is the decay constant for the radionuclide ($= \ln(2)/t_{1/2}$) and A_o is the concentration of the radionuclide at t = 0. This equation assumes no production in the sample. Then the age, t, can be calculated according to

$$t = \frac{1}{\lambda} \ln\left(\frac{A_o}{A}\right).$$

The crucial piece of information which is required for the use of radioisotopes as chronometers—information which is not always at hand—is the initial concentration of the nuclide, A_o. We can derive an age for a sample only if we know the ratio of the current nuclide concentration to the initial concentration. For those radionuclides which have long atmospheric residence times—^{14}C, ^{3}H, and noble

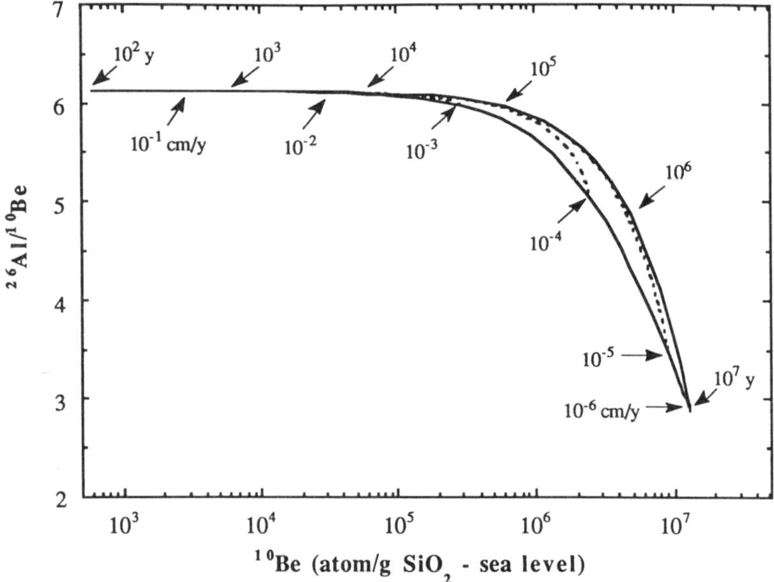

Figure 26. ^{10}Be concentration (normalized for atom/g quartz at sea level) plotted against $^{26}Al/^{10}Be$ ratio. The upper curve is calculated for zero erosion with different "exposure ages" as shown. The bottom curves is calculated for different "erosion rates." The erosion rates, 10^{-5}, 10^{-4}, and 10^{-3} cm/y with finite exposure age are shown with dashed line. Progress along the curve to the right represents increasing exposure time (upper curve) or decreasing erosion rate (lower curve) as indicated by the arrows (from Nishiizumi et al., 1991).

gases—the radionuclide mixes with the stable isotopes and a uniform equilibrium value is reached. If the reservoir is sufficiently large that short-term production variations are averaged, then the requirement of a known and approximately constant initial value can be attained. When the nuclide is transferred to a closed reservoir (sediment, biosphere, etc.) a well-defined initial concentration occurs which can be used to calculate an age. ^{14}C is the prime example of a nuclide which satisfies these requirements. ^{14}C remains in the atmosphere long enough to become homogeneously mixed with stable carbon. Therefore, the $^{14}C/C$ ratio can be used directly as a clock. The small isotopic fractionation which occurs as carbon moves between reservoirs can be corrected for by measuring changes in the stable isotope ratio, $^{13}C/^{12}C$. The production rate has changed in the past. This change has been determined by measurements in tree rings and in corals. These measurements can be used to produce a calibration curve of production rate versus time which is not available for any other isotope.

^{129}I is a special case. Although the atmospheric residence time is short, the production rate is so small that most of the ^{129}I in the atmosphere comes from remobilized marine iodine. The marine residence time of iodine and the half-life of ^{129}I are long enough that the $^{129}I/I$ in the ocean is expected to be approximately constant and, therefore, likewise the atmospheric ratio (Fabryka-Martin et al., 1985).

Double Dating

Other isotopes such as ^{10}Be and ^{36}Cl have such short atmospheric residence times that the ratio of radionuclide isotope to stable isotope is not constant in the atmosphere and changes as the isotope migrates from one reservoir to another. There are several techniques for deriving dates from such non-normalized species. These techniques include double-dating and profiling. Double-dating can be used when it may be assumed that one cosmogenic isotope can be used to normalize another because transport and production are similar enough that the ratio in the atmosphere remains constant. In this case the following relation applies:

$$\frac{A}{B} = \frac{A_o}{B_o} \exp[-(\lambda_A - \lambda_B)t].$$

For example, it was initially hoped that the production rate and subsequent transport of ^{10}Be and ^{36}Cl might be similar enough to allow the ratio to be used as a chronometer in ice core. It has since been shown that the ratio of ^{10}Be to ^{36}Cl can vary by a factor of five or more (Elmore et al., 1987; Suter et al., 1987). Clearly the atmospheric chemistry of chlorine is very different from that of Be. It may be that ^{26}Al can be used to normalize ^{10}Be to produce a chronometer ratio.

Profiling

Dating based on profiling requires that production, transport, and deposition remain constant. If one assumes, for example, constant sedimentation rate in a marine sediment and a constant flux of a radionuclide, then the concentration of the radionuclide versus depth can be used to date the layers of the sediment. The validity of the assumptions can be confirmed by checking to see whether the concentration decreases exponentially with depth. If the concentration profile is only partly exponential, it may still be possible to estimate deposition rates over the exponential interval, but deriving absolute ages will be more difficult. Similarly, ^{36}Cl concentrations can be determined along a hydrologic gradient in an acquifer. A smooth decrease in concentration, with no indication of input of chlorine (based on elemental chlorine concentrations), can be used to deduce an age (since contact with the atmosphere) for the water.

In those cases where the ideal of finding an absolute chronometer is frustrated, it may still be possible to use radioisotopes for generating relative dates. One example of the use of radionuclides as stratigraphic markers is the discovery in the Vostok core of two ^{10}Be spikes which occur at 35- and 60-kyr BP (Raisbeck and Yiou, 1988a). These spikes, so far of uncertain provenance, have been identified in several ice cores in the antarctic. Identification of these features in northern hemisphere ice would allow a good link between the northern and southern hemisphere time scales. More complex patterns, such as the climate-related ^{10}Be concentration in loess, which matches the oxygen isotope variation in marine sediments (Shen, 1986), and the ^{10}Be concentration in a sediment, which matches glacial-interglacial cycles (Eisenhauer et al., 1990), can also be used to determine relative dates.

In Situ

The situation is somewhat different in the case of nuclides produced in situ in terrestrial or extraterrestrial rocks. In these cases, the concentration increases with time after exposure in a way which depends on the half-life of the nuclide and on the rate of erosion of the surface until equilibrium is reached between production and decay. The higher the erosion rate the smaller this equilibrium value will be. In situ isotopes produced in rocks, either terrestrial or extraterrestrial, can be used to derive ages in two ways. In extraterrestrial samples, exposure times are often long relative to the nuclide half-life, and radioactive equilibrium is expected. That is to say the decay rate is expected to equal the production rate. Departures from equilibrium indicate either that the exposure age is short and equilibrium was not yet attained or that the sample was exposed at a time in the past and has begun to move away from production equilibrium due to radioactive decay. If external information about the history of the sample cannot provide the necessary information to distinguish the two cases, measuring a suite of isotopes with a succession of half-lives will often give the necessary information. For example, radionuclide concentrations in meteorites collected on the Earth can be used to calculate terrestrial ages: the length of time since the object fell to earth. Once shielded from the full interplanetary cosmic ray flux by the Earth's atmosphere, radionuclide concentrations begin to decrease. The movement from equilibrium can be used to deduce the length of time the object has sat on the Earth's surface.

The second manner in which in situ production can be used to derive ages primarily finds application in terrestrial samples. Erosion is so active on Earth that it must be included along with radioactive decay in calculating equilibrium isotopic concentrations for long-lived isotopes such as ^{10}Be and ^{26}Al. For exposures that are short relative to the isotope half-life, it is possible to use the very early part of the production curve (Figure 26), where the concentration grows linearly with time, to estimate an exposure age. For the surface of a rock with no inherited radionuclide from a previous exposure,

$$C(t) = \frac{Q}{\lambda + \mu\epsilon}\,[1\text{-}\exp\text{-}(\lambda + \mu\epsilon)t] \qquad \text{(Lal, 1990)}$$

where C(t) is the concentration of the radionuclide after time t, Q is the production rate of the radionuclide, $1/\mu$ is the absorption mean-free-path of cosmic rays in the rock being studied, and ϵ is the erosion rate in cm/yr. A given radionuclide, with decay constant λ, is suitable for determining an erosion rate ϵ (cm/yr) if λ is the same order as $\mu\epsilon$ ($\lambda \approx \mu\epsilon$), where $1/\mu$ is the absorption mean free path of cosmic rays in the rock, typically 50 cm (Lal, 1990). This equation, which is plotted in Figure 26 for various erosion rates, is the fundamental relation for using in situ radionuclides to study exposure and erosion rates of rocks. ^{10}Be, ^{26}Al, and ^{36}Cl have been measured in glacially polished rocks to date the time of glacial retreat, to date meteorite impacts, and, in young volcanic flows, to date the time of eruption. Since the rate of increase depends both on erosion and on the half-life of the nuclide, measurement of several nuclides can give information about erosion rates.

D. Caveats

It must be emphasized that, in most cases, measurement of a single nuclide concentration is not sufficient to derive geophysical information, because it is not possible to correct for the variability in production, transport, deposition, and postdeposition changes. Because it exists in well-mixed reservoirs, ^{14}C (and the noble gases, not discussed here) is the single exception to this rule. Double-dating using two nuclides can, in some cases, allow correction for disturbing parameters. However, this technique must be used with extreme care. Natural processes are often very adept at fractionating two nuclides which have very similar properties. Time profiles (depth profiles in sediments, areal profiles in aquifers) can also be used to get around the difficulty of unknown transport processes if the system can be shown to have been stable over the period under consideration.

VI. ISOTOPE ENCYCLOPEDIA

As already discussed, isotopic measurements in geophysics serve one of two main goals: *tracing* a geochemical pathway or *dating* a geophysical reservoir. An elegant example of tracing is the detection of ^{10}Be in young subduction zone volcanic rocks (Morris and Tera, 1989). This work demonstrated that there is a pathway connecting sea-floor sediments with volcanism occurring at active continental margins. A recent example of dating is the determination of the age of a glacial moraine in the Sierra Nevada mountains by measurement of ^{36}Cl produced directly in rocks making up the moraine (Phillips et al., 1990).

In order to provide a useful reference for researchers contemplating applying AMS to a geophysical problem, we have organized the discussion of the specific applications in terms of the isotopes which have been measured. This allows us to discuss the salient analytical parameters independently from the complexities of the geophysics involved in a particular investigation. The isotopes discussed are shown in Figure 27 with the range of isotope ratios found in nature. For each

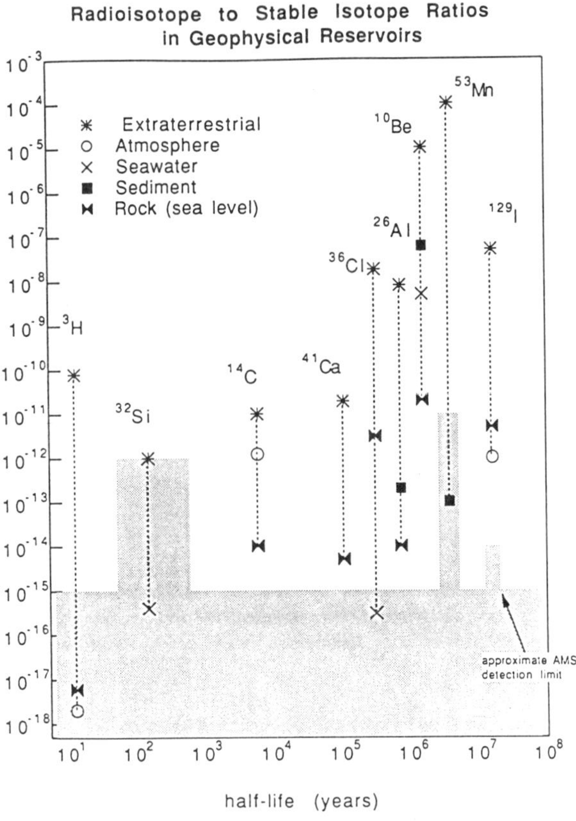

Figure 27. The ratio of a radioisotope to stable isotopes for various natural reservoirs is shown schematically. The ratios vary from 10^{-4} to 10^{-18}. The horizontal axis indicates the half-life of the radionuclide. Values shown are typical values. The range in actual samples may extend over several orders of magnitude. The AMS detection limit for many nuclides, currently about 10^{-15}, is indicated by the horizontal solid line. Extraterrestrial samples have the highest ratios because they are not shielded by the Earth's atmosphere and geomagnetic field. (Adapted from a figure by K. Nishiizumi)

isotope, we begin by discussing isotope production rates in the different geophysical reservoirs, and then discuss the analytical constraints associated with the measurement of each isotope. We conclude each section with mention of a few of the geophysical applications. The treatment is not meant to be, and could not hope to be, a complete catalog of geophysical applications, but it is intended to give a flavor of the kinds of systems which have been studied and suggest what parameters must be considered in new applications.

Several recent review articles and textbooks cover the use of radionuclides in the Earth sciences. These include (Brown, 1984; Faure, 1986; Elmore and Phillips, 1987; Bowen, 1988).

A. ^{10}Be ($t_{1/2}$ = 1.5×10^6 years[1]) and ^{7}Be ($t_{1/2}$ = 53.3 days)

Production and Distribution

Atmosphere. In the atmosphere the major production of ^{10}Be and ^{7}Be is from the interaction of high energy cosmic rays with atmospheric nitrogen and oxygen. The majority of the production is from neutrons with about 10% coming from protons (Lal and Peters, 1967). The production from capture of negative muons increases in importance with depth because the absorption length for muons is 270 g/cm^2, while that for neutrons is about 150 g/cm^2 (Nishiizumi et al., 1989d). The global production rate of ^{10}Be has been estimated to be about 4.9×10^{-2} at/cm^{2}sec (Somayajulu et al., 1984) and 3.8×10^{-2} at/cm^{2}sec (Monaghan et al., 1986). Assuming a ^{10}Be/^{7}Be production ratio of 0.6 (Lal, 1988b), the production rate of ^{7}Be is about 6×10^{-2} at/cm^{2}sec. Based on these production rates, the global inventory of ^{10}Be is about 170 tons. For ^{7}Be it is about 6 g. Concentrations of ^{10}Be in air are about 10^7 atoms/SCM (standard cubic meter) in the stratosphere, and 10^4 atoms/SCM in the troposphere (Raisbeck and Yiou, 1981).

Hydrosphere. Rainwater concentrations of ^{10}Be are on the order of 10^4 atoms/g (Brown et al., 1989). ^{10}Be concentrations in polar ice are between 2×10^4 and 2×10^5 atoms/g (Yiou et al., 1985; Beer et al., 1987). ^{10}Be found in marine and fresh water systems is primarily derived as input from the atmosphere, or is transferred along the erosion products from the continents. Concentrations in seawater are about 2000 atoms/g and generally show depth profiles similar to nutrient species, that is, higher levels in deep water than in surface water. ^{10}Be is about a factor of 2 higher in deep water than in surface water (Kusakabe et al., 1987; Ku et al., 1990). ^{10}Be/^{9}Be ratios in seawater are on the order of 10^{-7} (Ku et al., 1990).

Lithosphere. In rocks, ^{10}Be is primarily produced by spallation of oxygen and silicon. Production rates, which depend on both altitude and latitude (cf Sections

II-A and III-A), are about 6 atoms/g-yr in quartz (SiO_2) at sea level (Nishiizumi et al., 1989d). ^{10}Be concentrations in surface soils are about 10^8 atoms/g. ^{10}Be concentrations in sediments are on the order of 10^8 atoms/g.

Anthropogenic. There is no report of anthropogenic ^{10}Be or ^{7}Be in the environment.

Meteorites. In chondrites, ^{10}Be is produced by spallation reactions induced by GCRs. There is essentially no production from the lower energy SCRs (Nishiizumi et al., 1987). Concentrations are on the order of 10^{10} atoms/g, (10 dpm/kg in the units commonly used by meteoriticists) (Nishiizumi, 1987). Higher production rates are found in carbon phases in iron meteorites via the (n,2pn) reaction. Saturation values have been estimated to be about 10^{11} atoms/g-C (between 100 and 200 dpm/kg-C) (Nagai et al., 1990).

Sample Preparation

In most cases natural beryllium concentrations are so low that beryllium must be added to the sample as a carrier in order to obtain the approximately 1 mg BeO which is needed to prepare an AMS target. Beryllium can be separated from complex mixtures by a combination of cation exchange chromatography (using 50W-X8 resin) and acetylacetone extraction (e.g., Anderson et al., 1990). In the final step, $Be(OH)_2$ is ignited to BeO. The AMS target is usually prepared from the final BeO by mixing with Cu powder and pressing the mixture into the target holder.

Sample Purification

Because of the isobar ^{10}B (abundance 19.9%), boron is the major contaminant of concern in ^{10}Be analyses. Experience has shown that the critical steps to obtain low boron targets are the final washing of the $Be(OH)_2$ before ignition and the ignition itself. The $Be(OH)_2$ precipitate should be washed several times with pure water and should be ignited in a furnace which is constructed of an insulating material demonstrated not to evolve boron on heating. Several investigators use a quartz tube as a liner to isolate the BeO from the walls of the furnace. Others find that better results are obtained by using a gas burner for the final ignition of $Be(OH)_2$ to BeO.

Sensitivity and Special Techniques

Ordinary laboratory beryllium salts can contain ^{10}Be at the $^{10}Be/^{9}Be$ level of about 10^{-14}. Use of specially selected beryl as a source of Be carrier has been shown to reduce this blank to about 2×10^{-15} (Middleton et al., 1984). The detection limit for ^{10}Be is about 10^{-15} for samples with low boron content (cf. Section II-B).

Because of the significance of the $^{10}Be/^{7}Be$ ratio to many geophysical studies, it would be convenient to determine this ratio directly by an AMS measurement on a single sample. In spite of the short half-life of ^{7}Be, AMS can compete with decay counting using Ge(Li) well detectors in some cases. The detection limit for 2-day counting is about 10^5 atoms [J. Beer, personal communication], which is about an order of magnitude worse than the AMS detection limit of 10^4 atoms (Raisbeck and Yiou, 1988b). However, in many ways the decay counting method is simpler than AMS, and therefore the use of AMS for ^{7}Be must be evaluated on a case-by-case basis.

Illustrative Applications

General Information. Beryllium is both a very rare element and one which tends not to be very mobile in the environment; for example, the ^{10}Be residence time on soils under normal conditions is 10^5 yr (Pavich et al., 1986). These two properties stamp the geophysical uses of ^{10}Be. The rarity of stable beryllium means that ^{10}Be can be detected in many reservoirs where other nuclides are diluted by their stable isotope to such an extent that measurement is impossible. The lithophile character of beryllium means that ^{10}Be has a short atmospheric residence time (1–2 yr) because it becomes permanently bound to aerosols, a relatively short marine residence time (10^2–10^3 yr) (Anderson et al., 1990), and that it tends to remain fixed once it is deposited in rocks, sediments, soils, and ice cores. One disadvantage of this behavior of ^{10}Be is that the radioactive and stable isotopes do not become well enough mixed to reach a characteristic ratio. Thus, one cannot easily use stable Be to correct for the effects of dilution or removal as one can for ^{14}C. The best way to calculate concentrations depends on the application. In some cases, such as in open ocean sediments, ^{10}Be and ^{9}Be come close to equilibrium (Bourlès et al., 1989; Henken-Mellies et al., 1990). In other archives (ice, sediments, Mn-crusts, and nodules), the ^{10}Be concentration relative to the total sample weight (atoms/g) can be used. In other cases—for example, when studying production rate variations—the annual flux (atoms/cm^{2}y) may be a more useful parameter.

Recent reviews or research papers which provide a summary of current research are: Geoscience in general (Raisbeck and Yiou, 1984; Brown, 1987; McHargue and Damon, 1991; Bourlès, 1992); polar ice (Beer et al., 1990c); sediments (Anderson et al., 1990); soils (Brown, 1987); and meteorites (Nishiizumi, 1987).

Polar Ice. ^{10}Be concentrations in ice cores have proven to be extremely useful for deciphering the history of the cosmic ray flux in the atmosphere and give important information about the history of solar activity and of other causes of production rate variations such as changes in the geomagnetic field (Raisbeck et al., 1981; Beer et al., 1988; Beer et al., 1990). It has been shown in a series of papers by Beer et al. and by Raisbeck et al. that ^{10}Be concentrations in Greenland and antarctic ice cores are anticorrelated with sunspot numbers (Figure 28). This anticorrelation arises because sunspots are associated with increased solar magnetic activity which tends to shield the Earth from the cosmic rays which produce ^{10}Be in the atmosphere. The ice cores can be sampled in annual layers, determined by a number of methods including $\delta^{18}O$, δD, and H_2O_2 variations. Such sampling allows a continuous, high-resolution record of ^{10}Be concentrations to be measured. The concentrations observed in Greenland clearly vary in anticorrelation with the 11-year Schwabe solar sunspot cycle (Beer et al., 1990). Especially interesting are the periods during which solar sunspot activity remained at a near-zero level for many decades, the so-called Maunder minimum periods. During these periods ^{10}Be concentrations were generally elevated (Beer et al., 1991b). Similar periods of elevated ^{10}Be have been observed throughout the Holocene ice core record. The ^{14}C concentration in tree rings likewise shows variations with a characteristic time of centuries, called Suess Wiggles. In order to distinguish between production variations and variations caused by changes in precipitation rate or atmospheric circulation patterns, it has proven extremely useful to compare information from ^{14}C and ^{10}Be. For example, the agreement between ^{10}Be variations in ice cores and ^{14}C variations (Suess Wiggles) in tree rings gives good evidence that production rate variations and not meteorology, climate, or transport are the root causes of the observations (Beer et al., 1988) (Figure 29). The similar pattern observed for ^{10}Be variations with characteristic times on the order of 100 years in ice cores from both Greenland and the antarctic

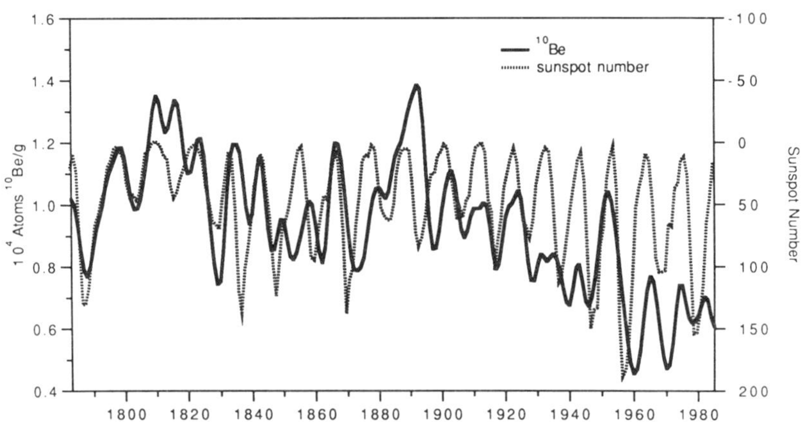

Figure 28. Comparison of the ^{10}Be concentrations at Dye 3, Greenland, with sunspot number form 1783 to 1985. The ^{10}Be concentrations have been smoothed. (Beer et al., 1990a)

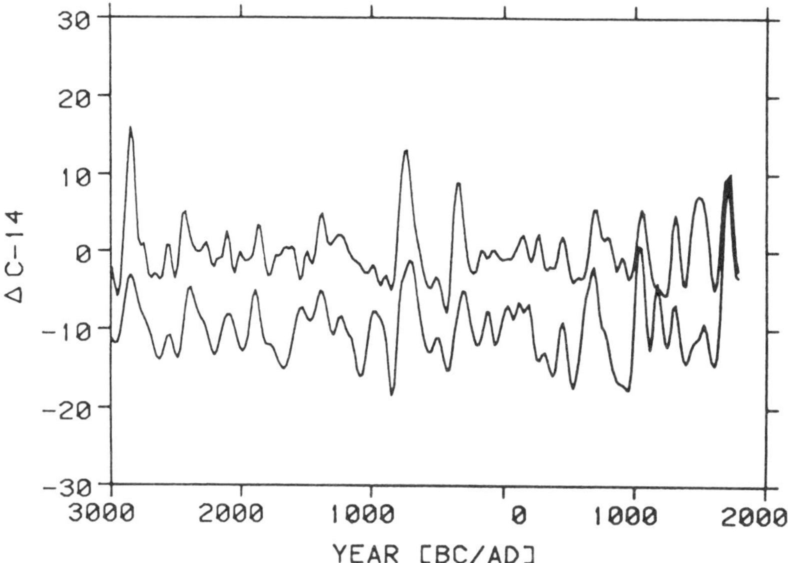

Figure 29. *Upper curve:* Detrended $\Delta^{14}C$ from tree rings (Stuiver et al., 1986a). *Lower curve:* Calculated $\Delta^{14}C$ using the ^{10}Be concentrations in the Camp Century ice core to model production rate changes and a carbon cycle model to take account of the transfer of carbon between atmosphere, ocean, and biosphere. The close similarity between the two curves indicates that the fluctuations of ^{14}C and ^{10}Be are both due to the same cause, i.e., changing production rate. (Siegenthaler and Beer, 1988)

likewise gives support to a production rate explanation for the observations (Beer et al., 1987).

Because ^{10}Be measurements in ice cores can be extended tens of thousands of years into the past, the ^{10}Be record can give information about solar activity for periods long before instrumental measurements came into existence. Understanding solar activity is important for a number of reasons, not the least of which is that recent satellite data has shown that solar magnetic activity, which controls the ^{10}Be production rate on Earth, and solar irradiance are positively correlated (Willson and Hudson, 1988). This correlation allows the use of ^{10}Be measurements to help understand the factors which control the terrestrial climate.

^{10}Be measurements in ice cores can also be used to study the history of the geomagnetic field by looking for long-term variations over centuries and thousands of years (Beer et al., 1988).

Marine Archives. The low mobility of beryllium, once it has become attached to mineral surfaces, has led to the use of ^{10}Be to study marine sedimentation rates, Mn-nodule growth rates, and subduction pathways in arc volcanism. In order to use ^{10}Be for these purposes, its geochemical behavior in seawater must first be considered. The residence time of Be in the deep sea is long enough for lateral mixing to make ^{10}Be in deep waters nearly constant within an ocean basin (Kusakabe et al., 1987), although differences between ocean basins have been observed (Ku et al., 1990). In the oceans, ^{10}Be concentrations increase with depth, being about a factor of 2 higher in deep water than in the mixed layer (Kusakabe et al., 1982; Ku et al., 1990) (Figure 30). This increase with depth is similar to that observed for nutrient elements which are important in the metabolism of marine microorganisms, and indicates that particle uptake of ^{10}Be is an important step in its ultimate removal from seawater. ^{9}Be also has a nutrient-like

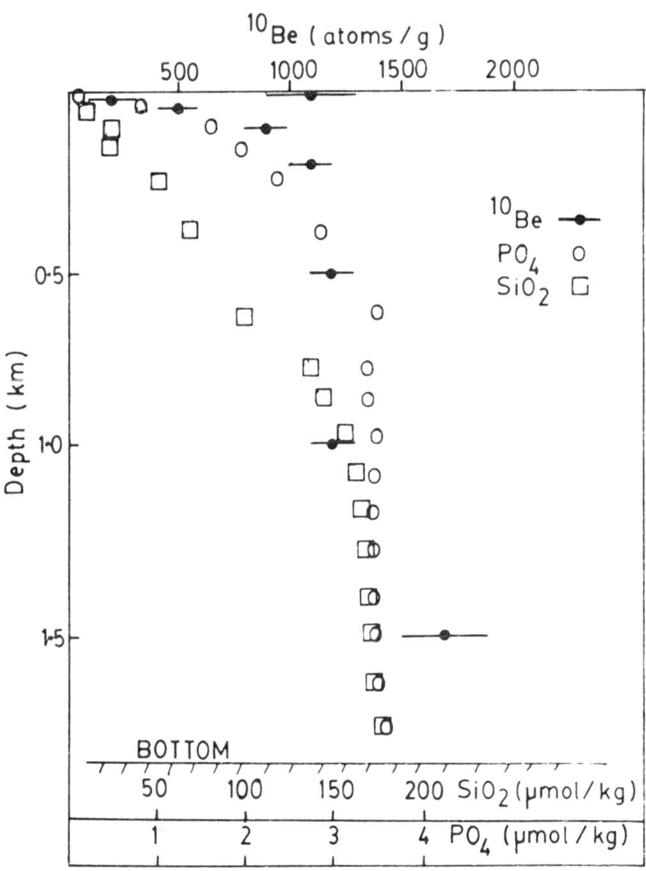

Figure 30. Water-column ^{10}Be in San Nicolas Basin. Comparison with the phosphate and silicate distributions (data provided by D. E. Hammond) suggests a biogenic origin for the ^{10}Be-carrying particles. (Kusakabe et al., 1987)

profile but the $^{10}Be/^{9}Be$ ratio is not constant. Anderson et al. (1990) have recently shown that enhanced scavenging at the boundary of the oceans leads to preferential deposition of ^{10}Be in ocean-margin sediments at rates at least an order of magnitude greater than in the open ocean (Figure 31). As much as 50% of the ^{10}Be supplied to the oceans is deposited at the margins (Anderson et al., 1990). The rate of ^{10}Be scavenging varies from region to region (Mangini et al., 1984; Eisenhauer et al., 1987; Anderson et al., 1990), making it necessary to consider regional residence times which vary from ~100 years at ocean margins to >1000 years in deep open-ocean regions. Beryllium seems to be primarily scavenged by aluminosilicates (Sharma et al., 1987). It is also to be expected that ^{10}Be deposition rates will change with time as continental erosion, marine chemistry, and oceanic circulation change.

Both sediments and authigenic minerals incorporate ^{10}Be in their structures as they are formed. The use of ^{10}Be for dating sediments requires that the concen-

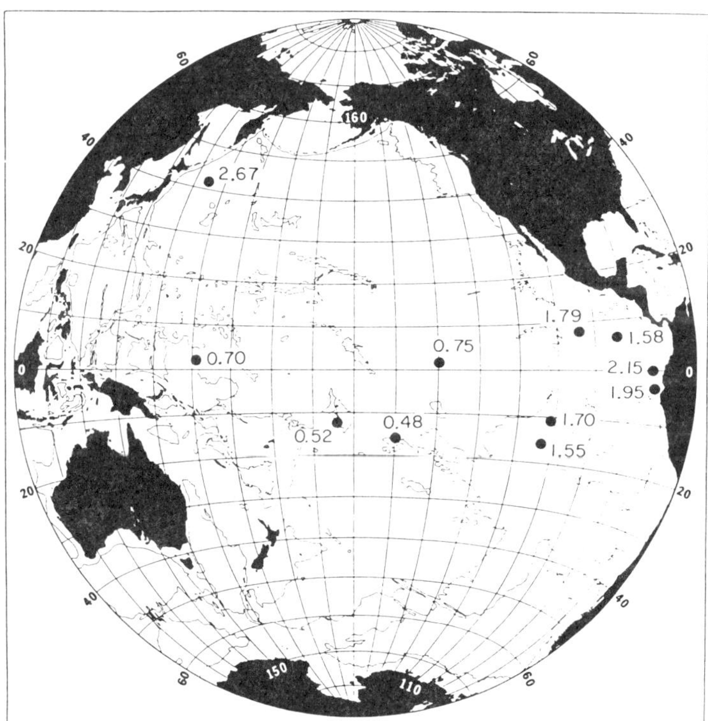

Figure 31. Deposition index values (relative $^{10}Be/^{230}Th$ ratios compared to the ocean-wide values) for ^{10}Be. The deposition index is proportional to the scavenging ratio of ^{10}Be at each site, with a value of 1.0 approximately representing the ocean-wide average ^{10}Be scavenging rate. (Anderson et al., 1990)

tration profile decreases exponentially with depth. Only under these conditions can one assume that production, transport, and sedimentation rate (or growth rate for Mn crusts) have remained constant with time and, therefore, that the ^{10}Be activity can be used to derive an age of the deposit, or, equivalently, to determine deposition rates. Changes in the exponential decrease of the ^{10}Be concentration versus depth curve indicate changes in accumulation rate. This technique has been utilized for dating marine sediments (Tanaka et al., 1977; Raisbeck and Yiou, 1984; Bourlès et al., 1989), manganese nodules (Segl et al., 1984; Mangini et al., 1986) (Figure 32), and phillipsite, an authigenic aluminosilicate, (Sharma et al., 1989). ^{10}Be concentrations in slowly accumulating marine deposits are high enough that in the past much work was done using decay counting (Tanaka et al., 1977). The emphasis of current work lies in understanding the factors which control the initial ^{10}Be concentration and $^{10}Be/^{9}Be$ ratio in marine deposits. The hope is that a range of conditions might be identified under which the $^{10}Be/^{9}Be$ initial ratio will be constant with time. Then the $^{10}Be/^{9}Be$ would immediately give an age. Conversely, if the sediment could be dated by other means (e.g., $^{230}Th/^{232}Th$ disequilibrium or magnetic reversal horizons), the $^{10}Be/^{9}Be$ concentrations could allow estimation of production rate changes. Observations to date indicate that the $^{10}Be/^{9}Be$ ratio in the surface of marine sediments is not constant over the world ocean, even for authigenic minerals. It is, rather, influenced by proximity to sources of continental input and varies between the Atlantic and Pacific basins. However, the relatively constant value observed in surface sediments from open ocean areas shows considerable promise for approximate dating of open ocean sediments (Bourlès et al., 1989). Because as much as 50% of the ^{10}Be supplied to the oceans is removed at the margins, a major change in boundary scavenging could lead to a significant redistribution of ^{10}Be burial within the ocean (Anderson et al., 1990). Much more needs to be understood about the factors which control ocean margin removal.

Because of their slow growth rates (mm/10^6 yr) ^{10}Be is very useful for determining growth rates of Mn nodules and crusts. There is some evidence that growth rates changed significantly at the times of quaternary and late tertiary paleoclimate events (Mangini et al., 1986). If this connection is confirmed by further work, Mn crust growth rates will be useful tools for understanding paleoclimate changes, and the relationship between climate and the chemistry of the oceans.

Sediments are subducted into the upper mantle, along active continental margins. The extent to which sediments are incorporated into the magmas produced along these subduction zones has long been a subject of inquiry. Because the origin of ^{10}Be is primarily in the atmosphere, its concentration is very low in mantle material, and ^{10}Be which becomes incorporated in surface sediments is a very useful tracer for the role of sediment in subduction zone magmatism (Monaghan et al., 1988; Ryan and Langmuir, 1988; Morris and

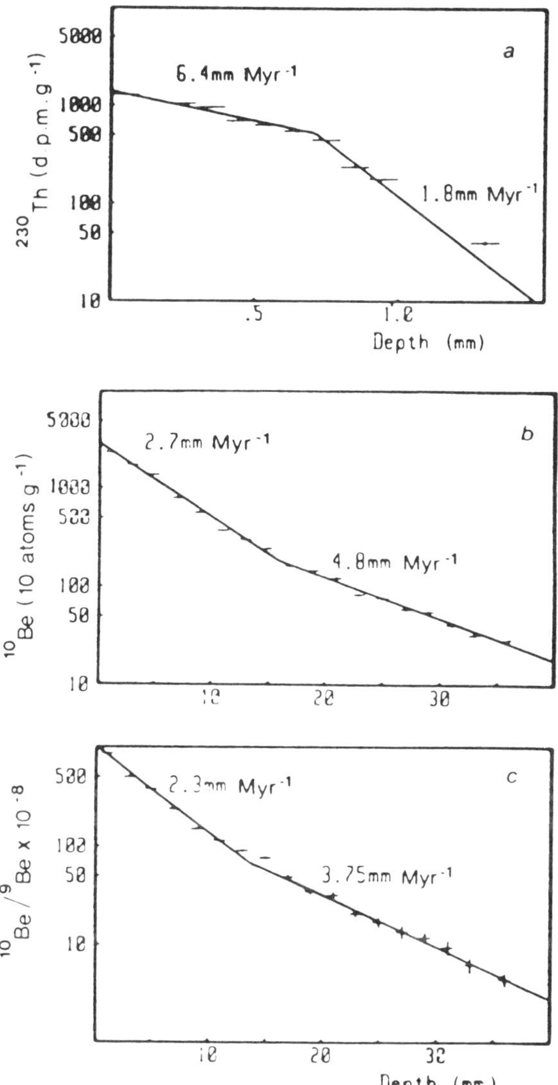

Figure 32. **a.** ^{230}Th profile for the uppermost 1.4 mm of the Mn crust. The exponential decrease yields growth rate of 6.4-mm Myr^{-1} for the uppermost 0.8 mm of the crust, and 1.8-mm Myr^{-1} between 0.8 and 1.4 mm. **b.** ^{10}Be profile for the top zone of the crust. For the section between 0 and 16 mm the growth rate amounts to 2.7-mm Myr^{-1}, and between 16 and 38 mm to 4.8-mm Myr^{-1}. **c.** ^{10}Be/^{9}Be ratio for the top zone. It yields growth rates of 2.3 mm Myr^{-1} for the upper-most 16 mm and 3.75-mm Myr^{-1} for the depth-range 16-38 mm. (Segl et al., 1984)

Tera, 1989). ^{10}Be in the sediments will be transported with the sediment and, if transit times are short enough, will appear in the lavas which form the subduction zone volcanoes. ^{10}Be has in fact been detected in fresh lavas from subduction zones, but in much smaller concentrations (or not at all) in lavas from ocean island and mid-ocean ridge basalts, in which magma genesis does not seem to incorporate sediments. ^{10}Be/^{9}Be ratios have been shown to be approximately constant in mineral separates from lavas from a number of volcanic eruptions in the Aleutians and in Central America (Monaghan et al., 1988; Morris and Tera, 1989) (Figure 33). The constancy of the ^{10}Be/^{9}Be ratio rules out incorporation of ^{10}Be after crystallization, and thus eliminates meteoric contamination as a source. Within volcanic arcs the range of values is generally large. Some of the differences between arcs can be attributed to differences in sediment subduction times, but the differences within arcs indicate real differences in the processes by which sediments are incorporated into magmas.

Soils. ^{10}Be incorporated in terrestrial soils as opposed to marine sediments is beginning to be used as a new tool to study Pleistocene surface processes (Brown, 1987). The affinity of beryllium for soil surfaces is sufficiently high that it is effectively immobilized on contact with the soil, thus ^{10}Be can be used as a tracer of sediment transport. This technique has been used to study erosion of soils in the eastern U.S. (Brown et al., 1988) and Taiwan (You et al., 1988). Under suitable conditions—no inherited ^{10}Be, well-defined starting time—the ^{10}Be inventory can be used to determine the age of a soil. This method has been used to estimate the age of river terraces along the Merced River in California (Pavich et al., 1986).

Rocks. The erosion rates of rocks (cf Section V-C), the timing of glacial stages, and the dating of young volcanic flows are all problems which have been difficult to explore because, before the high sensitivity of AMS made it possible to use cosmogenic isotopes, a suitable chronometer was lacking. The high sensitivity of AMS makes it possible to measure ^{10}Be produced directly in rocks by cosmic ray bombardment at the Earth's surface. As discussed in the section on production rates (cf. Section III-A), the production rate of ^{10}Be depends on the altitude and latitude of the sample. However, even at sea level, production rates are high enough that exposures of a few thousand years produce sufficient ^{10}Be to be measured. Once the in situ production rates are known (Nishiizumi et al., 1989c), the exposure history of a sample can be determined with useful accuracy by measurement of a few isotopes of different half-lives. Depending on the erosion rate, a measurement may suffice to constrain erosion rates, or, for short exposure times, to date the time of initial exposure. It is very useful to combine measurement of several isotopes to ensure the absence of meteoric contamination, and to be able to constrain erosion rates. This method has been applied to date meteorite impact craters (Nishiizumi et al., 1991a) and to determine erosion rates of antarctic rocks (Nishiizumi et al., 1991c).

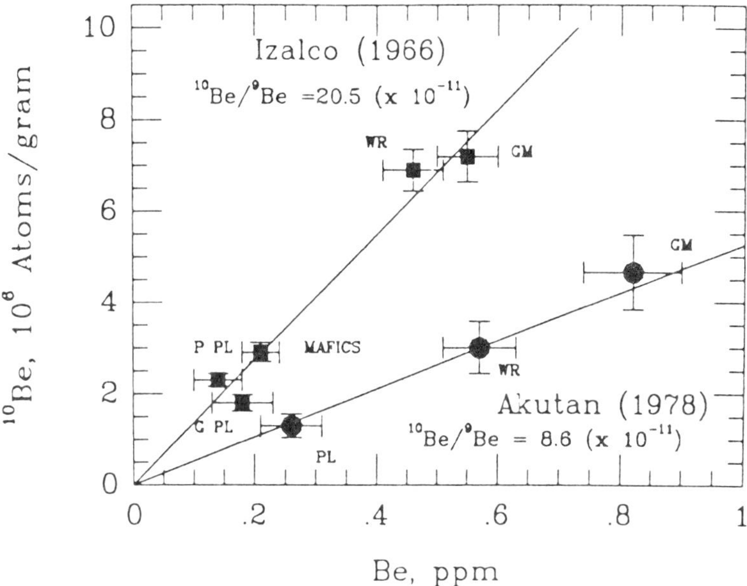

Figure 33. ^{10}Be vs. total Be (effectively ^{9}Be) for minerals separated from a 1966 eruption of Izalco Volcano in Central America and from a 1978 eruption of Akutan Volcano in the Aleutians. The straight lines indicate constant $^{10}Be/^{9}Be$ ratios in the two sets of whole rock (WR), groundmass (GM), Plagioclase (PL) separates. For Izalco, P PL indicates glomeroporphyritic phenocryst plagioclase 0.4 mm in size, while G PL indicates groundmass plagioclase (and broken P PL) $\leq -.20$ mm in size. "Mafics" are undifferentiated olivine and pyroxene combined to provide enough material to analyze. The constant $^{10}Be/^{9}Be$ ratio in each volcano indicates that meteoric contamination is not significant. (Morris and Tera, 1989)

Extraterrestrial Samples. In addition to its terrestrial geophysical uses, ^{10}Be was an important nuclide for meteorite studies even when decay counting was the only detection technique. ^{10}Be has been measured in stone and iron phases of meteorites (Nishiizumi, 1987), in lunar samples (Nishiizumi et al., 1987), and in interplanetary dust (Nishiizumi et al., 1991b). Because it is only produced by galactic cosmic rays in most cases (i.e., there is essentially no solar cosmic ray contribution), it can often be used as a simple indicator of exposure ages (Nishiizumi et al., 1987). In some cases exposure ages are short enough that the difference between measured ^{10}Be concentrations and saturation values can give important clues about the origin of the object. This has been proven to be true; for example, in studies of the set of meteorites from the MacAlpine Hills in Antarctica which are thought to have originated on the moon (Eugster et al., 1990).

Additional Applications. In addition, ^{10}Be been used for a number of other geophysical studies including:

1. Atmospheric circulation. Beryllium has been separated from air filters and precipitation and the $^{10}Be/^{7}Be$ ratio used to study atmospheric mixing, including stratosphere-troposphere exchange and seasonal variations (Raisbeck and Yiou, 1981; Monaghan et al., 1986; Brown et al., 1989; Beer et al., 1991a).
2. Hydrology. Because of its low solubility in nonacid waters, ^{10}Be has not been much investigated in groundwaters. A few measurements exist in hydrothermal waters (Valette-Silver et al., 1988).
3. Particle settling velocities in lakes (Wieland et al., 1990).
4. Study of the origin of diamonds and of the use of diamonds as probes of the upper mantle using ^{10}Be as an indicator of surface exposure. (Lal et al., 1987).
5. Loess. ^{10}Be has been measured in a Chinese loess deposit. Concentrations show a strong correlation with climate indicators (Shen, 1986).
6. Biogenic Reservoirs. ^{10}Be has been measured in several biogenic reservoirs including carbonate from mussels, clams, cockles, starfish and corals, plankton, and petroleum (Bourlès et al., 1984).

B. ^{14}C ($t_{1/2}$ = 5730 years[2])

Production and Distribution

Atmosphere. ^{14}C is produced primarily by the exothermic $^{14}N(n,p)^{14}C$ reaction on atmospheric nitrogen. ^{14}C, once produced, equilibrates with volatile carbon species so that to a good approximation the $^{14}C/C$ ratio in the atmosphere is homogeneous (Lerman et al., 1970). The natural value (derived from the measurement of 1860s' wood) is about 1.2×10^{-12} $^{14}C/C$ (13.6 dpm/g-C). The global production rate is about 2 atoms/cm^{2}min (Lal, 1988b). Based on this production rate, the global inventory is about 70 tons, of which about 90% resides in the ocean, about 8% resides in the biosphere and soils, and only about 2% remains in the atmosphere (Broecker and Peng, 1982). As is the case with all cosmogenic isotopes, the production rate of ^{14}C is not constant but can vary within roughly a factor of 2, depending on the state of the geomagnetic field and the intensity of solar modulation. Figure 34 shows the measured variation over the past 10,000 years. $\Delta^{14}C$ is the relative deviation from the NBS oxalic acid standard; decay corrected using the 5730-year physical half-life (Stuiver and Polach, 1977). The result is then further corrected for mass fractionation. The concentration of ^{14}C depends not only on the production rate but also on the rate of transfer between carbon reservoirs. Using measured ^{14}C concentrations in tree

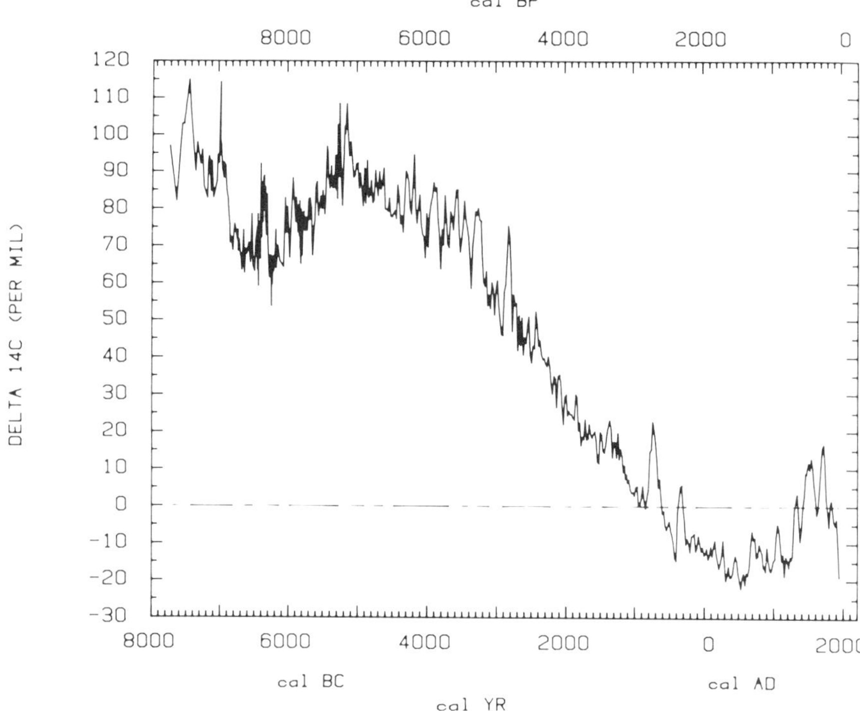

Figure 34. Atmospheric $\Delta^{14}C$ over the past 10,000 years. (Stuiver et al., 1986b)

rings of known age and a global carbon cycle model (Figure 35), the production rate can be estimated as a function of time in the past.

Hydrosphere. ^{14}C is transferred into marine and freshwater systems across the atmosphere water boundary. Concentrations in equilibrium with the atmosphere depend on the chemistry of the water body. Concentrations in the ocean also vary with depth and between oceans. The mean $^{14}C/C$ in the ocean is about 83% of that in the atmosphere, and the average total dissolved CO_2 in seawater is 2.2 mM/l (Broecker and Peng, 1982), making the ^{14}C content of seawater about 1.2×10^9 atoms/l. Concentrations in fresh water are comparable, but can depend heavily on the extent to which the chemistry of the water has been influenced by the addition of carbon from sources other than the atmosphere; for example, dead carbon from surrounding limestone deposits.

Lithosphere. The production rate of in situ ^{14}C was measured in a quartzite sample from the Allan Hills in the antarctic (Jull et al., 1989b). Based on this measured value Lal (Lal, 1991) estimated that ^{14}C production in quartz is about

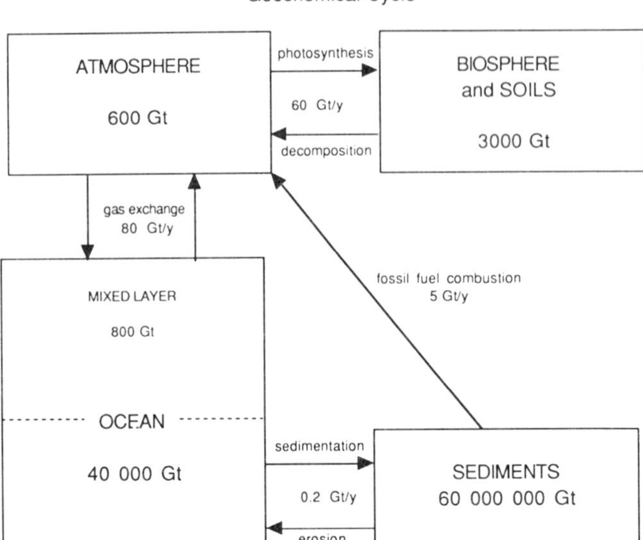

Figure 35. Carbon cycle model used for calculating ^{14}C concentrations. Reservoir sizes are in Gt C (gigatons of carbon, 1Gt = 10^9 t = 10^{15} g). Times are biospheric and atmospheric residence times. (Oeschger et al., 1987)

3.25 times ^{10}Be production. Thus, at sea level, the expected production rate of ^{14}C in quartz is about 20 atoms/g year.

Anthropogenic. An estimated 2.2 tons (9.6 MCi = 8.5×10^{17} Bq) of ^{14}C have been introduced into the atmosphere as the result of nuclear weapons tests (Anonymous, 1985). ^{14}C is produced in nuclear power reactors through absorption of neutrons by carbon, nitrogen, or oxygen which may be present as components of coolant, moderator, structural materials, fuel, or as impurities. By 1990, the total ^{14}C release from the nuclear power industry was estimated to be 0.05 tons (235 kCi = 8.7×10^{15} Bq) (Anonymous, 1985). The average release rate in the 1980s was about 40% of the natural production. The amount of ^{14}C incorporated into labeled compounds for use in medical or biological research is probably between 3.7×10^{12} and 1.8×10^{13} Bq annually (Anonymous, 1985). Locally, labeled compounds may be a significant source of ^{14}C either in field samples or as an inadvertent contaminant during sample preparation.

Meteorites. The highest production rates of ^{14}C occur in the silicate phase of chondrites because oxygen is the main target element. Saturation concentrations in H group chondrites are about 1.9×10^8 atoms/g meteorite (44 dpm/kg) or 2.4 × 10^8 atoms/g silicate (54 dpm/kg) (Jull et al., 1989a). In iron meteorites saturation concentrations are about 8.7 × 10^6 atoms/g Fe (2 dpm/kg) (Nishiizumi et al., 1989a).

Sample Preparation

Targets are made presently either in the form of filamentous graphite or as gaseous CO_2. In archaeological samples the starting materials may be almost any material containing carbon, such as bones, charcoal, textiles, blood, food remains on pottery, cast iron, seeds, and wood. In addition, geophysical samples include seawater, ice, meteorites, desert varnish, and sediments. The carbon is normally extracted as CO_2 by heating the sample in an oxidizing atmosphere. CO_2 can be extracted from water samples by flushing the acidified solution with a gas such as helium. Techniques are being developed for dating specific chemical compounds; for example, separated amino acids from bones (Currie et al., 1989), or identifiable microfossils. The CO_2 is either measured directly, or, more usually, converted to filamentous carbon by reduction with hydrogen in the presence of CO or Fe (Vogel et al., 1987), or by first converting to acetylene which is then dissociated in a high-voltage AC discharge (Kieser, 1989). For samples between 50 and 300 μg, iron carbide targets have been used (Verkouteren et al., 1987).

Sample Purification

There are no severe isobaric interferences since ^{14}N does not form stable negative ions. Therefore, the primary concern in sample preparation is to prepare a target which will deliver a good current in the ion source and which remains uncontaminated with modern carbon.

Sensitivity and Special Techniques

The sensitivity is determined by backgrounds, which are primarily limited by contamination during sample preparation. The best backgrounds (including sample preparation) currently attained are about 1.7 × 10^{-15}, which is equivalent to a ^{14}C age of 52 kyr (Kieser, 1989). High precision is required for many ^{14}C applications, especially oceanographic studies requiring water profiles. Techniques have been developed to give accuracies of 2 per mil (Kromer et al., 1987). In some cases, sample size rather than accuracy is of prime concern. This is the case when the ^{14}C content of CH_4 is used to determine the source of this greenhouse gas (Wahlen et al., 1989), or when biological remains from varved lakes are dated. ^{14}C has been determined in samples containing as little as 10 μg of carbon (Verkouteren et al., 1987).

Illustrative Applications

General Information. ^{14}C is a special nuclide in many respects. Its behavior has led to applications which differ strongly from those of the other long-lived nuclides we are discussing here. Its half-life and the end point energy of its beta spectrum make it possible to detect ^{14}C with good efficiency using conventional decay counting. Since Libby discovered its potential after 1947, ^{14}C has become the most studied cosmogenic nuclide.

The importance and wide range of application of ^{14}C is based on several facts. Once produced in the atmosphere, ^{14}C oxidizes rapidly to CO_2, which resides long enough in the atmosphere (about 7 yr) to mix well with CO_2 which is already present. Therefore, a constant $^{14}C/C$ is established throughout the atmosphere. Geographical variations are negligible. Due to the large atmospheric reservoir of CO_2, the effects of short-term variations in production are greatly reduced. The result is that $^{14}C/C$ varies only slowly with time. Carbon is of great importance in the biosphere, in which it is incorporated primarily by photosynthesis of CO_2 from the atmosphere or the oceans. This process defines a clear boundary after which no further carbon exchange takes place and only radioactive decay affects the ^{14}C level. All these facts are important for the use of ^{14}C in dating.

In the 50 years that ^{14}C has been studied, a very high level of sophistication has been reached, allowing measurement at high precision (0.2%) with conventional decay counting techniques (Polach, 1987). This level of precision is being approached by AMS (Kromer et al., 1987). High-precision ^{14}C measurements in tree rings have allowed calibration curves to be produced with 20-year resolution back to about 9000 BP (Figures 34, 36). At this precision proper normalization and data analysis become very important. In addition, mass fractionation which occurs on transfer of carbon from one reservoir to another must be considered. Fortunately carbon has two stable isotopes, ^{12}C and ^{13}C, which can be measured to monitor mass fractionation and to calculate a correction factor which can be applied to the $^{14}C/^{12}C$ ratio. It should be noted that "radiocarbon ages" are calculated using the Libby half-life of 5568 years. $\Delta^{14}C$ values, which are primarily used in geophysical studies, use the physical half-life of 5730 years. A detailed description of procedures for normalization, for applying mass fractionation corrections, and for reporting results has been given (Stuiver and Polach, 1977).

The major advantage of AMS over conventional decay counting for ^{14}C analysis is the enormous reduction in sample size. This reduction has opened many new fields of application. For materials containing little carbon the advantages are obvious. For ^{14}C dating of polar ice, conventional techniques would require tons of material. With AMS about 10 kilograms of ice suffices. For seawater, sample size has been reduced from 250 l to $<$500 ml, with clear advantages for determining depth profiles in the deep sea (Kromer et al., 1987).

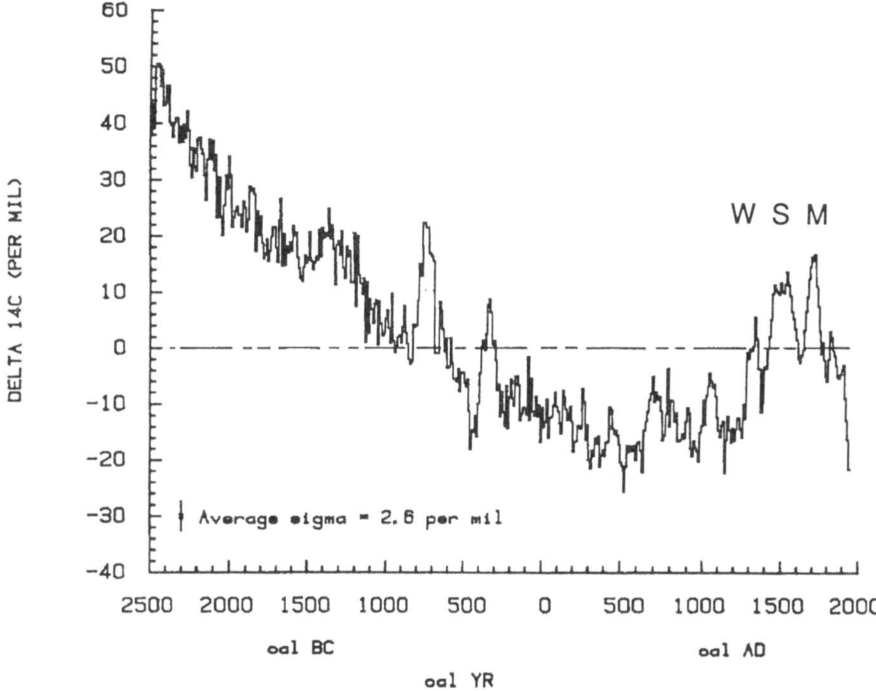

Figure 36. Atmospheric $\Delta^{14}C$ over the past 4500 years. M Maunder Minimum, S Spörer Minimum, W Wolff Minimum. (Stuiver et al., 1986b)

In other fields new techniques are being developed which allow extraction of specific chemical compounds to ensure that the object or time horizon under study is being dated and not contamination carbon which entered the sample at a later time. For example, amino acids can be extracted for bone dating (Verkouteren et al., 1987), or species-identifiable plant remains can be analyzed from lake sediments (Zbinden et al., 1989). The ability to date small samples brings with it a special set of problems. Contamination is more difficult to avoid. The problem of representativity must also be considered. Although one seed may be datable, one has to ensure, perhaps by dating several samples, that the seed is actually associated with the event being studied.

Recent reviews or research papers which provide a summary of current research are: Marine science (Broecker and Peng, 1982); archeology (Taylor, 1987); hydrology (Mook, 1980); ^{14}C calibration curve (Stuiver et al., 1990), and issues of the journal, *Radiocarbon*.

Atmosphere. Fundamental to the use of ^{14}C in all geophysical studies is an understanding of its concentration in the atmosphere and the way it has varied in the past. The best way to understand the variation in ^{14}C is through study of ^{14}C concentrations in tree rings. Because they are annual accumulations, tree rings can yield a precisely dated annual record of atmospheric ^{14}C concentrations. The tree ring record now extends to 9600-yr BP (Stuiver et al., 1990). The long-term trend visible in Figure 34 is probably the result of slow changes in the geomagnetic field. The variation with ~100 years characteristic time is thought to be the result of changes in solar magnetic activity which modulates the GCR flux. Figure 36 shows a blow-up of the recent part of the $\Delta^{14}C$ curve which makes the 100-year variations, the Suess Wiggles, more visible. The Maunder, Spörer, and Wolf minima are marked. These are periods during which the sun is thought to have had extremely low sunspot activity (Stuiver and Quay, 1980). The recent precipitous decrease in $\Delta^{14}C$ (the Suess effect) is the result of burning fossil fuel which contains no ^{14}C. Comparison of ^{14}C concentrations in corals raised off the island of Barbados with high-precision U-Th ages have recently provided very strong evidence that the ^{14}C concentration in the atmosphere at 20-kyr BP was 50% greater than today (Figure 37). This increase is most likely due to a lower geomagnetic field at that time (Bard et al., 1990).

Most of the ^{14}C produced in the atmosphere becomes incorporated into CO_2 shortly after production and then, by various processes of exchange and assimilation, ends up in organic material or in sedimentary carbonates. Some ^{14}C also ends up in atmospheric methane (CH_4). Depending on its source, CH_4 will contain different concentrations of ^{14}C. CH_4 is a strong infrared absorber and an important greenhouse gas; 15% of the greenhouse warming can be attributed to CH_4 (Rodhe, 1990). Its atmospheric concentration has more than doubled in the past 200 years and continues to increase at a rate of about 1% per year (Blake and Rowland, 1988). However, the source of this increase was difficult to determine. Recent measurements of ^{14}C in CH_4, when combined with ^{13}C results, have been interpreted to show that 21% of the annual input is derived from fossil fuel sources and the remainder from biogenic sources, including 25 to 26% from wetlands, peat bogs, and tundra, 19 to 22% from ruminants, 23 to 24% from rice production, and 9 to 10% from biomass burning (Wahlen et al., 1989).

Marine Archives. The largest single reservoir of ^{14}C is the sea. The distribution of ^{14}C in the oceans is a complicated function of biology, chemistry, exchange with the atmosphere, ocean circulation, anthropogenic activities, and nuclear decay. Since the circulation pattern and rate of circulation of the ocean are strongly coupled to climate, understanding of ocean circulation is of great practical importance. Understanding the effects of the increasing concentration of greenhouse gases (CO_2, CH_4, N_2O, O_3, and CFCs) requires that we understand the climate-ocean link. The present mode of circulation in the Atlantic Ocean is akin to a conveyor belt. Warm and salty upper waters flow northward to the vicinity of Iceland where they give up heat to the atmosphere. The consequent cooling densifies the water and causes it to sink to the abyss, creating new

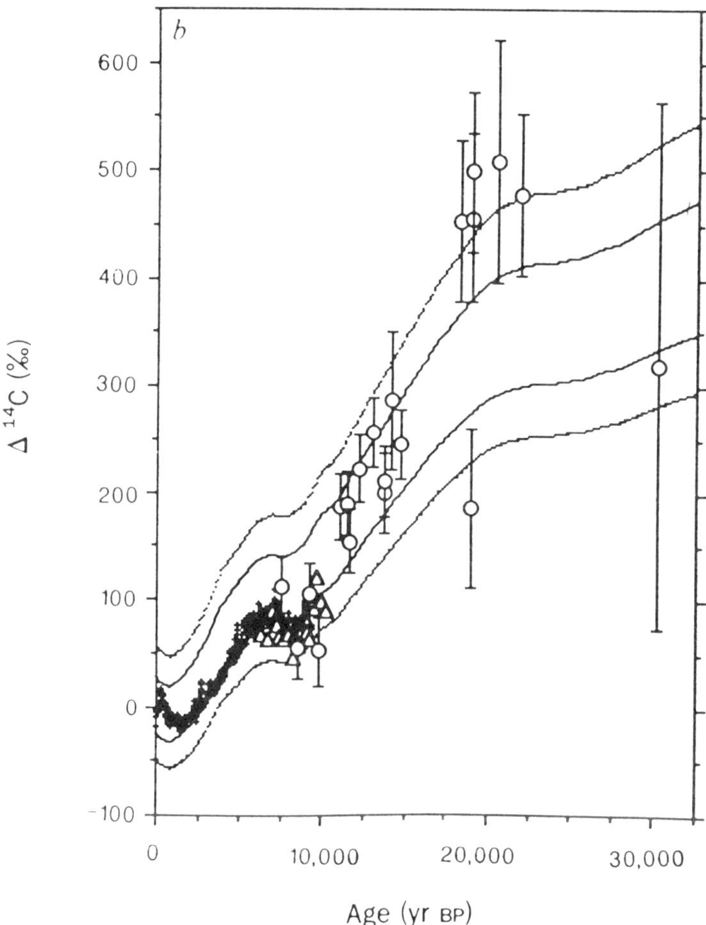

Figure 37. Atmospheric $\Delta^{14}C$ through time reconstructed by the U-Th/^{14}C age relationship (open symbols, the errors are quoted 2σ). The small crosses correspond to the tree-ring calibration and the open triangles to the Lake of the Cloud varved sediment calibration. The two sets of solid lines correspond to the envelopes of $\Delta^{14}C$ expected as a response to change of the Earth's magnetic field. (Bard et al., 1990)

deep water. ^{14}C profiles have proven to be a useful tool to understand the current pattern of ocean mixing by taking advantage of the fact that oceanic mixing times (on the order of 1000 yr) are comparable with the decay life time of ^{14}C (Broecker and Peng, 1982). Figure 38 shows the age difference between surface and deep water in the world ocean. The ^{14}C concentration measured in CO_2 extracted directly from seawater depends not only on ocean circulation, but also

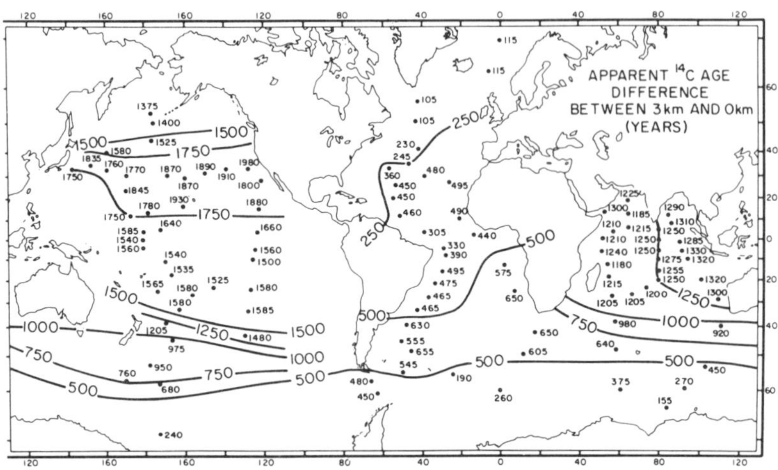

Figure 38. Map showing the ^{14}C age difference calculated from the Δ^{14}C values for pre-nuclear surface water and the Δ^{14}C values for water at 3-km depth as measuring during the GEOSECS program. See reference for details. (Andree et al., 1985)

on the transfer of ^{14}C by the remains of biologic organisms which can transport young carbon to the deep ocean more rapidly than might occur by water circulation alone. For these studies very high precision is needed. Pilot studies on small samples (<500 ml) measured by AMS have shown that nearly the same precision can be reached as with conventional counting on much larger samples (250 l) (Kromer et al., 1987). In connection with the World Ocean Circulation Experiment (WOCE) and the Global Ocean Flux Study (GOFS), two international geoscience programs, an AMS facility dedicated to ^{14}C measurements related to ocean studies has been constructed (Jones et al., 1990).

By examining ^{14}C contents in *foraminifera*, microscopic molluscs, different species of which live at different depths in the ocean, it has also been possible to learn about past circulation patterns. In a sediment sample, the ratio of ^{14}C/C in benthic (bottom dwelling) *foraminifera* relative to that for planktonic (free swimming) *foraminifera* provides a measure of the ^{14}C/C difference between surface and deep water at the time the sediment was deposited. This difference can be expressed as an age difference between surface and deep water. One conclusion of recent results using this technique is that Atlantic deep water was twice as old during glacial time as today. This implies slower circulation in the Atlantic during the glacial than today (Broecker et al., 1990). ^{14}C in *foraminifera* shells have also been used in other ways to date past climate events. For example Bard et al. (1987) have used ^{14}C dates on *foraminifera* at times of δ^{18}O change to determine the rate at which the polar front in the North Atlantic migrated during the last deglaciation.

Groundwater. ^{14}C is also found in fresh water and numerous investigations have been undertaken with the goal of using ^{14}C to date groundwater. The age of groundwater is the time since it was last in contact with the atmosphere. Although the ^{14}C method is the best developed technique for dating groundwater, many difficulties still exist with its application. Carbonate minerals are in a state of equilibrium with groundwater and even a slight change in the temperature of the acquifer can change the equilibrium and thus the ^{14}C content (Davis and Murphy, 1987). In addition, large amounts of organically derived carbon can enter groundwater from coal deposits, bringing significant amounts of dead carbon (Davis and Murphy, 1987). Subsurface production must also be considered in the case of very old water. Under certain conditions and within certain limits, absolute groundwater ages may be obtained by applying correction procedures using the chemical and stable carbon isotopic composition of the sample by techniques discussed by Mook (1980). Relative ages, which can be determined with more certainty, provide flow directions and velocities (Mook, 1980). This technique has been applied to the Carizzo aquifer in Texas (Pearson, Jr. et al., 1983) and to the central San Juan Basin in New Mexico (Phillips et al., 1989).

Archaeology. The importance of ^{14}C to archaeology is hard to overemphasize. Thousands of ^{14}C analyses are carried out each year, by both AMS and decay counting, in the service of a wide range of archaeological problems. Such studies as the dating of the migration of Homosapiens into the Western Hemisphere and the expansion of culture in Europe would be impossible without the time scales provided by ^{14}C. As discussed above, an important advantage offered by AMS is the ability to greatly reduce the sample size required for a date. This is especially important when working with valuable and unique archaeological specimens. Mook (1984) has summarized a number of the potential advantages to archaeology offered by small sample sizes. These include the ability to date: (1) a small collection of seeds found at a site; (2) plant remains in mortar; (3) traces of carbon found in iron objects; (4) paper, papyrus and textiles [e.g., the Shroud of Turin (Damon et al., 1989)], for which only a small mass of material is available; (5) plant remains in pottery; (6) specific biologic compounds in bone; (7) egg shells; and many other possibilities.

Additional Applications. ^{14}C measurements have, of course, contributed to many other fields of geophysical research, including:

1. Meteorites and in situ production. Recently Jull et al. (1989a) have begun to study terrestrial ages of antarctic meteorites using ^{14}C. This work has lead these authors to consider in situ production of ^{14}C on the Earth's surface, and to develop techniques for using in situ ^{14}C to study surface processes.

2. Ice core dating. The CO_2 content of ice is low, about 10–20 μg/kg of ice. Therefore on the order of 10 kg of ice is needed for a measurement. Because of the importance of ice cores as archives of climate history, much work has been expended on ^{14}C dating of ice (Andree et al., 1984; van de Wal et al., 1990; Wilson and Donahue, 1990).
3. Atmospheric chemistry. Recently ^{14}C has been used to help deduce the source of atmospheric carbonaceous particles (Currie et al., 1987). The low concentration of the particles in the atmosphere required the sensitivity of AMS.
4. Rock varnish dating. Rock varnish consists mostly of clay minerals, manganese, and iron oxides, but contains enough organic carbon for ^{14}C dating. This technique allows dating features which are otherwise undatable (Dorn et al., 1989).
5. Varved lake sediments. These sediments are tree analogs in that one can determine an age by counting the annual varves. Measurement of ^{14}C in identifiable biologic remains in these sediments offers the possibility to extend the ^{14}C record beyond the tree ring record (Zbinden et al., 1989).
6. Marine sediments. $\delta^{18}O$ values in carbonate sediments yield important climatic information. ^{14}C has been much used to determine the time scale in the sediments and the rates at which they are distributed by bioturbation (Berger et al., 1985).
7. Soil organic carbon. More carbon is sequestered in organic matter in soils than either in atmospheric CO_2 or as land biosphere. ^{14}C is an important tool in studies to understand the evolution of this soil organic carbon (Scharpenseel and Becker-Heidmann, 1989; Trumbore et al., 1989).

C. ^{26}Al ($t_{1/2}$ = 7.05 × 10^5 years[3])

Production and Distribution

Atmosphere. ^{26}Al is produced in the atmosphere by spallation of Ar. Because of the low atmospheric abundance of Ar (~1%) production rates are quite low. The $^{26}Al/^{10}Be$ production ratio has been measured to be about 4×10^{-3} (Raisbeck et al., 1983a). This ratio implies an ^{26}Al concentration of about 40 atoms/SCM in the troposphere.

Hydrosphere. The measured concentration in rainfall is about 7×10^4 atoms/1. The $^{26}Al/^{27}Al$ ratio varies because the amount of crustal Al varies considerably. The $^{26}Al/^{10}Be$ in rainfall has been measured in a few samples to be about 3×10^{-3} (Middleton and Klein, 1987). Two measurements exist for ^{26}Al in ice. These two samples, from Antarctica, have concentrations which differ by a factor of 6; that is, 6.8×10^4 and 4.13×10^5 atoms/kg respectively. In both cases, however the $^{26}Al/^{10}Be$ ratio is constant at $2.2 \pm 0.3 \times 10^{-3}$ (Middleton and Klein, 1987).

Lithosphere. In rocks ^{26}Al is primarily produced by spallation of Si. Production rates, which depend on both altitude and latitude, are about 37 atoms/g-yr in quartz (SiO_2) at sea level (Nishiizumi et al., 1989c). The high production rate from silicon implies that about 70% of the terrestrial inventory of ^{26}Al is produced in surface rocks. In comparison less than 1% of the ^{10}Be inventory is produced in rocks, the majority arising in the atmosphere (Sharma et al., 1989).

Anthropogenic. There is no report of anthropogenic ^{26}Al in the environment.

Meteorites. ^{26}Al is produced in chondrites is both by GCRs and SCRs. The concentration from GCR production is about 3.4×10^{-10} atoms/g-meteorite (60 dpm/kg) in H group chondrites at saturation (Middleton and Klein, 1987). Estimated SCR production in chondrites is about an order of magnitude greater, depending on the sample chemistry and irradiation conditions (Reedy, 1987; Nishiizumi et al., 1991b).

Sample Preparation

Targets are prepared by dissolution of the sample followed by ion exchange purification of Al, precipitation as $Al(OH)_3$ and oxidation to Al_2O_3 [e.g., (Sharma et al., 1989)]. In meteorites, where $^{26}Al/^{27}Al$ ratios are high, it is often convenient to add extra Al carrier. In most other settings, there is sufficient stable aluminum that carrier is not required.

Sample Purification

^{26}Al is measured using Al^- ion. Since ^{26}Mg does not form negative ions, there is no problem with isobaric interferences.

Sensitivity

Sensitivities for ^{26}Al detection are presently limited by beam currents, because one must use Al^- rather than AlO^-. The molecular ion yields higher beam currents but has a severe $^{26}MgO^-$ interference. $^{26}Mg^-$, on the other hand, is not a stable ion. The highest currents obtainable from Al_2O_3 have been on the order of 2 μA Al^- (Middleton and Klein, 1987). A sample with $^{26}Al/^{27}Al$ of 10^{-14} will give about 7 counts a minute at this current. Therefore rather long counting times are required for good precision.

Illustrative Applications

General Information. Like beryllium, aluminum is lithophile. However, unlike beryllium, it is an abundant element. This fact, coupled with the difficulty of obtaining high beam currents, makes measurement of ^{26}Al difficult because many samples which have sufficient ^{26}Al for measurement have very low $^{26}Al/^{27}Al$ ratios. Atmospheric production is a factor of 300 lower for ^{26}Al than for ^{10}Be, thus it is more difficult to measure in archives which derive their ^{26}Al from the atmosphere. In terrestrial rocks, however, ^{26}Al is produced at a rate about 6 times that of ^{10}Be. This is sufficiently high that ^{26}Al has become a very useful isotope for the study of terrestrial surface processes. In situ production in quartz is especially promising because of the low aluminum content of quartz.

A recent review which provides a summary of current research is: Geoscience in general (Middleton and Klein, 1987).

Marine Archives. A recurring problem in using cosmogenic isotopes in geophysics is the requirement of knowing the initial isotopic concentration in the reservoir. It is often difficult to keep track of changes which occur across reservoir boundaries; for example, during transfer of ^{26}Al from seawater to sediments or from the atmosphere to precipitation. Both isotope ratios $^{10}Be/^{9}Be$ and $^{26}Al/^{27}Al$ are not constant in the oceans or atmosphere because the residence times of both beryllium and aluminum are so short. It was once hoped that Be and Al would have similar enough geochemical behaviors that the ^{10}Be-^{26}Al pair would have constant initial ratios. This would allow the change in the $^{26}Al/^{10}Be$ ratio due to radioactive decay to be used directly as a clock. While further work remains to be done to ascertain whether this may be true for the atmospheric archive, which would allow the $^{26}Al/^{10}Be$ ratio to be used to date ice cores, it is now clear that the $^{10}Be/^{26}Al$ ratio is not constant in marine sediments. Recent work on marine phases shows that significant fractionation exists between ^{10}Be and ^{26}Al. In corals and Mn nodules, $^{26}Al/^{10}Be$ is about 5 times smaller than the atmospheric production value, and in phillipsites about an order of magnitude greater (Middleton and Klein, 1987; Sharma et al., 1989). The range of values for ^{26}Al in marine systems is indicated by the date in Table 2. Important topics for further work are the factors which control the relative uptakes of ^{10}Be and ^{26}Al, and how these factors may have changed with time; for example, due to changes in scavenging at the ocean margins, as discussed in the section on beryllium.

Rocks. By selecting appropriate mineral phases, it is possible to avoid many of the problems associated with measurement of ^{26}Al in the atmosphere and hydrosphere. Especially promising is the use of cosmogenic isotopes to study erosion and other terrestrial surface processes as is discussed in Section VI-A. ^{26}Al in quartz can be measured without much danger of meteoric contamination and production rates are about 37 atoms/g SiO_2-y at sea level.

Extraterrestrial Samples. In extraterrestrial material ^{26}Al is produced both by spallation reactions and by low energy SCR reactions, especially on Al as a target. Although the production rates are high enough to allow detection by

Table 2. ^{26}Al and ^{10}Be in Marine Sytems*

Sample	^{27}Al (%)	$^{26}Al/^{27}Al$ 10^{-14}	^{26}Al 10^{4} atoms g-1	^{10}Be 10^{7} atoms g-1	$^{26}Al/^{10}Be$ 10^{-3}
Authigenic					
coral†					
J56.F	1.7 ± 0.3 × 10^{-4}	37	1.40 ± 0.10	1.04 ± 0.10	1.35 ± 0.49
C3	1.5 ± 0.3 × 10^{-4}	7.6	0.97 ± 0.10	0.97 ± 0.10	0.27 ± 0.44
Mn-nodules					
Aries 15 D (0–3.8 mm)	0.255	6.3	360 ± 110	2350 ± 190	0.21 ± 0.06‡
Aries 12 D (0–4.7 mm)	0.374	6.7	560 ± 220	1850 ± 210	0.40 ± 0.17‡
Phillipsites					
Core AMPH-48 (5–7 cm)	5.96	6.1	8100 ± 1200	330 ± 33	31.8 ± 5.6§
Core AMPH-48 (16–17 cm)	2.90	6.2	4000 ± 460	317 ± 32	25.8 ± 3.9§
Sediments					
Calcareous					
MBC-3 site C (0–1.5 cm)	0.192	2.4	1200 ± 27	225 ± 16	1.53 ± 0.13
MBC-3 site C (1.5–3.5 cm)	0.77	8.8	4030 ± 65	384 ± 27	1.06 ± 0.19
Siliceous					
MBC-3 site S (0–1.0 cm)	3.20	2.6	2100 ± 320	583 ± 41	0.36 ± 0.07
MBC-3 site S (2.8–4.6 cm)	2.38	1.8	1070 ± 320	590 ± 42	0.18 ± 0.06

* From Middleton and Klein (1987).
† From Bourlès et al. (1984).
‡ Extrapolated to surface.
§ Extrapolated to surface from accumulation rate 0.21 mm ka^{-1} determined from ^{10}Be data.

decay counting in many cases, AMS has the advantage that small samples (~100 μg) can be measured, and that measurement times are short compared to decay counting times. The depth dependence of cosmogenic nuclides in extraterrestrial bodies has been studied by measurements on cores of lunar regolith from *Apollo 15* (Nishiizumi et al., 1984a,b). ^{26}Al showed significant surface enhancement while ^{10}Be showed no effect. This work was important for two reasons. First, it allowed determination of the pattern of ^{26}Al in extraterrestrial bodies, and second, it allowed study of the rate at which the lunar surface is stirred by meteorite impact, so-called lunar gardening.

The measurements of lunar samples, just discussed (Nishiizumi et al., 1984a,b), and of meteorites, summarized in Nishiizumi (1987), coupled with accelerator bombardment of model meteorites (Englert et al., 1984; Michel et al., 1986) and model calculations (Reedy and Arnold, 1972; Reedy, 1987), have laid the groundwork for using cosmogenic nuclides to help deduce the history and original of extraterrestrial material. The isotope content of a sample of extraterrestrial material found on the Earth depends on its exposure history. This history can be considered to consist of three episodes: (1) irradiation while a part of the parent-body; (2) irradiation while in transit from its origin to the Earth; and (3) decay while resident on the Earth. In these studies measurement of a suite of nuclides (and stable isotopes) with different half-lives and different production mechanisms is necessary to enable a reliable description of the exposure history of the object to be developed. This technique has been applied to antarctic meteorites (Nishiizumi et al., 1986), lunar meteorites (Nishiizumi et al., 1991d), eucrites (Herpers et al., 1990), and interplanetary particles (Raisbeck et al., 1983b; Nishiizumi et al., 1991b). As one example, Nishiizumi et al. (1991b) have studied stony cosmic spherules from deep sea sediments and from the Greenland ice cap. They deduced that these spherules show clear evidence of both GCR and SCR bombardment on time scales from a few times 10^5 to 10^7 years. The exposure took place in the inner solar system, not in highly eccentric orbits; therefore, these objects are not derived from comets. Although these objects were not much larger than their present size when they reached the Earth, it is not excluded that most of their cosmic ray exposure took place on the surface of an asteroidal body.

Additional Applications. In addition, ^{26}Al has been used for a few other geophysical studies including:

1. Preliminary work toward developing a dating method for polar ice (Middleton and Klein, 1987).
2. The study of the origin of exotic formations such as tektites and Libyan desert glass (Middleton and Klein, 1987).
3. Study of ^{10}Be and ^{26}Al produced in uranium and thorium ores (Middleton and Klein, 1987).

D. ^{36}Cl ($t_{1/2}$ = 3.01 × 10^5 years[4])

Production and Distribution

Atmosphere. ^{36}Cl is produced in the atmosphere by spallation of atmospheric ^{40}Ar. Production rates are estimated to be 1.1 × 10^{-3} atoms/cm^2 sec (Lal and Peters, 1967) and between 1.7 and 2.6 × 10^{-3} atoms/cm^2 sec (Oeschger et al., 1969). There are no direct measurements of global ^{36}Cl production rates; however, measured ^{10}Be/^{36}Cl ratios can provide some insight. ^{10}Be/^{36}Cl has been measured in Greenland snow. The average values are 8.3 [Camp Century (Elmore et al., 1987)] and 9 [Dye 3 (Suter et al., 1987)]. Using a ^{10}Be production rate of 4.9 × 10^{-2} atoms/cm^{2}sec, a production rate of 6.1 × 10^{-3} atoms/cm^2 sec is derived. This value is larger than any of the estimates. Clearly the value is only poorly known. Because the chloride concentration in the sample decreases with distance from the ocean, which is a source of chloride, ^{36}Cl/Cl ratios tend to increase. Isopleths of ^{36}Cl/Cl calculated for the United States vary between 100 and 1000 × 10^{-15} (Bentley et al., 1986a). The few measurements which exist lend support to the general conclusions of this model (Figure 39).

Hydrosphere. Concentrations in rainwater have not been much measured as of yet. Snow from Mt. Hermon in Israel had a concentration of 4 × 10^7 atoms ^{36}Cl/liter in 1987, which may be showing the effect of bomb-produced ^{36}Cl. Recent, but pre-bomb, North American groundwaters have concentrations in the vicinity of 10^7 atoms/liter (Bentley et al., 1986a). Concentrations in Greenland snow are about 10^6 atoms/kg at Dye 3 (Elmore et al., 1987; Suter et al., 1987). In the antarctic, where precipitation rates can be an order of magnitude lower, concentrations are on the order of 10^7 atoms/kg [Beer, personal communication].

Lithosphere. ^{36}Cl is produced by a wide range of reactions depending on the shielding depth (Phillips et al., 1986b; Fabryka-Martin, 1988). The dominant reaction in many rocks is thermal neutron activation of ^{35}Cl. A second mechanism of production which is more important in rocks with low amounts of chlorine is spallation of potassium and calcium. Negative muon capture by ^{40}Ca only becomes important at depth. Surface production rates of ^{36}Cl are on the order of 25 atoms/g-year (Fabryka-Martin, 1988). ^{36}Cl/Cl ratios at sea level are on the order of 10^{-12}, while production rates at depth due to primordial radionuclide decay are about two orders of magnitude less. Calculated equilibrium ^{36}Cl/Cl ratios from capture of neutrons deriving from the decay of primordial radionuclides commonly range from $<5 \times 10^{-15}$ to 3 × 10^{-14} (Fabryka-Martin, 1988).

Anthropogenic. ^{36}Cl was produced by oceanic nuclear weapons tests which resulted in the activation of ^{35}Cl in seawater. Fallout peaked around 1958 at

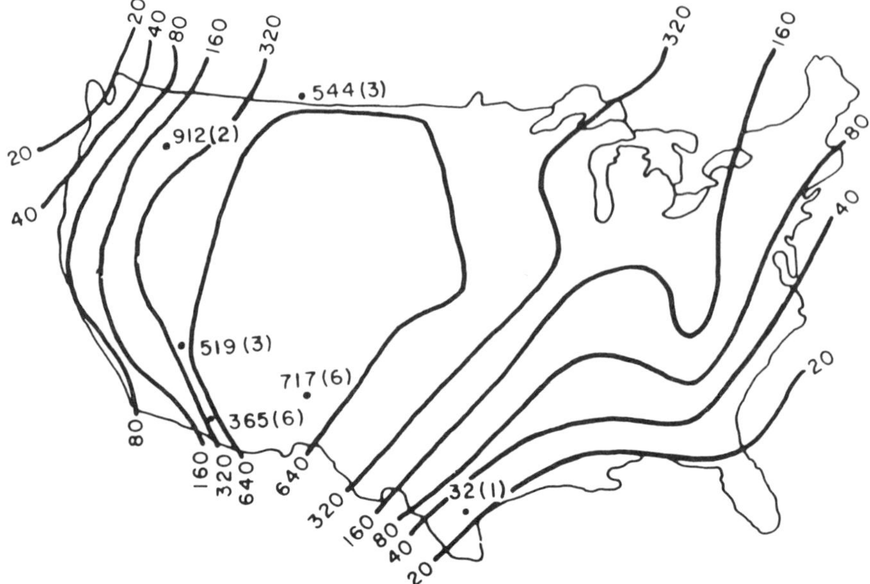

Figure 39. Estimated and measured ^{36}Cl/Cl ratios ($x10^{15}$) in pre-nuclear meteoric water and dry fallout in the continental U.S. Contour lines are estimated ^{36}Cl/Cl values (Bentley et al., 1986), while dots show the locations of field measurements. Numbers in parentheses indicate the number of samples measured at each site and included in the reported average ratio. (Fabryka-Martin et al., 1987)

levels about 700–1000 times natural (Bentley et al., 1986a; Synal et al., 1990) (see Figure 24).

Meteorites. In meteorites, ^{36}Cl can be produced by spallation or by neutron capture. Recently ^{36}Cl has been extensively measured in the iron phase of chondrites, where its production is solely by GCRs. Saturation concentrations are about 5.2×10^9 atoms/g-metal (22.8 $\pm$ 3.1 dpm/kg metal) (Nishiizumi et al., 1989).

Sample Preparation

^{36}Cl is prepared from aqueous samples by precipitation as AgCl. In most groundwater samples, chloride concentrations are high enough to allow direct precipitation of AgCl with $AgNO_3$. If sulfate levels are high, a preliminary precipitation of $BaSO_4$ using $Ba(NO_3)_2$ may be helpful. In rainfall and snow, a preconcentration step is usually necessary. This can be done either by evaporation or by ion exchange. In rock samples, meteoric chlorine is first removed by leaching with water. Afterwards, the rock is decomposed with acids and the in situ ^{36}Cl is separated by distillation. The final AgCl precipitate is dried and pressed directly into the sample holder to make the AMS target.

Sample Purification

^{36}S is the major impurity which causes problems. The AgCl precipitate can be cleaned by repeated dissolution in NH_4OH followed by filtration and reprecipitation with HNO_3 in a sulfur free laboratory.

Sensitivity

Currently sensitivity for ^{36}Cl is about 10^{-15}. A few mg of AgCl is required as target.

Illustrative Applications

General Information. The geochemical behavior of chlorine is different from most of the other nuclides discussed here in that chlorine has primarily atmophile and hydrophile properties. Like other nuclides, ^{36}Cl is produced primarily in the stratosphere. However, its atmospheric chemistry is less well understood than that of ^{14}C, which exists mostly as $^{14}CO_2$, or ^{10}Be, which is almost entirely associated with aerosols. ^{36}Cl occurs in the atmosphere partially as volatile HCl (perhaps greater than 90%) and partially as Cl^- on aerosols (Deck et al., 1990).

Recent reviews or research articles which provide a summary of current research are: Hydrology (Bentley et al., 1986a); meteorites (Nishiizumi et al., 1989); geoscience (Gove, 1987).

Hydrology. The greatest use of ^{36}Cl has been in hydrology, where it has been used in attempts to date groundwater (the time since the water fell as precipitation), to determine mixing of water from different sources, to delineate flow paths, to determine net infiltration in arid regions, to determine past climate conditions (Paul et al., 1986; Fabryka-Martin et al., 1987), and to determine evapotranspiration rates (Magaritz et al., 1990). ^{36}Cl *cannot* be used to date a groundwater in the same straightforward way in which ^{14}C can be used to date a bit of organic material. That is because one cannot assume that the $^{36}Cl/Cl$ ratio in the water is changed only by radioactive decay. Chloride concentrations almost always increase downgradient in an aquifer. In order to interpret the $^{36}Cl/Cl$ values, one needs to understand the reasons for this change. The $^{36}Cl/Cl$ ratio can be affected by many causes. These include: alterations in the distance from the coast in the past due to sea level changes; warmer climate in the past; solution of bedded salt; addition of Cl from rock leachate; and ion filtration (Fabryka-

Martin et al., 1987). Several studies in the Milk River Aquifer in Alberta, Canada (Phillips et al., 1986a) and in the Great Artesian Basin in Australia (Bentley et al., 1986b) have shown that while ^{36}Cl can provide useful information about groundwater ages, it cannot solve the difficult problem of determining a definitive age. Although ^{36}Cl will not revolutionize hydrology, it does seem on the way to becoming a reliable tool in the collection of tracers which hydrologists use to trace hydrological systems. It must be remembered (see above) that radiogenic production of ^{36}Cl can produce ^{36}Cl/Cl values as high as a few times 10^{-14}. This source of ^{36}Cl must always be considered in addition to the meteoric source in hydrological applications of ^{36}Cl. In very old groundwater, subsurface production can limit the useful dating range. Neutron capture by stable ^{35}Cl is the dominant production mechanism. Neutrons are produced by (α,n) reactions on light nuclei, and by spontaneous fission of ^{235}U. Calculated (Fabryka-Martin et al., 1987) and measured (Michelot et al., 1989) equilibrium ^{36}Cl/Cl ratios commonly range from $<5 \times 10^{-15}$ to 3×10^{-14} and, at least away from the coast where values are low due to the influence of marine chloride, are considerably lower than meteoric ratios.

^{36}Cl and chloride measurements have been used in the Mazowsze Basin in Poland to confirm recharge areas and to identify possible mixing of water from different formations (Dowgiallo et al., 1990). Similarly, studies in the Victorian and South Australian Mallee regions have been used to identify recharge regions and sources of salt to the aquifer (Davie et al., 1990), and in the Aquia Formation of southern Maryland to identify recharge (Purdy et al., 1987).

^{36}Cl can also be used to determine evapotranspiration rates and hydrological processes in arid regions. This technique has been used in the Jordan River/Dead Sea system to distinguish chloride content originating from young rainwater from that generated by the leaching of ancient rock chloride (Paul et al., 1986), and to determine evapotranspiration rates (Paul et al., 1986; Magaritz et al., 1990) (Figure 40).

In addition, the pulse of ^{36}Cl introduced by nuclear weapons tests has become a very important tracer. In New Mexico, Phillips et al. (1988) have used the bomb pulse of ^{36}Cl to trace the solute movement, and bomb ^{3}H to trace water flow in arid soils (Figure 41). ^{36}Cl and ^{3}H produced by specific nuclear tests has been used to trace migration from bomb cavities at the Nevada Test Site. These studies show that ^{36}Cl moves more slowly than ^{3}H, probably because of an ion filtration process (Ogard et al., 1988). In a related study, Buddemeier et al. (1991) compared the migration of ^{3}H, ^{36}Cl, ^{99}Tc, ^{125}Sb, ^{129}I, and ^{137}Cs in saturated and unsaturated conditions at the Nevada Test Site and concluded that the migration of these nuclides was very different in the two cases. Norris et al. (1987) used the bomb pulse of ^{36}Cl to study water infiltration at Yucca Mountain, Nevada, a potential nuclear waste storage site.

Atmosphere. The bomb pulse has been measured in Greenland snow and has been used to discuss the atmospheric residence time of chlorine. Earlier results

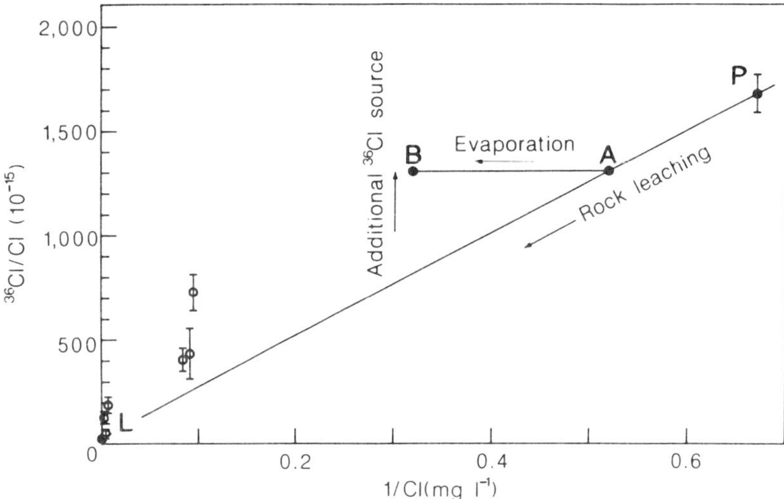

Figure 40. Variation of the $^{36}Cl/Cl$ isotopic ratio versus the reciprocal (1/Cl) of the stable chloride concentration and schematic representation of different effects on the isotopic ratio. The straight line PL represents a hypothetical closed system (no evaporation) in which all samples are admixtures of precipitation (P) and rock leachate (L); as the latter contribution increases, the result moves down along line PL(A). Line AB reflects the effect of evaporation, which does not change the $^{36}Cl/Cl$ ratio. Results may lie above the line also due to an additional source of ^{36}Cl (for example, the ^{36}Cl nuclear test pulse). Also shown are results of measurements (open circles and vertical bars) for the Jordan River/Dead Sea system. The fact that all data points lie above the mixing line is mostly explained by evaporation. (Paul et al., 1986)

suggested a residence time on the order of three to four years (Elmore et al., 1982; Suter et al., 1987), which was longer than that derived for other fallout isotopes. Recent work, with better time resolution, has allowed the detection of visible structure in the fallout pattern which is attributable to bomb tests which occurred after the peak of oceanic testing in the 1950s. By taking these later tests into account, a shorter atmospheric residence time of about two years was derived (Synal et al., 1990).

Recently Deck et al. (1990) have shown that ^{36}Cl exists to the extent of about 93% in a volatile form; probably HCl in stratospheric and only about 7% residues on particulates. This may be the main reason that the $^{36}Cl/^{10}Be$ ratio in ice cores is not constant enough to allow this ratio to be used as a clock (Suter et al., 1987).

Rocks. In addition to its main use as a hydrological tracer, ^{36}Cl is also being developed as a nuclide for rock dating using in situ-produced ^{36}Cl in silicates (Phillips et al., 1986b; Leavy et al., 1987) and in limestones (Kubik et al.,

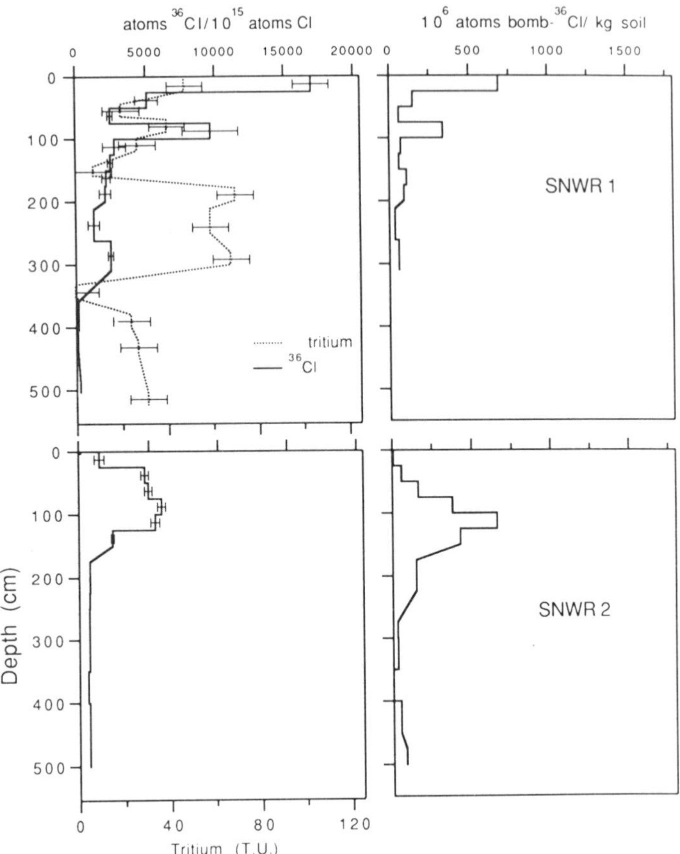

Figure 41. ^{3}H, ^{36}Cl/Cl and ^{36}Cl content as a function of depth in two desert soils from New Mexico. Note the different distribution of ^{3}H and ^{36}C. Natural ^{36}Cl (assumed ^{36}Cl/Cl ≈ 700 × 10^{-15}) has been subtracted to derive the bomb contribution. Such measurements can be used to investigate the transport of liquid and vapor in desert soils. (Data from Phillips et al., 1988)

1984). In situ ^{36}Cl is separated from meteoric ^{36}Cl by first leaching the crushed sample. The remaining ^{36}Cl, which is considered to be locked in the crystal matrix of the rock and therefore thought to be produced by in situ reactions on stable elements in the rock, is separated by distillation in a closed still while the rock is decomposed by attack with HF and HNO_3. Phillips have used in situ ^{36}Cl to date the Arizona meteorite impact crater and the ages of glacial moraines in the Sierra Nevada Mountains of California (Phillips et al., 1990). See Section V-C for a discussion of this method.

Extraterrestrial Samples. ^{36}Cl in the iron phase of meteorites is one of the most important isotopes for determining the terrestrial ages of meteorites. After its fall to Earth, a meteorite is well shielded from cosmic ray bombardment by the Earth's atmosphere and production effectively stops. The terrestrial age can be calculated using known saturation values. ^{36}Cl in the metal phase is produced only by high-energy nuclear reactions. A nearly constant production rate of ^{36}Cl is found since the rate does not depend strongly on the distance from the meteorite surface. Therefore the saturation value is easy to determine. The overall uncertainty in terrestrial ages due to small variations in the saturation value is about 60,000 years. The study of terrestrial ages has become especially important with the discovery that there are areas in the antarctic where ice is being ablated and sublimed faster than snow is accumulating. These areas can be used as giant natural meteorite collectors. Since the process was first discovered in 1969, thousands of meteorites have been recovered from the antarctic. As of 1989, 180 antarctic meteorites had been studied using ^{14}C, ^{81}Kr, or ^{36}Cl. Terrestrial ages as high as 1 million years have been determined. The terrestrial ages of these meteorites have been used to study the influx rates of meteorites and the history of the ice sheets. This work has been summarized by (Nishiizumi et al., 1989).

Additional Applications. In addition, ^{36}Cl has been used for a few other geophysical studies including:

1. Studies of evaporite deposits and brines (Phillips et al., 1983; Magaritz et al., 1986; Kaufman et al., 1990)
2. ^{36}Cl injected into oil fields has the potential as a tracer for understanding oil field structure to improve recovery.

E. ^{41}Ca ($t_{1/2} = 1.03 \times 10^5$ years[5])

Production and Distribution

Atmosphere. The atmospheric production rate is very small because the main target, Kr, has an abundance in air of only 1.14 ppm. Estimated global production is 4.6×10^{-6} atoms/cm^{2}sec from cosmic ray spallation of Kr (Kutschera, 1990). This production rate is about 30 times smaller than that of ^{26}Al, which is already very difficult to measure.

Hydrosphere. No significant production.

Lithosphere. The main production mechanism is neutron capture on ^{40}Ca. Calculated equilibrium values of $^{41}Ca/Ca$ are about 2×10^{-14} (Kutschera, 1990). Measured values in limestones indicate a sea level saturation value of $^{41}Ca/Ca$ of 8×10^{-15} (Fink et al., 1990c; Kutschera, 1990). Erosion would lower the actual value observed in surface rocks below this value. Mineral calcium becomes incorporated in the food chain and eventually into the bones of animals. Concentrations in bone have been measured at values which range from 5×10^{-16} (Zebra, Kenya) to 5×10^{-14} (Elk, Los Angeles) (summarized in (Kutschera, 1990)).

Anthropogenic. Estimate of global inventory is 550 kg ^{41}Ca, compared to a natural inventory of between 20 and 200 kg (Kutschera, 1990). This anthropogenic source is primarily from nuclear weapons testing and is clearly non-negligible relative to the natural inventory. Anthropogenic ^{41}Ca must be considered as a potential contaminant in any environmental use of ^{41}Ca.

Meteorites. ^{41}Ca has not been much studied in meteorites because of the difficulty of its detection prior to AMS. Saturation values in small iron meteorites have been measured to be 1.8×10^9 at/g Fe (24 ± 1 dpm/kg Fe) (Fink et al., 1990b).

Sample Preparation

Ca can be separated using wet chemistry and precipitation as calcium oxalate, which can then be oxidized as CaO. The CaO is then reduced to Ca metal using Zr powder and distilled at 1500 °C. The Ca metal is collected on a water-cooled Cu cup and then converted to CaH_2 by heating to about 650 °C in H_2 (Fink et al., 1990c). Yields are on the order of 90%.

Sample Purification

^{41}K is the potential isobaric interference. By using CaH_3^- produced from CaH_2 it is possible to effectively reduce the contribution from this isotope, since KH_3^- is not stable (Fink et al., 1990c).

Sensitivity

The sensitivity is limited by the background, which is about $(6 \pm 3) \times 10^{-16}$ (Fink et al., 1990c).

Illustrative Applications

General Information. ^{41}Ca is primarily produced in the terrestrial environment by neutron capture on stable Ca. Because of the abundance of stable Ca, ^{41}Ca measurements are difficult, and the applications, both actual and proposed, have been somewhat limited to date. The potential applications of ^{41}Ca are dating bones, determining the chronology of terrestrial surface processes, and determining terrestrial ages of meteorites. The variability of the $^{41}Ca/Ca$ observed in terrestrial samples has frustrated the hope that ^{41}Ca would be useful for bone dating, although the problem continues to be investigated. Typical ^{41}Ca values in terrestrial samples are given in Table 3.

Recent reviews or research articles which provide a summary of current research are: Geosciences (Kutschera, 1990); meteorites (Fink et al., 1990b).

Bone Dating. The most significant use of ^{41}Ca would be the dating of bones in the several hundred thousand year range. This would be of tremendous importance to archaeology by extending the chronology of human evolution beyond the limit set by ^{14}C. ^{41}Ca found in bones is incorporated along with dietary calcium and would be expected to have the $^{41}Ca/Ca$ ratio found in the mineral calcium in the environment in which the organism lived. From this point of view, measurements of $^{41}Ca/Ca$ in rocks and bones have given disappointingly variable values. Ratios in terrestrial surface rocks vary by about two orders of magnitude in the 10^{-15}–10^{-13} range (Kutschera, 1990). Ratios in bone vary about the same. If these measurements reflect true natural variability, then the hope of using ^{41}Ca for bone dating is not very great since the requirement of knowing the initial ratio will be very difficult to satisfy. It is, however, possible that some of the variation is due to laboratory or global contamination with man-made ^{41}Ca, which was extensively produced in nuclear weapons tests. Further work will be required to answer this question (Kutschera, 1990).

Extraterrestrial Samples. Saturation production of ^{41}Ca has been measured in four small iron meteorites and found to be 23.8 ± 0.7 dpm/kg (Fink et al.,

Table 3. ^{41}Ca Results in Bone and Terrestrial Rocks[a]

Sample Description			$^{41}Ca/Ca$
Type	*Location*	*Elevat. (m)*	(10^{-15})
Bones			
Human[b]	Penna., USA	~100	1.9 ± 0.7
Sheep[b]	Rocky Mts. USA	~1300	1.9 ± 0.5
Sheep[b]	Cork, Ireland	~200	2.6 ± 0.7
Sheep	Lake Turkana, Kenya	~400	4.2 ± 1.8
Lion	Lake Turkana, Kenya	~400	1.3 ± 0.6
Zebra	Lake Turkana, Kenya	~400	0.6 ± 0.3
Terrestrial Rocks			
Dolomite top	Heart Mt., WY USA	2200	5.9 ± 1.3
Tufa (upper plateau)	Ain Amûr, Egypt	525	7.0 ± 1.2
Tufa (lower plateau)	Ain Amûr, Egypt	505	3.0 ± 0.3
Coral	Efaté Is., New Hebrides	100	1.4 ± 0.5
Blank[c]			
CaH_2	Aldrich Chem. Co.		0.6 ± 0.2

[a] Data from Fink et al., 1990c.

[b] Samples pre-date nuclear weapons testing.

[c] No blank subtraction was applied to the above results.

1990c); in good agreement with earlier estimates based on the measured $Fe(p,x)^{41}Ca$ spallation cross section (Fink et al., 1987). Further work on chondrites will be required to establish the utility of ^{41}Ca as a useful isotope for determining terrestrial ages of meteorites in a manner similar to that done using ^{3} ^{6}Cl. In silicate phases, ^{41}Ca is produced by neutron capture on ^{40}Ca and may be a sensitive monitor of shielding at depths where another widely used shielding indicator, $^{22}Ne/^{21}Ne$, varies only marginally. ^{41}Ca may also be useful as a size indicator for small objects (Fink et al., 1990b).

^{41}Ca can also be produced by low-energy SCRs. A Ti-rich lunar basalt was analyzed with the goal of detecting SCR-produced ^{41}Ca from Ti (Fink et al., 1987). Clear evidence for SCR production was found. There is some evidence that the SCR flux in the past 10^4 years (inferred from ^{22}Na, ^{55}Fe, and ^{14}C) may have been a factor of four greater than during the previous 10^6 years (inferred from ^{53}Mn and ^{26}Al). The ^{41}Ca half-life falls between these ranges and may help define the timing of this increase and its reality. Deconvolution of the various production mechanisms responsible for the SCR effect and interpreting them in terms of a flux of SCR protons is currently underway (Klein et al., 1990).

F. ^{129}I ($t_{1/2} = 1.6 \times 10^7$ years[6])

Production and Distribution

Atmosphere. ^{129}I in the atmosphere comes from several sources: spallation of atmospheric Xe by cosmic rays in the atmosphere, spontaneous fission of ^{238}U, and neutron induced reactions on ^{128}Te and ^{130}Te (Anonymous, 1983). These sources produce an equilibrium value in the atmosphere and oceans of about $^{129}I/I \approx 1 \times 10^{-12}$ (Fabryka-Martin et al., 1985).

Hydrosphere. Pre-1945 concentrations in rainwater are estimated to be about 2×10^4 atoms/liter. The total inventory is estimated to be 1.1×10^{12} Bq, or 7 tons (Anonymous, 1983). The residence time of iodine is long enough that the oceans should be isotopically well mixed but too short for ^{129}I decay to be important (Fabryka-Martin et al., 1985). The result of large amounts of marine iodine entering the atmosphere is that the atmospheric ^{129}I concentration should be buffered at about $^{129}I/I = 1 \times 10^{-12}$ (Fabryka-Martin et al., 1985). The small amount of data on marine sediments and groundwater supports this conclusion.

Lithosphere. ^{129}I is produced by spontaneous fission of ^{238}U and neutron-induced fission of ^{235}U. Concentrations vary between $1–100 \times 10^{-13}$ $^{129}I/U$. Fission ^{129}I is the dominant source; however, in surface layers of ultramafic rocks (low U concentration) and Te ores (high-target element abundance) μ^- and neutron capture can lead to surface excesses of a factor of 10 or more (Takagi et al., 1974; Fabryka-Martin, 1988). $^{129}I/I$ in rocks is generally on the order of 1–10 $\times$ 10^{-12}. The total inventory is estimated to be 3.7 $\times$ 10^{11} Bq (Anonymous, 1983).

Anthropogenic. Since 1945, $^{129}I/I$ has been elevated above natural values in the environment due to contributions from nuclear weapons tests, nuclear power plant operation, and nuclear fuel reprocessing (Fabryka-Martin et al., 1989). Anthropogenic ^{129}I derives primarily as a fission product of ^{235}U and ^{239}Pu. Peak values reached the $^{129}I/I$ ratio of 10^{-7} in the 1960s and 1970s. The total estimated release due to nuclear weapons testing is about 3.7 $\times$ 10^{11} Bq (Anonymous, 1983). In 1983 it was estimated that by 1990 the total amount of ^{129}I stored in spent nuclear fuel would be about 3.7 $\times$ 10^{13} Bq. Although this value could be significantly lower, it is still clear that the potential for release is large compared to the natural inventory.

Meteorites. ^{129}I is produced in meteorites by cosmic-ray secondary neutron reactions on Te and in lunar surface materials mainly by spallation on Ba and rare earth elements. Concentrations are a few times 10^6 atoms/g in meteorites and about an order of magnitude higher in lunar surface material (Nishiizumi et al., 1983).

Sample Preparation

Iodine can be separated from water samples by anion exchange extraction (Dowex 1-x8) followed by stripping from the resin using sodium hypochlorite (Fabryka-Martin et al., 1989; Wilkins, 1989). Iodine can also be extracted after evaporative concentration with no loss of iodine (Fabryka-Martin et al., 1989). In both cases the iodine is purified after extraction by solvent extraction following oxidation to I_2 followed by reduction and back extraction (Kleinberg and Cowan, 1960).

Sample Purification

Since ^{129}Xe doesn't form stable negative ions, there are no isobaric interferences. The target is prepared by precipitating AgI and may be mixed with Ag metal to increase the heat conductivity of the target.

Sensitivity

Reagent KI was measured to be 1.33 $\pm$ 0.09 $\times$ 10^{-13}. The lowest blanks that have been measured using Woodward iodine (obtained from the high grade brines in Woodward County, Oklahoma) are 0.39 $\pm$ 0.07 $\times$ 10^{-13} (Boaretto et al., 1990).

Illustrative Applications

General Information. Unlike most of the other long-lived isotopes discussed here, ^{129}I has important noncosmogenic sources in the terrestrial environment. ^{129}I is somewhat like ^{14}C in that it should have a more or less uniform $^{129}I/^{127}I$ ratio in the atmosphere, hydrosphere, and biosphere prior to 1950 when anthropogenic inputs started to become important. This is because its long half-life and relatively long marine residence time produce an approximately constant $^{129}I/^{127}I$ in the marine reservoir. Because atmospheric iodine is primarily of marine origin, oceanic iodine should continue to be a good buffer of concentration changes even in remote, inland locations (Fabryka-Martin et al., 1985; Davis and Murphy, 1987). It has similar chemistry to chlorine in that it has both atmosphile and hydrophile tendencies. However, iodine has more complicated redox properties and is a much larger ion; thus its distribution and transport will follow that of chloride only in the most general sense. Its primary use has been in hydrology, studies of uranium ores, and studies of extraterrestrial material.

Most of the limited work on ^{129}I in environmental science has been concerned with laying the foundation for further studies. Measurements in meteorite and lunar samples (Nishiizumi et al., 1983), in groundwater (Fabryka-Martin et al., 1989), in rocks (Boaretto et al., 1990), in sea sediments (Fehn et al., 1986), and in crude oil (Fehn et al., 1987) have all been made. However, these studies are primarily of a feasibility nature. The results are promising, but further work will be needed to establish ^{129}I as a really useful geophysical tracer (Fabryka-Martin et al., 1987). More field samples will be needed to establish whether the initial ratio of $^{129}I/I$ in surface reservoirs has been constant in time and place. Better quantification is needed of natural production mechanisms. The role of volcanic emissions needs to be understood. The extent to which fissiogenic ^{129}I escapes into the water are needed. Perhaps the most straightforward use of ^{129}I is as a monitor of fission product release to the atmosphere resulting from man's use of nuclear power.

A recent review which provides a summary of current research is: Geosciences (Fabryka-Martin et al., 1985).

Hydrology. ^{129}I has been studied in groundwater mostly to understand the phenomena associated with fissiogenic ^{129}I generation and its distribution between solid and liquid phases (Fabryka-Martin et al., 1989; Boaretto et al., 1990). At this time the systems seem not to be well enough understood to explain the observations. ^{129}I was also studied in the Milk River Aquifer and in the Stripa granite (Fabryka-Martin et al., 1987).

Anthropogenic. Because of the important fission source of ^{129}I, it has been much studied in relation to release from nuclear weapons tests and nuclear fuel cycle operation. ^{129}I was detected by AMS after the Chernobyl reactor accident in April 1986 (Kutschera et al., 1988). These authors concluded that the spike of Chernobyl fallout may be a useful tracer for hydrological and other studies in the future.

Extraterrestrial Samples. ^{129}I offers the possibility of extending the time span over which cosmogenic radioisotopes can be used to study extraterrestrial phenomena by about a factor of 5 over ^{53}Mn ($t_{1/2} = 3.7 \times 10^6$ yr). Target elements for ^{129}I ($Z > 28$) are normally present in low concentration in natural material making detection difficult. ^{129}I has been measured in small number of meteorites and lunar samples (Nishiizumi et al., 1983, 1985, 1989b). Samples on the order of 1 to 10 g are required and the measurements are not difficult at the levels encountered.

G. Isotopes Less Studied by AMS

^{3}H ($t_{1/3}$ = 12.3 years)

Tritium finds its primary geochemical use in hydrology and physical oceanography. Its short half-life, and therefore small saturation concentration, typically ^{3}H/H $\approx 10^{-18}$, imply that AMS will be useful only in special circumstances. AMS has not yet actually been used as an analytical tool for ^{3}H analyses in a real application. However, AMS is potentially competitive with scintillation counting where small sample sizes are required (Glagola et al., 1984; Jiang et al., 1984). The most significant potential applications probably lie not in geophysics, but in medical research where AMS-based studies would allow tracer experiments to be performed with very low dose rates.

^{32}Si ($t_{1/2}$~130 years)

The primary production is in atmosphere by spallation of Ar. Silicon is an important nutrient in seawater and the determination of ^{32}Si is useful in studying the cycle of silica and ocean circulation (Lal and Somayajulu, 1984; Somayajulu et al., 1987). Because silicon is such a common element, ^{32}Si/Si ratios are often very low. This poses no fundamental problem for decay counting since ^{32}Si can be determined by counting its short-lived daughter ^{32}P. Therefore, decay counting may be the best technique for studying ^{32}Si except in ice cores which contain very low silica content. ^{32}Si has been used to date polar ice (Clausen, 1973). Much AMS work has gone toward determining the ^{32}Si half-life, which, however, remains uncertain. Determinations range from 108 to 172 years (Elmore et al., 1980; Kutschera et al., 1980; Alburger et al., 1986; Hofmann et al., 1990). The best geochemical estimate, 276 ± 32 years, is based on ^{32}Si measurements in a varved core from the Gulf of California dated both by varve counting and the excess ^{210}Pb method (Demater, 1980).

^{53}Mn ($t_{1/2} = 3.7 \times 10^6$ years)[7]

There is very little production of ^{53}Mn on Earth. The prime terrestrial source is cosmic dust and meteorite ablation. ^{53}Mn/Mn has been measured with a sensitivity of 3×10^{-11} ^{53}Mn/Mn (Korschinek et al., 1987). The ^{53}Mn concentration in iron meteorites and the metal phase of chondrites is typically about 1.1×10^{12} atom/g Fe (~400 dpm/kg Fe) (Nishiizumi, 1987), or 10^{-4} ^{53}Mn/Mn. In deep-sea sediments the ratio is $\sim 10^{-13}$ (Arnold, 1987). The detection limit for determination of ^{53}Mn by neutron activation, which is set by the ^{55}Mn(n,2n)^{54}Mn interference, is about 10^{-11} for a well-moderated reactor (Arnold, 1987).

^{60}Fe ($t_{1/2} = 1.5 \times 10^6$ years)

The half-life of ^{60}Fe has been measured by AMS to be $(1.49 \pm 0.27) \times 10^6$ years (Kutschera et al., 1984), significantly longer than the one previous estimate of 3×10^5 years. ^{60}Fe has been determined with a sensitivity of about ^{60}Fe/Fe $= 10^{-12}$. Even in meteorites, the concentrations are expected to be on the order of 10^{-14}. Therefore, further work is needed before this isotope can be used for actual applications.

Others

AMS has also been considered for ^{22}Na (Heinemeier et al., 1987), ^{59}Ni (Faestermann et al., 1990), ^{126}Sn (Kutschera et al., 1989; Paul et al., 1989), and ^{205}Pb (Ernst et al., 1984; Henning, 1987; Henning and Schuell, 1988). Only preliminary studies have been done, however, on these isotopes.

SI-AMS (Secondary Ion AMS)

The application of AMS techniques to stable isotope mass spectrometry (secondary ion accelerator mass spectrometry) offers increased sensitivity primarily because of the ability to remove interferences from molecular isobars. AMS has been used to measure osmium isotope concentrations and Re/Os ratios to study mantle-crust differentiation in the Earth (Teng et al., 1987). AMS has also been used to analyze isotopic composition of Pt, Ir, and Os in clays from the Cretaceous-Tertiary boundary, and Pt and Os ratios in meteorites (Chew et al., 1984). Recently an AMS system has been used as an ion probe to measure Au distributions in minerals from sulphide-rich ores (Rucklidge et al., 1990). This technique has currently been extended to the measurement of Pt, Ir, Os, Ag, Pd, Rh, and Ru within polished mineral grains within the host rock. Extension of these studies will help understand the formation of ore bodies (Kilius et al., 1990). Investigations are also being undertaken to assess the applicability of AMS to actinide determination; for example, for measuring ^{230}Th/^{232}Th ratios for dating (Kilius et al., 1990). Work in stable SIAMS is still mostly at the feasibility stage. It remains to be seen whether the technique will prove to be a useful tool for geochemistry.

Secondary ion mass spectrometry (SIMS) is limited in sensitivity because of molecular interferences. Application of AMS techniques, coupled with mass analysis of the Cs sputter beam, can yield sensitivities which are several orders of magnitude greater than can be reached with SIMS. Recent work at the University of North Texas has indicated that SIAMS may be very useful in materials science where the production of very pure materials are required (Matteson et al., 1990).

VII. CONCLUSIONS

AMS facilities worldwide contribute to geophysical research by measuring thousands of samples yearly. The situation is rapidly being reached in which analytically useful data can be obtained on samples containing 10^5 to 10^6 atoms of most radionuclides which have geophysical applications.

AMS, however, does remain a complex, costly technique which requires a well-trained staff for operation, data analysis, and maintenance. AMS facilities rely either on older tandem accelerators ($\sim$20 facilities) no longer needed full-time for nuclear physics, or new accelerators of the Tandetron type ($\sim$8 facilities) designed specifically for AMS. The older accelerators can reach terminal voltages between 6 and 11 MV. Although requiring extensive upgrading to attain good AMS performance, and being costly to maintain, they are very flexible in the number of nuclides which can be studied. The Tandetron facilities operate at a lower terminal voltage, around 2 MV. These systems are simpler and cheaper to operate, but have limited capability when analyses require separation of strong isobaric relationships.

Most facilities have not been fully optimized. Significant improvements can be expected in efficiency, with a resultant improvement in statistical uncertainty and throughput as high-intensity sources become more common and better optics and stripper designs are utilized. If demand for more and cheaper analyses continues, new designs can be expected to result in smaller, simpler systems, perhaps dedicated to single nuclides.

The range of studies in which AMS plays an important role continues to grow. Current concern with climate change and the factors which control the terrestrial environment have created a huge demand for ^{14}C measurements in seawater, lake sediments, tree rings, atmospheric gases, and ice cores. ^{10}Be, after languishing for years in the era of relatively low-sensitivity decay counting, has become the second most important radionuclide for studying Pleistocene and Holocene processes. ^{36}Cl has become an important hydrological tool. ^{26}Al, ^{41}Ca, ^{129}I are being actively utilized and new applications discovered as understanding of their distribution in nature grows. The sensitivity of AMS has allowed the development of totally new dating (e.g., in situ production) and tracing (e.g., bomb pulse ^{36}Cl) techniques. Experience has shown that once an isotope can be measured by AMS, its range of application increases dramatically. Further developments may bring ^{32}Si, ^{53}Mn, ^{60}Fe, and thorium isotopes into the stable of AMS nuclides. In addition, progress can be expected in the almost untapped field of coupled AMS and SIMS systems.

ACKNOWLEDGMENTS

This review would not have been possible without the continued stimulating interaction we have had with our many colleagues involved in the exciting effort of applying accelerator techniques to solve geophysical and geochemical problems. Many authors provided preprints of works in progress so that our coverage could be as up to date as possible. We are especially indebted to our co-workers at the ETH/PSI AMS Facility in Zurich and the Center for Accelerator Mass Spectrometry in Livermore for stimulating discussions and hard questions. The manuscript was much improved by the thoughtful comments of Jürg Beer, Georges Bonani, Dittrich, Köbi Hofmann, Arno Synal, and Wölfli. The authors, of course, remain responsible for any errors in fact and judgment. RCF is grateful to the Paul Scherrer Institute and Professor Wölfli, head of the ETH/PSI AMS facility, for hospitality during a sabbatical year in which this manuscript was completed.

This article was prepared under the auspices of the U.S. Department of Energy by Lawrence Livermore National Laboratory under contract W-7405-Eng-48.

NOTES

1. Hofmann et al., 1987a
2. Godwin, 1962
3. Norris et al., 1983
4. Bartholomew et al., 1955; Goldstein, 1966
5. Mabuchi et al., 1974
6. Emery et al., 1972
7. Honda and Imamura, 1971

REFERENCES

Alburger, D. E., Harbottle, G. and Norton, E. F. (1986). Half-life of ^{32}Si. *Earth Planet Sci. Lett.*, **78**, 168–176.

Alton, G. D. (1990). Recent advancements in high intensity heavy negative ion sources based on sputter principle. *Nucl. Instr. Meth.*, **A287**, 139–149.

Alvarez, L. W. and Cornog, R. (1939). ^{3}He in helium. *Phys. Rev.*, **56**, 379.

Anderson, R. F., Lao, Y., Broecker, W. S., Trumbore, S. E., Hofmann, H. J. and Wölfli, W. (1990). Boundary scavenging in the Pacific Ocean: a comparison of ^{10}Be and ^{231}Pa. *Earth Planet. Sci. Lett.*, **96**, 287–304.

Andree, M., Moor, E., Beer, J., Oeschger, H., Stauffer, B., Bonani, G., Hofmann, H. J., Morenzoni, E., Nessi, M., Suter, M. and Wölfli, W. (1984). ^{14}C dating of polar ice. *Nucl. Instr. Meth.*, **B5**, 385–388.

Andree, M., Beer, J., Oeschger, H., Mix, A., Broecker, W., Ragano, N., O'Hara, P., Bonani, G., Hofmann, H. J., Morenzoni, E., Nessi, M., Suter, M. and Wölfli, W. (1985). Accelerator radiocarbon ages on foraminfera separated from deep-sea sediments. In: *The Carbon Cycle and Atmospheric CO_2: Natural Variations Archean to Present* (Sundquist, E. T. and Broecker, W. S., Eds.). Geophysical Monograph 32. American Geophysical Union, Washington, D.C. pp. 143–153.

Andrews, J. N., Davis, S. N., Fabryka-Martin, J., Fontes, J.-C., Lehmann, B. E., Loosli, H. H., Michelot, J.-L., Moser, H., Smith, B. and Wolf, M. (1989). The in situ production of radioisotopes in rock matrices with particular reference to the Stripa granite. *Geochim. et Cosmochim. Acta*, **53**, 1803–1815.

Anonymous (1983). *Iodine-129: Evaluation of Releases from Nuclear Power Generation.* NCRP Report No. 75, National Council on Radiation Protection and Measurement (NCRP, Washington, D.C.).

Anonymous (1985). *Carbon-14 in the Environment.* NCRP Report No. 81, National Council on Radiation Protection and Measurement (NCRP, Washington, D.C.).

Anonymous (1989). *Zurich Sunspot Number Series.* Computer diskette, National Geophysical Data Center, Boulder, CO, USA.

Arnold, J. R. (1987). Decay counting in the age of AMS. *Nucl. Instr. Meth.*, **B29**, 424–426.

Balzer, R. (1984). Medium power pulse sources for deflection chamber modulation. *Nucl. Instr. Meth.*, **B5**, 247–250.

Balzer, R., Bonani, G., Nessi, M., Stoller, C., Suter, M. and Wölfli, W. (1984). Optimizing the geometry of a negative ion sputter source. *Nucl. Instr. Meth.*, **B5**, 204–207.

Bard, E., Arnold, M., Maurice, P., Duprat, J., Moyes, J. and Duplessy, J.-C. (1987). Retreat velocity of the North Atlantic polar front during the last deglaciation determined by ^{14}C accelerator mass spectrometry. *Nature*, **328**, 791–794.

Bard, E., Hamelin, B., Fairbanks, R. G. and Zindler, A. (1990). Calibration of the ^{14}C timescales over the past 30,000 years using mass spectrometric U-Th ages from Barbados corals. *Nature*, **345**, 405–410.

Bartholomew, R. M., Boyd, A. W., Brown, F., Hawkings, R. C., Lounsbury, M. and Merritt, W. F. (1955). The half-life of ^{36}Cl. *Can. J. Phys.*, **33**, 43–48.

Beer, J., Bonani, G., Hofmann, H. J., Suter, M., Synal, A., Wölfli, W., Oeschger, H., Siegenthaler, U. and Finkel, R. C. (1987). ^{10}Be measurements on polar ice: Comparison of the arctic and Antarctic records. *Nucl. Instr. Meth.*, **B29**, 203–206.

Beer, J., Siegenthaler, U., Bonani, G., Finkel, R. C., Oeschger, H., Suter, M. and Wölfli, W. (1988). Information on past solar activity and geomagnetism from ^{10}Be in the Camp Century ice core. *Nature*, **331**, 675–679.

Beer, J., Blinov, A., Bonani, G., Finkel, R. C., Hofmann, H. J., Lehmann, B., Oeschger, H., Sigg, A., Schwander, J., Staffelbach, T., Stauffer, B., Suter, M. and Wölfli, W. (1990). Use of ^{10}Be in polar ice to trace the 11-year cycle of solar activity. *Nature*, **347**, 164–166.

Beer, J., Finkel, R. C., Bonani, G., Gäggeler, H., Görlach, U., Jacob, P., Klockow, D., Langway Jr., C. C. Neftel, A., Oeschger, H., Schotterer, U., Schwander, J., Siegenthaler, U., Suter, M., Wagenbach, D. and Wölfli, W. (1991a). Seasonal Variations in the Concentration of ^{10}Be, Cl^-, NO_{3-}, SO_4^{2-}, H_2O_2, ^{210}Pb, ^3H, Mineral Dust, and $\delta^{18}O$ in Greenland Snow. *Atm. Env.* **25a**, 899–904.

Beer, J., Raisbeck, G. M. and Yiou, F. (1991b). *Time Variations of ^{10}Be and Solar Activity.* The Sun in Time. University of Arizona, Tucson, AZ. pp. 343–359.

Bennet, C. L., Beukens, R. P., Clover, M. R., Gove, H. E., Lievert, R. B., Litherland, A. E., Purser, K. H. and Sondheim, W. E. (1977). Radiocarbon dating using accelerators: Negative ions provide the key. *Science*, **198**, 508–509.

Bentley, H. W., Phillips, F. M. and Davis, S. N. (1986a). "Chlorine-36 in the terrestrial environment." In: *Handbook of Environmental Isotope Geochemistry* (Fritz, P. and Fontes, J. C., Eds.). Elsevier Scientific Publishing, Amsterdam, Vol. 2, pp. 427–480.

Bentley, H. W., Phillips, F. M., Davis, S. N., Habermehl, M. A., Airey, P. L., Calf, G. E.,

Elmore, D., Gove, H. E. and Torgersen, T. (1986b). Chlorine-36 dating of very old groundwater 1. The Great Artesian Basin, Australia. *Water Res. Res.*, **22**, 1991–2001.

Berger, W. H., Killingley, J. S., Metzler, C. V. and Vincent, E. (1985). Two-step deglaciation: ^{14}C-dated high-resolution δ^{18}O records from the tropical Atlantic Ocean. *Qatternary Res.*, **23**, 258–271.

Berkovits, D., Boaretto, E., Hollos, G., Kutschera, W., Naaman, R., Paul, M. and Vager, Z. (1989). Selective suppression of negative ions by lasers. *Nucl. Instr. Meth.*, **A281**, 663–666.

Berkovits, D., Boaretto, E., Hollos, G., Kutschera, W., Naaman, R., Paul, M. and Vager, Z. (1990). Study of laser interaction with negative ions. *Nucl. Instr. and Meth.*, (AMS 5).

Bertsche, K. L., Friedman, P. G., Morris, D. E., Muller, R. A. and Welch, J. J. (1987). Status of the Berkeley small cyclotron AMS project. *Nucl. Instr. Meth.*, **B29**, 105–109.

Bertsche, K. J., Karadi, C. and Muller, R. A. (1990). First detection of radiocarbon in the cyclotrino. *Nucl. Instr. and Meth.*, (AMS 5).

Betz, H. D. (1972). Charge states and charge-changing cross sections of fast heavy ions penetrating through gaseous and solid media. *Rev. of Mod. Phys.*, **44**, 465–539.

Betz, H. D. (1983). Heavy ion charge states. In: *Applied Atomic Collision Physics* (Datz, S., Ed.). Academic Press, New York, Vol. 4, pp. 1–42.

Beukens, R. P. (1990). *Radiocarbon Sample Research.* Annual Report 1990, Isotrace Laboratory, Accelerator Mass Spectrometry Facility at the University of Toronto, pp. 5–7.

Beukens, R. P., Gurfinkel, D. M. and Lee, H. W. (1986). Progress at the IsoTrace radiocarbon facility. *Radiocarbon*, **28**, 229–236.

Blake, D. R. and Rowland, F. S. (1988). Continuing worldwide increase in tropospheric methane, 1978 to 1987. *Science*, **239**, 1129–1134.

Blinov, A. (1988). The dependence of cosmogenic isotope production rate on solar activity and geomagnetic field variations. In: *Secular Solar and Geomagnetic Variations in the last 10,000 Years* (Stephenson, F. R. and Wolfendale, A. W., Eds.). Kluwer Academic Publishers, Dordrecht, The Netherlands, Vol. 236, pp. 329–340.

Boaretto, E., Berkovits, D., Hollos, G. and Paul, M. (1990). Measurements of natural concentrations of ^{129}I in uranium ores by accelerator mass spectrometry. *Nucl. Instr. Meth.*, **B50**, 280–285.

Bonani, G., Beer, J., Hofmann, H. J., Synal, H. A., Suter, M., Wölfli, W., Pfleiderer, C., Kromer, B., Junghans, C. and Münnich, K. O. (1987). Fractionation, precision and accuracy in ^{14}C and ^{13}C measurements. *Nucl. Instr. Meth.*, **B29**, 87–90.

Bonani, G., Eberhardt, P., Hofmann, H. J., Niklaus, T. R., Suter, M., Synal, H. A. and Wölfli, W. (1990). Efficiency improvements with a new stripper arrangement. *Nucl. Instr. and Meth.* (AMS 5).

Bourlès, D.L. (1992). Beryllium isotopes in the Earth's environment. In: *Encyclopedia of Earth System Science*. Academic Press, New York, pp. 337–352.

Bourlès, D., Raisbeck, G. M., Yiou, F., Loiseaux, J. M., Lieuvin, M., Klein, J. and Middleton, R. (1984). Investigation of the possible association of ^{10}Be and ^{26}Al with biogenic matter in the marine environment. *Nucl. Instr. Meth.*, **B5**, 365–370.

Bourlès, D., Raisbeck, G. M. and Yiou, F. (1989). ^{10}Be and ^{9}Be in marine sediments and their potential for dating. *Geochim. Cosmochim. Acta*, **53**, 443–452.

Bowen, R. (1988). *Isotopes in the Earth Sciences.* Elsevier Applied Science Publishers, Essex, England.

Broecker, W. S. and Peng, T.-H. (1982). *Tracers in the Sea.* ELDIGIO Press, Palisades, NY.

Broecker, W. S., Andree, M., Bonani, G., Wölfli, W., Oeschger, H., Klas, M., Mix, A. and Curry, W. (1988). Preliminary estimates for the radiocarbon age of deep water in the glacial ocean. *Paleoceanography*, **3**, 659–669.

Broecker, W. S., Peng, T.-H., Trumbore, S., Bonani, G. and Wölfli, W. (1990). The distribution of radiocarbon in the glacial ocean. *Biogeochemical Cycles.* Submitted.

Bronk, C. R. and Hedges, R. E. M. (1987). A gas ion source for radiocarbon dating. *Nucl. Instr. Meth.*, **B29**, 45–49.

Bronk, C. R. and Hedges, R. E. M. (1990). A gaseous ion source for routine AMS radiocarbon dating. *Nucl. Instr. and Meth.*, (AMS 5).

Brown, L. (1984). Applications of accelerator mass spectrometry. *Ann. Rev. Earth Plant. Sci.*, **12**, 39–59.

Brown, L. (1987). ^{10}Be: recent applications in Earth sciences. *Phil. Trans. R. Soc. Lond.*, **A323**, 75–86.

Brown, L., Pavich, M. J., Hickman, R. E., Klein, J. and Middleton, R. (1988). Erosion of the eastern United States observed with ^{10}Be. *Earth Surf. Processes Landforms*, **13**, 441–457.

Brown, L., Stensland, G. J., Klein, J. and Middleton, R. (1989). Atmospheric deposition of ^{7}Be and ^{10}Be. *Geochim. et Cosmochim. Acta*, **53**, 135–142.

Brownlee, D. E., Tomandl, D. A. and Olszewski, E. (1977). Interplanetary dust; A new source for extraterrestrial material for laboratory studies. *Proc. Lunar Sci. Conf. 8th*, 149–160.

Buddemeier, R. W., Finkel, R. C., Marsh, K. V., Ruggieri, M. R., Rego, J. H. and Silva, R. J. 1991. Hydrology and radionuclide migration at the Nevada Test Site. *Rad. Acta* **52/53**, 275–282.

Castagnoli, G. and Lal, D. (1980). Solar modulation effects in terrestrial production of Carbon-14. *Radiocarbon*, **22**, 133–158.

Chen, M.-B., Gao, W.-Z. and Li, D.-M. (1989a). Design criteria on a minicyclotron as a mass spectrometer for dating. *Nucl. Instr. Meth.*, **A278**, 402–408.

Chen, M.-B., Gao, W.-Z., Zhang, D.-M. L. and Zhou, R.-F. (1989b). Adapting a non-sinusoidal wave dee voltage in minicyclotrons for mass spectrometers for dating. *Nucl. Instr. Meth.*, **A278**, 409–416.

Chew, S. H., Greenway, T. J. L. and Allen, K. W. (1984). Accelerator mass spectrometry for heavy isotopes at Oxford (OSIRIS). *Nucl. Instr. Meth.*, **B5**, 179–184.

Clausen, H. B. (1973). Dating of polar ice by ^{32}Si. *J. Glaciol.*, **66**, 411–416.

Creer, K. M. (1988). Geomagnetic field and radiocarbon activity through Holocene time. In: *Secular Solar and Geomagnetic Variations in the Last 10,000 Years* (Stephenson, F. R. and Wolfendale, A. W., Eds.). Kluwer Academic Publishers, Dordrecht, The Netherlands, Vol. 236, pp. 381–397.

Currie, L. A., Fletcher, R. A. and Klouda, G. A. (1987). On the identification of carbonaceous aerosols via ^{14}C accelerator mass spectrometry and laser microprobe mass spectrometry. *Nucl. Instr. Meth.*, **B29**, 346–354.

Currie, L. A., Stafford, T. W., Sheffield, A. E., Klouda, G. A., Wise, S. A., Fletcher, R. A., Donahue, D. J., Jull, A. J. T. and Linick, T. W. (1989). Microchemical and molecular dating. *Radiocarbon*, **31**, 448–463.

Damon, P. E., Donahue, D. J., Gore, B. H., Hatheway, A. L., Jull, A. J. T., Linick, T. W., Sercel, P. J., Toolin, L. J., Bronk, C. R., Hall, E. T., Hedges, R. E. M., Housley, R., Law, I. A., Perry, C., Bonani, G., Trumbore, S., Wölfli, W., Ambers, J. C., Bowman, S. G. E., Leese, M. N. and Tite, M. S. (1989). Radiocarbon dating of the shroud of Turin. *Nature*, **337**, 611–615.

Davie, R. F., Kellet, J. R., Fifield, L. K., Evans, W. R., Calf, G. E., Bird, J. R., Topham, S. and Ophel, T. T. (1990). ^{36}Cl measurements in the Murray Basin: Preliminary results from the Victorian and South Australian Mallee region. *J. Aust. Geol. & Geophys.* In press.

Davis, S. N. and Murphy, E. (1987). *Dating Ground Water and the Evaluation of Repositories for Radioactive Waste.* Report, NUREG/CR-4912 or NRC FIN B6628, Division of Engineering, Office of Nuclear Regulatory Research, U.S. Nuclear Regulatory Commission.

Deck, B., Wahlen, M., Weyer, H., Kubik, P.W., Sharma, P., and Gove, H. (1990). ^{36}Cl in the stratosphere. Initial results. *EOS*, **71**, 1249.

de Haas, A. P., Langerak, J. J., Oskamp, C. J., de Vries, H. and van der Borg, K. (1987). Pulsed beam measurement system. *Nucl. Instr. Meth.*, **B29**, 91–93.

Demaster, D. J. (1980). The half-life of ^{32}Si determined from a varved Gulf of California sediment core. *Earth Planet. Sci. Lett.*, **48**, 209–217.

Dorn, R. I., Jull, A. J. T., Donahue, D. J., Linick, T. W. and Toolin, L. J. (1989). Accelerator mass spectrometry radiocarbon dating of rock varnish. *Geol. Soc. Am. Bull.*, **101**, 1363–1372.

Dowgiallo, J., Nowicki, Z., Beer, J., Bonani, G., Suter, M., Synal, H. A. and Wölfli, W. (1990). ^{36}Cl in groundwater of the Mazowsze Basin (Poland). *J. Hydrol.* **111B**, 373–385.

Eddy, J. A. (1976). The Maunder Minimum. *Science*, **192**, 1189–1201.

Eisenhauer, A., Mangini, A., Segl, M., Beer, J., Bonani, G., Suter, M. and Wölfli, W. (1987). High Resolution ^{10}Be and ^{230}Th profiles in DSDP site 580. *Nucl. Instr. Meth.*, **B29**, 326–331.

Eisenhauer, A., Mangini, A., Botz, R., Walter, P., Beer, J., Bonani, G., Hofmann, H. J., Suter, M. and Wölfli, W. (1990). High resolution ^{230}Th and ^{10}Be stratigraphy of late quaternary sediments from the From Strait (core 23235). In: *Geologic History of the Polar Oceans: Arctic versus Antarctic* (Bleil, V. and Thiede, J., Eds.). Kluwer Academic Publishers, Dordrecht, The Netherlands, pp. 475–487.

Elmore, D. and Phillips, F. M. (1987). Accelerator mass spectrometry for measurement of long-lived radioisotopes. *Science*, **236**, 543–550.

Elmore, D., Anantaraman, N., Fulbright, H. W., Gove, H. E., Hans, H. S., Nishiizumi, K., Murell, M. T. and Honda, M. (1980). Half-life of ^{32}Si from tandem accelerator mass spectrometry. *Phys. Rev. Lett.*, **45**, 589–592.

Elmore, D., Tubbs, L. E., Newman, D., Ma, X. Z., Finkel, R., Nishiizumi, K., Beer, J., Oeschger, H. and Andree, M. (1982). ^{36}Cl bomb pulse measured in a shallow ice core from Dye 3, Greenland. *Nature*, **300**, 735–737.

Elmore, D., Conard, N. J., Kubik, P. W., Gove, H. E., Wahlen, M., Beer, J. and Suter, M. (1987). ^{36}Cl and ^{10}Be profiles in Greenland ice: Dating and production rate variations: *Nucl. Inst. Meth.*, **B29**, 217–210.

Emery, J. F., Reynolds, S. A., Wyatt, E. I. and Gleason, G. I. (1972). Half-lives of radionuclides—IV. *Nucl. Sci. Eng.*, **48**, 319–323.

Englert, P., Theis, S., Michel, R., Tuniz, C., Moniot, R. K., Vadja, S., Kruse, T. H., Pal, D. K. and Herzog, G. F. (1984). Production of ^{7}Be, ^{22}Na, ^{24}Na, and ^{10}Be from Al in a 4π-irradiated meteorite model. *Nucl. Instr. Meth.*, **B5**, 415–419.

Ernst, H., Korschinek, G., Kubik, P., Mayer, W., Morinaga, H., Nolte, E., Ratzinger, U., Henning, W., Kutschera, W., Mueller, M. and Schuell, D. (1984). ^{205}Pb: accelerator mass spectrometry of a very heavy radioisotope and the solar neutrino problem. *Nucl. Instr. Meth.*, **B5**, 426–429.

Eugster, O., Beer, J., Burger, M., Finkel, R. C., Hofmann, H. J., Krähenbühl, U., Michel, T., Synal, H. A. and Wölfli, W. (1990). History of paired lunar meteorites MAC88104 and MAC88105 derived from noble gas isotopes and radionuclides. Submitted.

Fabryka-Martin, J. T. (1988). *Production of Radionuclides in the Earth and their Hydrologic Significance, with Emphasis on Chlorine-36 and Iodine-129*. PhD. Thesis, The University of Arizona.

Fabryka-Martin, J., Bentley, H., Elmore, D. and Airey, P. L. (1985). ^{129}I as an environmental tracer. *Geochim. Cosmochim. Acta*, **49**, 337–347.

Fabryka-Martin, J., Davis, S. N. and Elmore, D. (1987). Applications of ^{129}I and ^{36}Cl in hydrology. *Nucl. Instr. Meth.*, **B29**, 361–371.

Fabryka-Martin, J. T., Davis, S. N., Elmore, D. and P. W. Kubik (1989). In situ production and migration of ^{129}I in the Stripa granite, Sweden. *Geochim. et Cosmochim. Acta*, **53**, 1817–1823.

Faestermann, H., Kato, K., Korschinek, G., Krauthan, P., Nolte, E., Rühm, W. and Zerle, L. (1990). Accelerator mass spectrometry with fully stripped ^{26}Al, ^{36}Cl, ^{41}Ca, and ^{59}Ni ions. *Nucl. Instr. Meth.*, **B50**, 275–279.

Faure, G. (1986). *Principles of Isotope Geology*. John Wiley & Sons, New York/Chichester/Brisbane/Toronto/Singapore.

Fehn, U., Holdren, G. R., Elmore, D., Brunelle, T., Teng, R. and Kubik, P. W. (1986). Determination of natural and anthropogenic ^{129}I in marine sediments. *Geophys. Res. Lett.*, **13**, 137–139.

Fehn, U., Tullai, S., Teng, R. T. D., Elmore, D. and Kubik, P. W. (1987). Determination of ^{129}I in heavy residues of two crude oils. *Nucl. Instr. Meth.*, **B29**, 380–382.

Fink, D., Paul, M., Hollos, G., Theis, S., Vogt, S., Stueck, R., Englert, P. and Michel, R. (1987).

Measurements of ^{41}Ca spallation cross sections and ^{41}Ca concentrations in the Grant meteorite by accelerator mass spectrometry. *Nucl. Instr. Meth.*, **B29**, 275–280.

Fink, D., Klein, J. and Middleton, R. (1990a). AMS studies with ^{41}Ca. *Nucl. Instr. and Meth.*, (AMS 5). In press.

Fink, D., Klein, J., Middleton, R., Vogt, S. and Herzog, G. F. (1990b). ^{41}Ca in iron falls, Grant and Estherville: production rates and related exposure age calculations. *Earth Planet. Sci. Lett.* **107**, 115–128.

Fink, D., Middleton, R., Klein, J. and Sharma, P. (1990c). ^{41}Ca measurement be accelerator mass spectrometry and applications. *Nucl. Instr. Meth.*, **B47**, 79–96.

Fowler, M. M., Lysaght, P. and Wilhelmy, J. B. (1987). The accelerator-based mass spectrometry of methane. *Nucl. Instr. Meth.*, **B24/25**, 672–675.

Glagola, B. G., Phillips, G. W., Marlow, K. W., Myers, L. T. and Omohundro, R. J. (1984). Low level tritium detection using accelerator mass spectrometry. *Nucl. Instr. Meth.*, **B5**, 221–225.

Godwin, H. (1962). Half-life of radiocarbon. *Nature*, **195**, 984.

Goldstein, G. (1966). Partial half-life for β decay of ^{36}Cl. *J. Inorg. Nucl. Chem.*, **28**, 937–939.

Gove, H. E. (1978). Proc. First Conf. on Radiocarbon Dating with Accelerators, University of Rochester, USA, pp. 1–401.

Gove, H. (1987). Tandem-accelerator mass-spectrometry measurements of ^{36}Cl, ^{129}I and osmium isotopes in diverse natural samples. *Phil. Trans. R. Soc. Lond.*, **A323**, 103–119.

Gove, H. E., Elmore, D., Ferraro, R., Beukens, R. P., Chang, K. H., Kilius, L. R., Lee, H. W., Litherland, A. E. and Purser, K. H. (1980). Radioisotope detection with tandem electrostatic accelerators. *Nucl. Instr. Meth.*, **168**, 425–433.

Heinemeier, J., Hornshøj, P., Nielson, H. L., Rud, N. and Thomson, M. S. (1987). Accelerator mass spectrometry applied to $^{22,24}Na$, $^{31,32}Si$ and ^{14}C. *Nucl. Instr. Meth.*, **B29**, 110–113.

Henken-Mellies, W. U., Beer, J., Heller, F., Hsü, K. J., Shen, C., Bonani, G., Hofmann, H. J., Suter, M. and Wölfli, W. (1990). ^{10}Be and ^{9}Be in South Atlantic DSDP Site 519: Relation to geomagnetic reversals and to sediment composition. *Earth Planet. Sci. Lett.*, **98**, 267–276.

Henning, W. (1987). Accelerator mass spectrometry of heavy elements: ^{36}Cl to ^{205}Pb. *Phil. Trans. R. Soc. Lond.*, **A323**, 87–99.

Henning, W., Kutschera, W., Paul, M., Smither, R. K., Stephenson, E. J. and Yntema, J. L. (1981). Accelerator mass spectrometry and radioisotope detection at the Argonne FN tandem facility. *Nucl. Instr. Meth.*, **184**, 247–268.

Henning, W., Bell, W. A., Billquist, P. J., Glagola, B., Kutschera, W., Liu, Z., Lucas, H. F., M. Paul, Rehm, K. E. and Yntema, J. L. (1987). Calcium-41 concentrations in terrestrial materials: prospects for dating Pleistocene samples. *Science*, **236**, 725–727.

Henning, W. and Schuell, D. (1988). On the separation and accelerator mass spectrometry of ^{205}Pb. *Nucl. Instr. Meth.*, **A271**, 324–327.

Herpers, U., Vogt, S., Bremer, K., Hofmann, H. J., Wölfli, W., Bobe, K., Stoffler, D., Wieler, R., Signer, P., Michel, R., Dragovitsch, P. and Filges, D. (1990). Cosmogenic nuclides in eucrites. *Nucl. Instr. Meth.*, (AMS 5).

Hofmann, H. J. (1987). *Nachweis seltener kosmogener Radionuklide mit der Methode der Beschleuniger-massenspektrometrie*. PhD., ETH Zürich.

Hofmann, H. J., Bonani, G., Morenzoni, E., Nessi, M., Suter, M. and Wölfli, W. (1984). Charge state distributions and resulting isotopic fractionation effects of carbon and chlorine in the 1-7 MeV energy range. *Nucl. Instr. Meth.*, **B5**, 254–258.

Hofmann, H. J., Beer, J., Bonani, G., von Gunten, H. R., Raman, S., Suter, M., Walker, R. L., Wölfli, W. and Zimmerman, D. (1987a). ^{10}Be: half-life and AMS-standards. *Nucl. Instr. Meth.*, **B29**, 32–36.

Hofmann, H. J., Bonani, G., Suter, M. and Wölfli, W. (1987b). Charge state distributions and isotopic fractionation. *Nucl. Instr. Meth.*, **B29**, 100–104.

Hofmann, H. J., Bonani, G., Suter, M., Wölfli, W., Zimmermann, D. and Gunten, H. R. v. (1990). A new determination of the half-life of ^{32}Si. *Nucl. Instr. and Meth.*, **B52**, 544–551.

Honda, M. and Imamura, M. (1971). Half-life of ^{53}Mn. *Phys. Rev.*, **C4**, 1182–1188.

Husain, L., Dutkiewicz, V. and Rusheed, A. (1979). *Origin of tropospheric ozone*. International Union of Geodesy and Geophysics, XVII General Assembly. Canberra, 3–15 Dec. 1979, 582.

Imamura, M., Hashimoto, Y., Yoshida, K., Yamane, I., Yamashita, H., Inoue, T., Tanaka, S., Nagai, H., Honda, M., Kobayashi, K., Takaoka, N. and Ohba, Y. (1984). Tandem accelerator mass spectrometry of $^{10}Be/^{9}Be$ with internal beam monitor method. *Nucl. Instr. Meth.*, **B5**, 211–216.

Jiang, S., Yu, A., Cui, Y., Zhou, Z., Luo, D. and Li, Q. (1984). Determination of tritium using a small van de Graaff accelerator. *Nucl. Instr. Meth.*, **B5**, 226–229.

Jones, G. A., Schneider, R. J., von Reden, K. F. and McNichol, A. P. (1990). The National Ocean Sciences AMS facility at Woods Hole Oceanographic Institution: objectives. *Nucl. Instr. Meth.*, **B52**, 278–284.

Jull, A. J. T., Donahue, D. J. and Linick, T. W. (1989a). ^{14}C activities in recently fallen meteorites and Antarctic meteorites. *Geochim. Cosmochim. Acta*, **53**, 2095–2100.

Jull, A. J. T., Donahue, D. J., Linick, T. W. and Wilson, G. C. (1989b). Spallogenic ^{14}C in high-altitude rocks and in Antarctic meteorites. *Radiocarbon*, **31**, 719–724.

Kaufman, A., Magaritz, M., Paul, M., Hillaire-Marcel, C., Hollos, G., Boaretto, E. and Taieb, M. (1990). The ^{36}Cl ages of the brines in the Magadi-Natron Basin, East Africa. *Geochim. Cosmochim. Acta*. **54**, 2827–2833.

Kieser, W. E. (1989). *Isotrace Laboratory 1989 Annual Report*. Annual Report, University of Toronto.

Kilius, L. R. and Litherland, A. E. (1987). High-resolution accelerator-based mass spectrometry at IsoTrace. *Nucl. Instr. Meth.*, **B26**, 371–376.

Kilius, L. R., Baba, N., Garan, M. A., Litherland, A. E., Nadeau, M. J., Rucklidge, J. C. and Zhao, X. L. (1990). AMS of heavy ions with small accelerators. *Nucl. Instr. and Meth.*, **B52**, 357–365.

Klein, J., Middleton, R. and Tang, H. Q. (1982). Instrumentation of an FN tandem for the detection of ^{10}Be. *Nucl. Instr. Meth.*, **193**, 601–616.

Klein, J. and Middleton, R. (1984). Accelerator mass spectrometry at the University of Pennsylvania. *Nucl. Instr. Meth.*, **B5**, 129–133.

Klein, J., Fink, D., Middleton, R., Vogt, S., Herzog, G. F., Reedy, R. C., Sisterson, J. M., Koehler, A. M. and Magliss, A. (1990). *Average SCR Flux during Past 10^5 Years: Inference from ^{41}Ca in Lunar Rock 74275*. Lunar and Planetary Science Conference, XXI. Lunar and Planetary Science Institute, Houston, Texas. pp. 635–636.

Kleinberg, J. and Cowan, G. A. (1960). *The Radiochemistry of Fluorine, Chlorine, Bromine and Iodine*. NAS-NS-3005, Subcommittee on Radiochemistry, National Academy of Sciences—National Research Council.

Korschinek, G., Morinaga, H., Nolte, E., Preisenberger, E., Ratzinger, U., Urban, A., Dragovitsch, R. and Vogt, S. (1987). Accelerator mass spectrometry with completely stripped ^{41}Ca and ^{53}Mn ions at the Munich tandem accelerator. *Nucl. Instr. Meth.*, **B29**, 67–71.

Kromer, B., Pfleiderer, C., Schlosser, P., Levin, I., Münnich, K. O., Bonani, G., Suter, M. and Wölfli, W. (1987). AMS ^{14}C measurement of small volume oceanic water samples: experimental procedure and comparison with low-level counting technique. *Nucl. Instr. Meth.*, **B29**, 302–305.

Ku, T.-L., Kusakabe, M., Measures, C. I., Southon, J. R., Cusimano, G., Vogel, J. S., Nelson, D. E. and Nakaya, S. (1990). Beryllium isotope distribution in the western North Atlantic: a comparison to the Pacific. *Deep Sea Res.*, **37**, 795–808.

Kubik, P. W., Korschinek, G., Nolte, E., Ratzinger, U., Ernst, H., Teichmann, S., Morinaga, H., Wild, E. and Hille, P. (1984). Accelerator mass spectrometry of ^{36}Cl in limestone and some paleontological samples using completely stripped ions. *Nucl. Instr. Meth.*, **B5**, 326–330.

Kubik, P. W., Elmore, D., Hemmick, T. K. and Kutschera, W. (1989). The gas-filled magnet: an isobar separator for accelerator mass spectrometry. *Nucl. Instr. Meth.*, **B40/41**, 741–744.

Kusakabe, M., Ku, T. L., Vogel, J., Southon, J. R., Nelson, D. E. and Richards, G. (1982). ^{10}Be profiles in seawater. *Nature*, **299**, 712–714.

Kusakabe, M., Ku, T. L., Southon, J. R., Vogel, J. S., Nelson, D. E., Measures, C. I. and Nozaki,

Y. (1987). Distribution of ^{10}Be and ^{9}Be in the Pacific Ocean. *Earth Planet Sci. Lett.*, **82**, 231–240.

Kutschera, W. (1990). Accelerator mass spectrometry: a versatile tool for research. *Nucl. Instr. Meth.*, **B50**, 252–261.

Kutschera, W., Henning, W., Paul, M., Smither, R. K., Stephenson, E. J., Yntema, J. L., Alburger, D. E., Cumming, J. B. and Harbottle, G. (1980). Measurement of the ^{32}Si half-life via accelerator mass spectrometry. *Phys. Rev. Lett.*, **45**, 592–596.

Kutschera, W., Billquist, P. J., Frekers, D., Henning, W., Jensen, K. J., Xiuzeng, M., Pardo, R., Paul, M., Rehm, K. E., Smither, R. K. and Yntema, J. L. (1984). Half-life of ^{60}Fe. *Nucl. Instr. Meth.*, **B5**, 430–435.

Kutschera, W., Fink, D., Paul, M., Hollos, G. and Kaufman, A. (1988). Measurement of the ^{129}I/^{131}I ratio in Chernobyl fallout. *Phys. Scr.*, **37**, 310–313.

Kutschera, W., Ahmad, I., Billquist, P. J., Glagola, B. G., Pardo, R. C., Paul, M., Rehm, R. E. and Yntema, J. L. (1989). Accelerator mass spectrometry at ATLAS. *Nucl. Instr. Meth.*, **B42**, 101–108.

Kvale, T. J., Alton, G. D., Compton, R. N., Pegg, D. J. and Thompson, J. S. (1985). Experimental determination of the energy level of Be^- $(1s^22s2p^2)^4P$. *Phys. Rev. Lett.*, **55**, 484–487.

Lal, D. (1987). ^{10}Be in polar ice: Data reflect changes in cosmic ray flux or polar meteorology. *Gephys. Res. Lett.* **14**, 785–788.

Lal, D. (1988a). In situ-produced cosmogenic isotopes in terrestrial rocks. *Ann. Rev. Earth Planet. Sci.*, **16**, 355–388.

Lal, D. (1988b). Theoretically expected variations in the terrestrial cosmic-ray production rates of isotopes. In: *Solar Terrestrial Relationships and the Earth Environment in the last Millennia* (Castagnoli, G. C., Ed.). North-Holland, Amsterdam, Vol. XCV, 216–233.

Lal, D. (1991). Cosmic ray tagging of erosion surfaces: in situ production rates and erosion models. *Earth Planet. Sci. Lett.*, **104**, 424–439.

Lal, D. and Peters, B. (1967). Cosmic ray produced radioactivity on the Earth. In: *Handbuch der Physik* (Flugge, S., Ed.), Vol. 4612, pp. 551–612.

Lal, D. and Somayajulu, B. L. K. (1984). Some aspects of the geochemistry of silicon isotopes. *Tectonophysics*, **105**, 383–397.

Lal, D., Nishiizumi, K., Klein, J., Middleton, R. and Craig, H. (1987). Cosmogenic ^{10}Be in Zaire allumial diamonds: implications for ^{3}He contents in diamonds. *Nature*, **328**, 139–141.

Leavy, B. D., Phillips, F. M., Elmore, D., Kubik, P. W. and Gladney, E. (1987). Measurement of cosmogenic ^{36}Cl/Cl in young volcanic rocks: an application of accelerator mass spectrometry in geochronology. *Nucl. Instr. Meth.*, **B29**, 246–250.

Lerman, J. C., Mook, W. G. and Vogel, J. C. (1970). ^{14}C in tree rings from different localities. In: *Radiocarbon Variations and Absolute Chronology* (Olsson, I. U., Ed.). Almquist and Wiksell, Stockholm, pp. 275–299.

Leskovar, B. (1977). Microchannel plates. *Physics Today*, **30**, 42–49.

Litherland, A. E. (1980). Ultrasensitive mass spectrometry with accelerators. *Ann. Rev. Nucl. Part. Sci.*, **30**, 437–473.

Litherland, A. E. (1984). Accelerator mass spectrometry. *Nucl. Instr. Meth.*, **B5**, 100–108.

Litherland, A. E. and Kilius, L. R. (1990). Recombinator for accelerator mass spectrometry. *Nucl. Instr. Meth.*, (AMS 5).

Lobb, D. E., Southon, J. R., Nelson, D. E., Wiesehahn, W. and Korteling, R. G. (1981). An ion injection system for use in tandem accelerator radioisotope dating devices. *Nucl. Instr. Meth.*, **179**, 171–180.

Mabuchi, H., Takahashi, H., Nakamura, Y., Notsu, K. and Hamaguchi, H. (1974). Half-life of ^{41}Ca. *J. Inorg. Nucl. Chem.*, **36**, 1687–1688.

Magaritz, M., Kaufman, A., Levy, Y., Fink, D., Meirav, O. and Paul, M. (1986). ^{36}Cl in a halite layer from the bottom of the Dead Sea. *Nature*, **320**, 256–257.

Magaritz, M., Kaufman, A., Paul, M., Boaretto, E. and Hollos, G. (1990). A new method to determine regional evapotranspiration. *Water Res. Res.*, **26**, 1759–1762.

Mangini, A., Segl, M., Bonani, G., Hofmann, H. J., Morenzoni, E., Nessi, M., Suter, M., Wölfli,

W. and Turekian, K. K. (1984). Mass-spectrometric ^{10}Be dating of deep-sea sediments applying the Zürich tandem accelerator. *Nucl. Instr. Meth.*, **B5**, 353–358.

Mangini, A., Segl, M., Kudrass, H., Wiedicke, M., Bonani, G., Hofmann, H. J., Morenzoni, E., Nessi, M., Suter, M. and Wölfli, W. (1986). Diffusion and supply rates of ^{10}Be and ^{230}Th radioisotopes in two manganese encrustations from the South China Sea. *Geochim. Cosmochim. Acta*, **50**, 149–156.

Matteson, S., Marble, D. K., Hodges, L. S., Hajaleh, J. Y., Arrale, A. M., McNeir, M. R. and Duggan, J. L. (1990). Molecular-interference-free accelerator mass spectrometry. *Nucl. Instr. Meth.*, **B45**, 575–579.

Maurette, M., Hammer, C., Brownlee, D. E., Reeh, N. and Thomsen, H. H. (1986). Placers of cosmic dust in the blue lakes of Greenland. *Science*, **233**, 869–872.

McElhinny, M. W. and Senanyake, W. E. (1982). *J. Geomag. Geoelect.*, **34**, 39–51.

McHargue, L.R. and Damon, P.E. (1991). The global beryllium-10 cycle. *Rev. of Geophys.* **29**, 141–158.

Michel, R., Dragovitsch, P., Englert, P., Pfeiffer, F., Stück, R., Theis, S., Begemann, F., Weber, H., Signer, P., Wieler, R., Filges, D. and Cloth, P. (1986). On the depth dependence of spallation reactions in a spherical thick diorite target homogeneously irradiated by 600 MeV protons: Simulation of production of cosmogenic nuclides in small meteorites. *Nucl. Instr. Meth.*, **B16**, 61–82.

Michelot, J. L., Fontes, J. C., Soreau, S., Lehmann, B. E., Loosli, H. H., Balderer, W., Elmore, D., Kubik, P. W., Wölfli, W., Beer, J. and Synal, A. (1989). Chlorine-36 in deep groundwaters and host-rocks of Northern Switzerland: sources, evolution and hydrological implications. In: *Water-Rock Interactions* (Miles, Ed.). Balkema, Rotterdam, pp. 483–486.

Middleton, R. (1983). A versatile high intensity negative ion source. *Nucl. Instr. Meth.*, **214**, 139–150.

Middleton, R. (1984). A review of ion source for accelerator mass spectrometry. *Nucl. Instr. Meth.*, **B5**, 193–199.

Middleton, R. (1989). *A Negative-Ion Cookbook.* University of Pennsylvania.

Middleton, R., Klein, J., Brown, L. and Tera, F. (1984). ^{10}Be in commercial beryllium. *Nucl. Instr. Meth.*, **B5**, 511–513.

Middleton, R. and Klein, J. (1987). ^{26}Al: Measurement and applications. *Phil. Trans. R. Soc. Lond.*, **A323**, 121–143.

Middleton, R., Klein, J. and Fink, D. (1989). A CO_2 negative ion source for ^{14}C dating. *Nucl. Instr. Meth.*, **B43**, 231–239.

Millard, H. T. and Findelman, R. B. (1970). Chemical and mineralogical compositions of cosmic and terrestrial spherules from a marine sediment. *J. Geophys. Res.*, **75**, 2125–2134.

Monaghan, M. C., Krishnaswami, S. and Turekian, K. (1986). The global-average production of ^{10}Be. *Earth Planet. Sci. Lett.* **76**, 279–287.

Monaghan, M. C., Klein, J. and Measures, C. J. (1988). The origin of ^{10}Be in island-arc volcanic rocks. *Earth Planet. Sci. Lett.* **89**, 288–298.

Mook, W. G. (1980). Carbon-14 in hydrological studies. In: *Handbook of Environmental Isotope Geochemistry* (Fritz, P. and Fontes, J. C., Eds.). Elsevier Scientific Publishing Co., Amsterdam, Vol. 1, pp. 49–74.

Mook, W. G. (1984). Archaeological and geological interest in applying ^{14}C to small samples. *Nucl. Instr. Meth.*, **B5**, 297–302.

Morris, J. and Tera, F. (1989). ^{10}Be and ^{9}Be in mineral separates and whole rocks from volcanic arcs: Implications for sediment subduction. *Geochim. et Cosmochim. Acta*, **53**, 3197–3206.

Muller, R. A. (1977). Radioisotope dating with a cyclotron. *Science*, **196**, 489–494.

Nadeau, M.-J., Kieser, W. E., Beukens, R. P. and Litherland, A. E. (1984). Quantum mechanical effects on sputter source isotope fractionation. *Nucl. Instr. Meth.*, **B29**, 83–86.

Nagai, H., Imamura, M., Kobayashi, K., Yoshida, K., Ohashi, H. and Honda, M. (1990). High ^{10}Be production rate found in meteoritic carbons. *Nucl. Instr. and Meth.*, **B52**, 568–571.

Nelson, D. E., Korteling, R. G. and Stott, W. R. (1977). Carbon-14: Direct detection at natural concentrations. *Science*, **198**, 507–508.

Nishiizumi, K. (1987). ^{53}Mn, ^{26}Al, ^{10}Be, and ^{36}Cl in meteorites: data compilation. *Nucl. Track Radiation Meas.*, **13**, 209–273.

Nishiizumi, K., Elmore, D., Honda, M., Arnold, J. R. and Gove, H. E. (1983). Measurements of ^{129}I in meteorites and a lunar rock by tandem accelerator mass spectrometry. *Nature*, **305**, 611–612.

Nishiizumi, K., Elmore, D., Ma, X. Z. and Arnold, J. R. (1984a). ^{10}Be and ^{36}Cl depth profile in *Apollo 15* drill core. *Earth Planet. Sci. Lett.*, **70**, 157–163.

Nishiizumi, K., Klein, J., Middleton, R. and Arnold, J. R. (1984b). ^{26}Al depth profile in *Apollo 15* drill core. *Earth Planet. Sci. Lett.*, **70**, 164–168.

Nishiizumi, K., Arnold, J. R., Elmore, D. and Kubik, P. W. (1985). Cosmogenic ^{129}I in iron meteorites. *Meteoritics*, **20**, 719–720.

Nishiizumi, K., Klein, J., Middleton, R., Elmore, D., Kubik, P. W. and Arnold, J. R. (1986). Exposure history of Shergottites. *Geochim. Cosmochim. Acta*, **50**, 1017–1021.

Nishiizumi, K., Imamura, M., Kohl, C. P., Nagai, H., Kobayashi, K., Yoshida, K., Yamashita, H., Reedy, R. C., Honda, M. and Arnold, J. R. (1987). ^{10}Be profiles in lunar surface rock 68815. *Proc. Lunar Planet. Sci. Conf. 18th*, 79–85.

Nishiizumi, K., Elmore, D. and Kubik, P. W. (1989a). Update on terrestrial ages of Antarctic meteorites. *Earth Planet. Sci. Lett.*, **93**, 299–313.

Nishiizumi, K., Kubik, P. W., Sharma, P. and Arnold, J. R. (1989b). ^{129}I depth profiles in cores from Jilin and the moon. *Meteoritics*, **24**, 310.

Nishiizumi, K., Winterer, E. L., Kohl, C. P., Klein, J., Middleton, R., Lal, D. and Arnold, J. R. (1989c). Cosmic ray production rates of ^{10}Be and ^{26}Al in quartz from glacially polished rocks. *J. Geophys. Res.*, **94**, 17907–17915.

Nishiizumi, K., Kohl, C. P., Shoemaker, E. M., Arnold, J. R., Lal, D., Klein, J., Fink, D. and Middleton, R. (1991a). In situ ^{10}Be-^{26}Al exposure ages at Meteor Crater, Arizona. *Geochim. Cosmochim. Acta*, **55**, 2699–2703.

Nishiizumi, K., Arnold, J. R., Fink, D., Klein, J., Middleton, R., Brownlee, D. E. and Maurette, M. (1991b). Exposure history of individual cosmic particles. *Earth Planet. Sci. Lett.*, **104**, 315–342.

Nishiizumi, K., Kohl, C. P., Arnold, J. R., Klein, J., Fink, D. and Middleton, R. (1991c). Cosmic ray produced ^{10}Be and ^{26}Al in Antarctic rocks: exposure and erosion history. *Earth Planet. Sci. Lett.*, **104**, 440–454.

Nishiizumi, K., Arnold, J.R., Klein, J., Fink, D., Middleton, R., Kubik, P.W., Sharma, P., Elmore, D. and Reedy, R.C. (1991d). Exposure histories of lunar meteorites: ALHA 81005, MAC88104, MAC88105, and Yamato 791197. *Geochim. Cosmochim. Acta*, **55**, 3149–3155.

Norris, A. E., Wolfsberg, K., Gifford, S. K., Bentley, H. W. and Elmore, D. (1987). Infiltration at Yucca Mountain, Nevada, traced by ^{36}Cl. *Nucl. Instr. Meth.*, **B29**, 376–379.

Norris, T. L., Gancarz, A. J., Rokop, D. J. and Thomas, K. W. (1983). Half-life of ^{26}Al. *Proc. 14th Lunar Planet. Sci. Conf., J. Geophys. Res.*, **88 B**, 331–333.

Northcliffe, L. C. and Schilling, R. F. (1970). Range and stopping-power tables for heavy ions. *Nucl. Data Tab.*, **A7**, 233–463.

O'Brien, K. (1979). Secular variations in the production of cosmogenic isotopes in the Earth's atmosphere. *J. Geophys. Res.*, **84**, 423–431.

Oeschger, H., Houtermans, J., Loosli, H. and Wahlen, M. (1969). The constancy of radiation from isotope studies of meteorites and on the Earth. In: *Radiocarbon Variations and Absolute Chronology, 12th Nobel Symposium* (Olssun, I. U., Ed.). Wiley-Interscience, New York, pp. 471–496.

Oeschger, H., Beer, J. and Andree, M. (1987). ^{10}Be and ^{14}C in the earth system. *Phil. Trans. R. Soc. Lond.*, **A323**, 45–56.

Ogard, A. E., Thompson, J. L., Rundberg, R. S., Wolfsberg, K., Kubik, P. W. and Elmore, D. (1988). Migration of chlorine-36 and tritium from an underground nuclear test. *Radiochimica Acta*, **44/45**, 213–217.

Pal, Y. and Peters, B. (1964). Meson production at high energies and the propogation of cosmic rays through the atmosphere. *Kgl. danske Vidensk. Selsk. Mat.-fys. Medd.*, **33**, 1–53.

Paul, M. (1990). Separation of isobars with a gas filled magnet. *Nucl. Instr. and Meth.*, (AMS 5).

Paul, M., Kaufman, A., Magaritz, M., Fink, D., Henning, W., Kaim, R., Kutschera, W. and Meirav, O. (1986). A new ^{36}Cl hydrological model and ^{36}Cl systematics in the Jordan River/Dead Sea system. *Nature*, **321**, 511–515.

Paul, M., Glagola, B. G., Henning, W., Keller, J. G., Kutschera, W., Lia, Z., Rehm, K. E., Schneck, B. and Siemsson, R. H. (1989). Heavy ion separation with a gas-filled magnetic spectrograph. *Nucl. Instr. Meth.*, **A277**, 418–430.

Pavich, M. J., Brown, L., Harden, J., Klein, J. and Middleton, R. (1986). ^{10}Be distribution in soils from Merced River terraces, California. *Geochim. Cosmochim. Acta*, **50**, 1727–1735.

Pearson, Jr., F. J., Noronha, C. J. and Andrews, R. W. (1983). Mathematical modeling of the distribution of natural ^{14}C, ^{234}U, and ^{238}U in a regional groundwater system. *Radiocarbon*, **25**, 291–300.

Phillips, F. M., Smith, G. I., Bentley, H. W., Elmore, D. and Gove, H. E. (1983). Chlorine-36 dating of saline sediments: Preliminary results from Searles Lake, California. *Science*, **222**, 925–927.

Phillips, F. M., Bentley, H. W., Davis, S. N., Elmore, D. and Swanick, G. B. (1986a). Chlorine 36 dating of very old groundwater 2. Milk River Aquifer, Alberta, Canada. *Water Res. Res.*, **22**, 2003–2016.

Phillips, F. M., Leavy, B. D., Jannik, N. O., Elmore, D. and Kubik, P. W. (1986b). The accumulation of cosmogenic chlorine-36 in rocks: A method for surface exposure dating. *Science*, **231**, 41–43.

Phillips, F. M., Mattick, J. L., Duval, T. A., Elmore, D. and Kubik, P. W. (1988). Chlorine 36 and tritium from nuclear weapons fallout as tracers for long-term liquid and vapor movement in desert soils. *Water Res. Res.*, **24**, 1877–1891.

Phillips, F. M., Tansey, M. K., Peeters, L. A., Cheng, S. and Long, A. (1989). An isotopic investigation of groundwater in the Central San Juan Basin, New Mexico: Carbon 14 dating as a basis for numerical flow modeling. *Water Res. Res.*, **25**, 2259–2273.

Phillips, F. M., Zreda, M. G., Smith, S. S., Elmore, D., Kubik, P. W. and Sharma, P. (1990). Cosmogenic chlorine-36 chronology for glacial deposits at Bloody Canyon, Eastern Sierra Nevada. *Science*, **248**, 1529–1532.

Polach, H. A. (1987). Perspectives in radiocarbon dating by radiometry. *Nucl. Instr. Meth.*, **B29**, 415–423.

Pomerantz, M. A. (1971). *Cosmic Rays*. Van Nostrand Reinhold Co., New York.

Proctor, I. D., Southon, J. R., Roberts, M. L., Davis, J. C., Heikkinen, W., Moore, T. L., Garibaldi, J. L. and Zimmerman, T. A. (1990). The LLNL ion source- past, present and future. *Nucl. Instr. and Meth.*, (AMS 5).

Purdy, C. B., Mignerey, A. C., Helz, G. R., Drummond, D. D., Kubik, P. W., Elmore, D. and Hemmick, T. (1987). ^{36}Cl: A tracer in groundwater in the Aquia formation of Southern Maryland. *Nucl. Instr. Meth.*, **B29**, 372–375.

Purser, K. H., Liebert, R. B., Litherland, A. E., Beukens, R. P., Gove, H. E., Bennet, C. L., Clover, M. R. and Sondheim, W. E. (1977). An attempt to detect stable N^- ions from a sputter ion source and some implications of the results for the design of tandems for ultra-sensitive carbon analysis. *Rev. Phys. App.*, **12**, 1487–1492.

Purser, K. H., Schneider, R. J., Dobbs, J. G. and R. Post (1981). *A preliminary description of a dedicated commercial untra-sensitive mass spectrometer for direct atom counting of ^{14}C*. Symp. on Accelerator Mass Spectrometry, Argonne, ANL/PHY-81-1. pp. 431–462

Purser, K. H., Smick, T., Litherland, A. E., Beukens, R. P., Kieser, W. E. and Kilius, L. R. (1988). A third generation ^{14}C accelerator mass spectrometer. *Nucl. Instr. Meth.*, **B35**, 284–291.

Purser, K. H. and Smick, T. H. (1990). A high throughput ^{14}C accelerator mass spectrometer. *Nucl. Instr. and Meth.*, (AMS 5).

Raisbeck, G. M. and F. Yiou (1979). ^{26}Al measurement with a cyclotron. *J. Phys.*, **40**, 241–244.

Raisbeck, G. M. and Yiou, F. (1981). Cosmogenic ^{10}Be/^{7}Be as a probe of atmospheric transport processes. *Geophys. Res. Lett.*, **8**, 1015–1018.

Raisbeck, G. M. and Yiou, F. (1984). Production of long-lived cosmogenic nuclei and their applications. *Nucl. Instr. Meth.*, **B5**, 91–99.

Raisbeck, G. M. and Yiou, F. (1988a). ^{10}Be as a proxy indicator of variations in solar activity and geomagnetic field intensity during the last 10,000 Years. In: *Secular Solar and geomagnetic Variations in the last 10,000 Years* (Stephenson, F. R. and Wolfendale, A. W., Eds.). Kluwer Academic Publishers, Durham. pp. 287–296.

Raisbeck, G. M. and Yiou, F. (1988b). Measurement of ^{7}Be by accelerator mass spectrometry. *Earth and Planet. Sci. Lett.*, **89**, 103–108.

Raisbeck, G. M., Yiou, F., Fruneau, M., Loiseaux, J. M., Lieuvin, M., Ravel, J. C. and Hays, J. D. (1979). A search in a marine sediment core for ^{10}Be concentration variations during a geomagnetic field reversal. *Geophys. Res. Lett.*, **6**, 717–719.

Raisbeck, G. M., Yiou, F., Fruneau, M., Loiseaux, J. M., Lieuvin, M., Ravel, J. C. and Lorius, C. (1981). Cosmogenic ^{10}Be concentrations in antarctic ice during the past 30,000 years. *Nature*, **292**, 825–826.

Raisbeck, G. M., Yiou, F., Klein, J. and Middleton, R. (1983a). Accelerator mass spectrometry measurement of cosmogenic ^{26}Al in terrestrial and extraterrestrial matter. *Nature*, **301**, 690–692.

Raisbeck, G. M., Yiou, F., Klein, J., Middleton, R., Yamakoshi, Y. and Brownlee, D. E. (1983b). ^{26}Al and ^{10}Be in deep sea stony spherules: evidence for small parent bodies (abstract). *Lunar Planet. Sci. XIV*, 622–623.

Raisbeck, G. M., Yiou, F., Bourlès, D., Lestringuez, J. and Deboffle, D. (1984). Measurement of ^{10}Be with a tandetron accelerator operating at 2 MeV. *Nucl. Instr. Meth.*, **B5**, 175–178.

Reedy, R. C. (1987). Nuclide production by cosmic-ray protons. *J. Geophys. Res.*, **92**, 697–702.

Reedy, R. C. and Arnold, J. R. (1972). Interaction of solar and galactic cosmic-ray particles with the Moon. *J. Geophys. Res.*, **77**, 537–555.

Reedy, R. C., Arnold, J. R. and Lal, D. (1983). Cosmic-ray record in solar system matter. *Ann. Rev. Nucl. Part. Sci.*, **33**, 505–537.

Reedy, R. C., Nishiizumi, K. and Arnold, J. R. (1990). *Solar cosmic rays: Fluxes and reaction cross sections.* Lunar and Planetary Science Conference, XX. Lunar and Planetary Science Institute, Houston, Texas, pp. 890–891.

Reiter, E. (1975). Stratospheric-tropospheric exchange processes. *Rev. of Geophys. and Space Phys.*, **13**, 459–474.

Rodhe, H. (1990). A comparison of the contribution of various gases to the greenhouse effect. *Science*, **248**, 1217–1219.

Rucklidge, J. C., Wilson, G. C. and Kilius, L. R. (1990). AMS advances in the geosciences and heavy-element analysis. *Nucl. Instr. Meth.*, **B45**, 565–569.

Ryan, J. G. and Langmuir, C. H. (1988). Beryllium systematics in young volcanic rocks: Implications for beryllium-10. *Geochim. Cosmochim. Acta*, **52**, 237–244.

Sarmiento, J. and Gwinn, E. (1986). Strontium 90 fallout prediction. *J. Geophys. Res.*, **91**, 7631–7646.

Scharpenseel, H.-w. and Becker-Heidmann, P. (1989). Shifts in ^{14}C patterns of soil profiles due to bomb carbon, including effects of morphogenetic and turbation processes. *Radiocarbon*, **31**, 627–636.

Schmidt-Böcking, H. (1978). Penetration of heavy ions through matter. In: *Experimental Methods in Heavy Ion Physics* (Ehlers, J., Hepp, K., Kippenhahn, R., Weidmüller, H. A., and Zittartz, J., Eds.). Springer-Verlag, Berlin, Vol. 83, pp. 81–149.

Schmidt-Böcking, H. and H. Hornung. (1978). Energy straggling of Cl ions in gases. *Z. für Physik*, **A286**, 253–261.

Segl, M., Mangini, A., Bonani, G., Hofmann, H. J., Nessi, M., Suter, M., Wölfli, W., Friedrich, G., Plüger, W. L., Wiechowski, A. and Beer, J. (1984). ^{10}Be-dating of a manganese crust from Central North Pacific and implications for ocean palaeocirculation. *Nature*, **309**, 540–543.

Sharma, P., Mahannah, R., Moore, W. S., Ku, T. L. and Southon, J. R. (1987). Transport of ^{10}Be and ^{9}Be in the ocean. *Earth Planet. Sci. Lett.*, **86**, 69–76.

Sharma, P., Church, T. M. and Bernat, M. (1989). Use of cosmogenic ^{10}Be and ^{26}Al in phillipsite for the dating of marine sediments in the South Pacific Ocean. *Chem. Geol.*, **73**, 279–288.

Shen, C. (1986). *^{10}Be in Chinese Loess*. PhD Thesis, Universität Bern.

Shima, K., Mikumo, T. and Tawara, H. (1986). Equilibrium charge state distributions of ions ($Z_1 \geq 4$) after passage through foils: compilation of data after 1972. *Atomic Data and Nuclear Data Tables*, **34**, 357–391.

Siegenthaler, U. and Beer, J. (1988). Model comparison of ^{14}C and ^{10}Be isotope records. In: *Secular Solar and Geomagnetic Variations in the Last 10,000 Years* (Stephenson, F. R., and Wolfendale, A. W., Eds.). Kluwer Academic Publishers, Dordrecht, The Netherlands, Vol. 236, pp. 315–328.

Somayajulu, B. L. K., Sharma, P., Beer, J., Bonani, G., Hofmann, H. J., Morenzoni, E., Nessi, M., Suter, M. and Wölfli, W. (1984). ^{10}Be annual fallout in rains in India. *Nucl. Instr. Meth.*, **B5**, 398–403.

Somayajulu, B. L. K., Rengarajan, R., Lal, D., Weiss, R. F. and Craig, H. (1987). GEOSECS Atlantic ^{32}Si profiles. *Earth Planet. Sci.*, **85**, 329–342.

Southon, J. R., Vogel, J. S. and Nelson, D. E. (1986). *An ion injection system for simultaneous AMS measurements*. Workshop on Techniques in Accelerator Mass Spectrometry. Oxford. pp. 40–43

Steinhof, A. (1990). Use of pilot beams in accelerator mass spectrometry. *Nucl. Instr. Meth.*, **B51**, 133–139.

Steinhof, A., Henning, W., Mueller, M., Roeckl, E., Schuell, D., Korschinek, G., Nolte, E. and Paul, M. (1987). Accelerator mass spectrometry of ^{41}Ca with a positive-ion source and the UNILAC accelerator. *Nucl. Instr. Meth.*, **B29**, 59–62.

Steinhof, A., Folger, H., Henning, W., Mueller, M., Paic, A., Roeckl, E. and Schuell, D. (1989). Charge state distributions of potassium and calcium ions, and full-stripping probabilities in carbon foils for medium-heavy fast ions. *Nucl. Instr. Meth. Nuclear Instruments and Methods in Physics Research*, **B36**, 93–95.

Stoller, C., Suter, M., Himmel, R., Bonani, G., Nessi, M. and Wölfli, W. (1983). Charge state distributions of 1 to 7 MeV and Be ions stripped in thin foils. *IEEE Trans. Nucl. Sci.*, **NS-30**, 1074–1075.

Stuiver, M. and Polach, H. (1977). Reporting of ^{14}C data. *Radiocarbon*, **19**, 355–363.

Stuiver, M. and Quay, P. (1980). Changes in atmospheric carbon-14 attributed to a variable sun. *Science*, **207**, 11–19.

Stuiver, M., Kromer, B., Becker, B. and Ferguson, C. W. (1986a). Radiocarbon age calibration back to 13,300 years BP and the ^{14}C age matching of the German oak and US bristlecone pine chronologies. *Radiocarbon*, **28**, 969–979.

Stuiver, M., Pearson, G. W. and Braziunas, T. (1986b). Radiocarbon age calibration of marine samples back to 9000 cal yr BP. *Radiocarbon*, **28**, 980–1021.

Stuiver, M., Braziunas, T. F., Becker, B. and Kromer, B. (1990). Late-glacial and Holocene atmospheric $^{14}C/^{12}C$ change: Climate, solar, oceanic and geomagnetic influences. *Quat. Res.*, **35**, 1–24.

Subotic, K. M., Novokic, D., Stojanovic, M. S. and Milinkovic, L. S. (1990). Superconducting minicyclotrons in AMS. *Nucl. Instr. Meth.*, **B50**, 267–270.

Suter, M. (1990). Accelerator mass spectrometry: the state of the art in 1990. *Nucl. Instr. Meth.*, (AMS 5).

Suter, M., Balzer, R., Bonani, G., Hofmann, H. J., Morenzoni, E., Nessi, M., Wölfli, W., Andree, M., Beer, J. and Oeschger, H. (1984a). Precision measurements of ^{14}C in AMS—some results and prospects. *Nucl. Instr. Meth.*, **B5**, 117–122.

Suter, M., Balzer, R., Bonani, G. and Wölfli, W. (1984b). A fast beam pulsing system for isotope ratio measurements. *Nucl. Instr. Meth.*, **B5**, 242–246.

Suter, M., Bonani, G., Stoller, C. and Wölfli, W. (1984c). An accelerator voltage regulating system using a position sensitive faraday cup. *Nucl. Instr. Meth.*, **B5**, 251–253.

Suter, M., Balzer, R., Bonani, G., H. J. Hofmann, Morenzoni, E., Nessi, M. and W. Wölfli. (1985). Efficiency parameters of the ETH AMS facility. *Nucl. Instr. Meth. Nuclear Instruments and Methods in Physics Research*, **B10/11**, 877–880.

Suter, M., Beer, J., Bonani, G., Hofmann, H. J., Michel, D., Oeschger, H., Synal, H. A. and Wölfli, W. (1987). ^{36}Cl Studies at the ETH/SIN-AMS facility. *Nucl. Instr. Meth.*, **B29**, 211–215.

Suter, M., Beer, J., Billeter, D., Bonani, G., Hofmann, H. J., Synal, H. A. and Wölfli, W. (1989). Advances in AMS at Zurich. *Nucl. Instr. Meth.*, **B40/41**, 734–740.

Synal, H. A. (1989). *Beschleunigermassenspektrometrie mit ^{36}Cl.* PhD. Thesis, ETH Zürich.

Synal, H. A., Beer, J., Bonani, G., Hofmann, H. J., Suter, M. and Wölfli, W. (1987). Detection of ^{32}Si and ^{36}Cl with the ETH/SIN EN-tandem. *Nucl. Instr. Meth.*, **B29**, 146–150.

Synal, H.-A., Beer, J., Bonani, G., Suter, M. and Wölfli, W. (1990). Atmospheric transport of bomb-produced ^{36}Cl. *Nucl. Instr. Meth.*, **B52**, 483–488.

Synal, H. A., Bonani, G., Wölfli, W., Finkel, R. C. and Suter, M. (1991). The heavy ion injector at the Zürich AMS facility. *Nucl. Instr. Meth.*, (Denton 1990).

Takagi, J., Hampel, W. and Kirsten, T. (1974). Cosmic-ray muon-induced ^{129}I in tellurium ores. *Earth Planet. Sci. Lett.* **24**, 141–150.

Tanaka, S., Inoue, T. and Imamura, M. (1977). The ^{10}Be method of dating marine sediments—comparison with the paleomagnetic method. *Earth Planet. Sci. Lett.*, **37**, 55–60.

Tarling, D. H. (1988). Secular Variations of the Geomagnetic Field—the Archaeomagnetic Record. In: *Secular Solar and Geomagnetic Variations in the Last 10,000 Years* (Stephenson, F. R. and Wolfendale, A. W., Eds.). Kluwer Academic Publishers, Dordrecht, The Netherlands, Vol. 236, pp. 349–365.

Taylor, R. E. (1987). *Radiocarbon Dating An Archaeological Perspective.* Harcourt Brace Jovanovich, Orlando.

Teng, R. T. D., Fehn, U., Elmore, D., Hemmick, T. K. and Kubik, P. W. (1987). Determination of Os isotopes and Re/Os ratios using AMS. *Nucl. Instr. Meth.*, **B29**, 281–285.

Trumbore, S. E., Vogel, J. S. and Southon, J. R. (1989). AMS ^{14}C measurements of fractionated soil organic matter: an approach of deciphering the soil carbon cycle. *Radiocarbon*, **31**, 644–654.

Turner, N., Purser, K. H. and Sieradzki, M. (1987). Design considerations of VLSI compatible production MeV ion implantation system. *Nucl. Instr. Meth.*, **B21**, 285–295.

Valette-Silver, N. J., Tera, F., Klein, J. and Middleton, R. (1988). ^{10}Be/^{9}Be in the Salton Sea geothermal system (U.S.A.). *Trans. Geotherm. Res. Council*, **12**, 143–150.

Van de Wal, R. S. W., de Jong, A. F. M., van der Borg, K. and Oerlemans, J. (1990). ^{14}C dating of small CO_2 samples for dating ice cores. *Nucl. Instr. Meth.*, **B52**, 469–472.

Verkouteren, R. M., Klouda, G. A., Currie, L. A., Donahue, D. J., Jull, A. J. T. and Linick, T. W. (1987). Preparation of microgram samples on iron wool for radiocarbon analysis via accelerator mass spectrometry: a closed system approach. *Nucl. Instr. Meth.*, **B29**, 41–44.

Vogel, J. S., Nelson, D. E. and Southon, J. R. (1987). ^{14}C background levels in an accelerator mass spectrometry system. *Radiocarbon*, **29**, 323–333.

Wahlen, M., Tanaka, N., Henry, R., Deck, B., Zeglen, J., Vogel, J. S., Southon, J., Shemesh, A., Fairbanks, R. and Broecker, W. (1989). ^{14}C in methane sources and in atmospheric methane: the contribution from fossil carbon. *Science*, **245**, 286–290.

White, N. R., Hedges, R. E. M., Wand, J. O. and Hall, E. T. (1981). *The radiocarbon facility at the Research Laboratory for Archeology in Oxford—A Review.* Symp. on Accelerator Mass Spectrometry, Argonne National Laboratory, ANL/PHY-81-1. Argonne National Laboratory. pp. 472–483

Wieland, E., Santschi, P. H., Beer, J., Bonani, G., Hofmann, H. J., Suter, M. and Wölfli, W. (1990). A multi-tracer study of radionuclides in Lake Zurich, Switzerland: II. Residence times, removal processes, and sediment focusing. *J. Geophys. Res.*, **96**, 17076–17080.

Wilkins, B. T. (1989). *Investigations of Iodine-129 in the natural environment: results and implications.* Report, NRPB-R225. National Radiological Protection Board, Oxon OX11 ORQ, England.

Willson, R. C. and Hudson, H. S. (1988). Solar luminosity variations in solar cycle 21. *Nature*, **332**, 810–812.

Wilson, A. T. and Donahue, D. J. (1990). AMS carbon-14 dating of ice—progress and future prospects. *Nucl. Instr. Meth.*, AMS 5. In press.

Wiza, J. L. (1979). Microchannel plate detectors. *Nucl. Instr. Meth.*, **162**, 587–601.

Wölfli, W. (1987). Advances in accelerator mass spectrometry. *Nucl. Instr. Meth.*, **B29**, 1–13.

Yiou, F., Raisbeck, G. M., Bourlès, D., Lorius, C. and Barkov, N. I. (1985). ^{10}Be in ice at Vostok, Antarctica, during the past climatic cycle. *Nature*, **316**, 616–617.

You, C. F., Lee, T., Brown, L. and Shen, J. J. (1988). ^{10}Be study of rapid erosion in Taiwan. *Geochim. Cosmochim. Acta*, **52**, 2687–2691.

Zbinden, H., Andree, M., Oeschger, H., Ammann, B., Lotter, A., Bonani, G. and Wölfli, W. (1989). Atmospheric radiocarbon at the end of the last glacial: An estimate based on AMS radiocarbon dates on terrestrial macrofossils from lake sediments. *Radiocarbon*, **31**, 795–804.

Ziegler, J. F., Biersack, J. and Littmark, U. (1985). *The Stopping and Range of Ions in Solids.* Pergamon Press, New York.

RESONANCE IONIZATION MASS SPECTROMETRY FOR GEOCHEMICAL APPLICATIONS

B. L. Fearey, C. M. Miller, and N. S. Nogar

I. Introduction 116
A. General 116
B. Mass Spectrometry 116
C. Resonance Ionization Mass Spectrometry 117
II. Modeling and Experimental Considerations 120
A. Modeling 120
B. Mass Spectrometry 128
C. Pulsed RIMS 132
D. Continuous Wave (cw) RIMS 136
III. Applications 138
A. Thorium RIMS 138
B. Re/Os 143
C. Technetium 146
D. Krypton 147
References 149

Advances in Analytical Geochemistry
Volume 1, pages 115–154

ISBN: 1-55938-332-1

I. INTRODUCTION

A. General

With the renewed interest in understanding global environmental and geological processes, new instrumentation will be critical in reducing the required sample sizes, as well as the data-gathering and interpretation process. Much of what we know about the geochemistry and environmental chemistry of the atmosphere, water and ice movements, and paleogeology is inferred from isotope ratio measurements. In this chapter, we describe the implementation and application of resonance ionization mass spectrometry (RIMS) to isotope ratio measurements. This technique combines multistep photoionization with mass spectral sorting and detection of the resultant ions. High spectral brightness lasers are typically used to effect ionization. When used in conjunction with high detection efficiency mass spectrometers, this combination offers the possibility of unparalleled sensitivity and selectivity.

This chapter is divided into several sections. Initially we present introductory material on applications, including isotope geochemistry. This is followed by a brief survey of conventional mass spectrometry and a description of the similarities and differences with respect to RIMS. We then present a modeling study of the RIMS process, followed by a discussion of experimental techniques and apparatus. Lastly, we discuss in some detail the application of RIMS to selected problems in geochemical analysis.

B. Mass Spectrometry

The science of mass spectrometry is over a half-century old and has been applied to a wide variety of problems. These range from basic research to routine measurements in all areas of science and technology. Mass spectrometry is one of the most comprehensive and versatile instrumental techniques available, with general applicability to a wide range of problems.

Mass spectral analysis, both qualitative and quantitative, can be used for a variety of systems, ranging from simple atoms and diatomic molecules to large organic molecules (Zenobi et al., 1989) and very large proteins. In general, analytical mass spectrometry, for the purposes of this chapter, will involve measurements of atomic isotopes for both overall concentration and relative abundances in geologic materials, with samples ranging from carefully chemically prepared and separated atomic species to whole rock analysis (Smith, 1988). Some aspects speak to the thrust of this chapter in that the "ultimate" wish for many isotope geochemists is to analyze a virgin sample, thereby eliminating any potential for biasing (and hence altering) the specimen by sample preparation. RIMS is a technique which has potential to directly address this problem.

A mass spectrometer system is required to separate, either in space, time, or frequency, the isotopes of interest for quantitation. Many types of spectrometers may find application in isotope geochemistry: magnetic sector, quadrupole, time-of-flight, and potentially others, such as ion cyclotron resonance.

The forms of analysis most frequently needed are the *concentration* and *isotopic distribution* of the elements of interest. The latter capability is where mass spectrometry excels. Some important applications include isotope abundance measurements for dating rocks and their relative fractionations (e.g., Rb-Sr, K-Ar, Re-Os, Lu-Hf, K-Ca, U, and Th-Pb) (Faure, 1986). It is not the purpose of this chapter to review the plethora of isotope geochemistry applications utilizing mass spectrometry, but simply to indicate its wide use.

Although mass spectrometry has opened entirely new frontiers for geochemistry and geology, this technique by itself has limitations and difficulties which reduce its overall applicability. Some of the major difficulties in making isotope measurements with mass spectrometers are isobaric interferences, high background levels, abundance sensitivity, and ionization efficiency. Isobars, or same mass interferences, are particularly difficult when the different elements have very similar chemistry (i.e., chemical separation is difficult) and when the abundance of the isotope of interest is small relative to the background level of the isobar. Obviously these isobars, whether they are elemental or organic in nature, can lead to troublesome (and potentially uncorrectable) background levels. Another source of background is ion up- or down-scattering because of high atomic densities near the ionization source. This is most common for analysis of refractory elements where high ionization temperatures are required. In addition, ion scattering typically leads to a reduction in the effective abundance sensitivity of the mass spectrometer, and thus limits the ability to make high dynamic range isotope ratio measurements.

Another limitation of conventional mass spectrometry is that elements which have high ionization potentials are often very difficult or impossible to analyze by standard methods. Alternative ionization techniques (e.g., sputtering or ion impact) can sometimes solve this problem but often with accompanying matrix effects. Often frustrating are atomic complexation or speciation difficulties which generate undesired molecular species reducing overall sensitivity.

C. Resonance Ionization Mass Spectrometry

Many of these problems can be reduced or eliminated by using laser resonance ionization in combination with a mass spectrometer. Resonance ionization spectroscopy was initially developed for applications in atomic physics, but it was soon apparent that there was considerable potential for RIMS applications in geochemistry and geophysics. RIMS combines multistep photoionization with mass spectral sorting and detection of the the resultant ions. Gas phase atoms are irradiated by laser pulses tuned to a resonant electronic transition (Paisner and

Figure 1. Photograph of a cw-RIMS apparatus currently in use for isotope ratio measurements at Los Alamos National Laboratory.

Solarz, 1987), thus insuring a high degree of selectivity in the excitation process (Hurst et al., 1979). Absorption of subsequent photons results in further excitation, and eventually, ionization. Figure 1 is a photograph of one RIMS apparatus currently in use at the Los Alamos National Laboratory, while Figure 2 depicts a generic RIMS apparatus, with key features identified. Isobars are reduced or eliminated because of elemental selectivity, typically $\geq 10^8$. Typical operating temperatures are quite moderate thereby minimizing ion scattering and potentially increasing abundance sensitivity. In addition, high ionization potential elements can be analyzed (Walker and Fassett, 1986), albeit often with complicated ionization schemes (Blum et al., 1990a).

Most analytically useful resonance ionization processes use a relatively small number of photons, typically 2–4, to generate ions. Figure 3 depicts two of the most common ionization processes. This figure suggests the use of a single laser to produce an excited electronic state(s), by either a one- or a two-photon process, and to ionize the atom from that excited state (Downey et al., 1984a; Nogar et al., 1985a). More complex schemes, utilizing three or more lasers are also possible. The particular ionization scheme chosen will depend on a number of factors: the electronic energy level structure of both the analyte element of interest, and potential interfering species (either atomic or molecular); the anticipated concentration or size of the sample; and the available laser and mass spectral hardware. The information on electronic structure is available from a number of sources (e.g., see Moore, 1971). Rather than discuss the choice of ionization processes in detail here, a number of examples are provided later in this chapter.

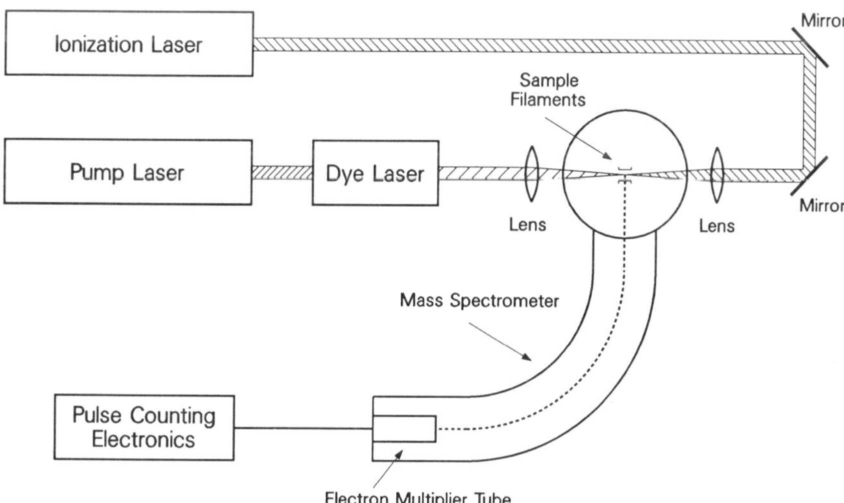

Figure 2. Schematic of a generic RIMS apparatus.

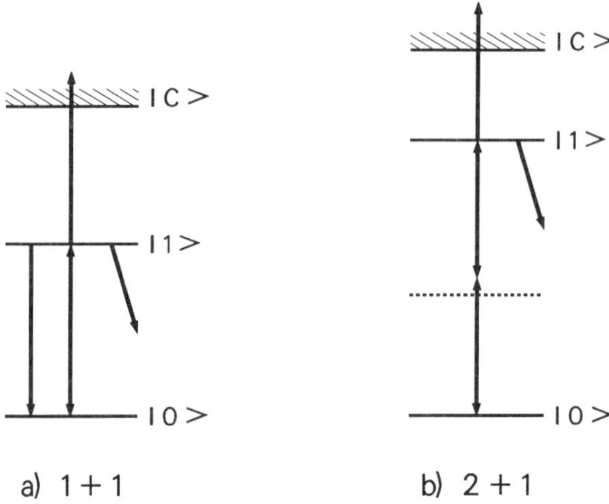

Figure 3. Schematic energy level diagrams for two simple resonance ionization schemes. The one on the left is a "1 + 1" process ("photons to resonance + photons to ionize"), while the one on the right is a "2 + 1" process.

High spectral brightness (photons/cm^2/sec/Hz) lasers are typically used for the excitation and ionization processes. In some cases, it is possible to saturate the ionization; that is, to ionize all analyte atoms within the focal volume of the ionization laser. When used in conjunction with high detection efficiency mass spectrometers, resonance ionization thus offers the possibility of unparalleled sensitivity and selectivity. Early reports of resonance ionization (Hurst et al., 1979) stressed the ability to detect a single analyte atom in the presence of as many as 10^{16} other atoms.

II. MODELING AND EXPERIMENTAL CONSIDERATIONS

A. Modeling

Before undertaking the investment in time and money to build an experimental RIMS apparatus, it is useful to understand the parameters relevant to efficient ionization, separation, and detection. Such understanding will allow the experimentalist to choose the most appropriate laser system (i.e., pulsed or cw) and mass spectrometer for the target element and matrix. A certain amount of numerical modeling of the resonance ionization process is useful in gaining insight into reasonable expectations in the laboratory.

A rate equations method for modeling ionization is one that many find quite useful, and an attractive alternative to the density matrix approach (Ackerhalt and Shore, 1977; Agostini et al., 1978; Swain, 1980). In essence, one establishes a set of equations which express the rate at which an atom moves from one level or state to another. Laser-driven processes will reflect the relevant absorption cross section and laser intensity; spontaneous processes (such as relaxation of an intermediate electronic excited state) will be independent of the laser. In total, these equations comprise a set of coupled differential equations which can be solved either numerically or in closed form. With literature values (or reasonable estimates) of the relevant atomic parameters, the virtues of various laser types and characteristics can be explored (Miller and Nogar, 1983a).

The detail with which the resonance ionization process is modeled can be quite great, limited practically by the computing time available. Additional characteristics of the particular experiment can be included, such as the spatial intensity distribution of the laser and atom source (Nogar and Estler, 1990a), motion of the atoms with respect to the laser, and spectral profiles of the atom and laser. Generally, zeroth-order approximations suffice for initial investigations of the "best" laser for a particular resonance ionization scheme. Higher order detail must be included for a thorough understanding of a particular system, especially when considering the measurement of isotope ratios. For the latter, features such as laser lineshape, isotope shifts, and hyperfine structure must also be included (Young et al., 1989b).

To demonstrate the utility of the rate equations approach, we present two simple calculational examples. One is based upon pulsed laser irradiation, as might be applicable to time-of-flight analysis, while the second will reflect cw laser conditions, which would be applicable with magnetic sector analysis.

For both examples, we will use the same basic atomic system, as illustrated in Figure 3a, that of a two-level atom. The electronic excited state, $|1\rangle$, is chosen to correspond to the energy of a 420-nm photon above the ground state, $|0\rangle$, ($\sim$ 3 eV), while the ionization continuum, $|C\rangle$, may be reached by a second photon (i.e., $\leq$ 6 eV). The excited electronic state is taken to have a lifetime of 10 nsec, with 90% of the fluorescence returning the atom to the ground state, while 10% decays to another electronic level not communicating (on the time scale of the experiment) with the ground state. Additional terms are defined in Table 1.

Rate equations can be written which reflect the system's evolution under the influence of the photon field as a function of time:

$$dn_0/dt = \{(k_0 + k_{-1})n_1\} - k_1 n_0 \quad (1a)$$

$$dn_1/dt = k_1 n_0 - (k_{-1} + k_2 + k_0 + k_p)n_1 \quad (1b)$$

$$dn_c/dt = k_2 n_1 \quad (1c)$$

We note that as the rates k_1, k_{-1}, and k_2 reflect the laser intensity at a particular location and time, the ionization efficiency will necessarily also vary with location and time. Atomic motion may also influence the exposure, particularly

Table 1. Definition of Symbols

$\|0\rangle\ \|1\rangle,\ \|C\rangle$	ground and excited states, and continuum
n_0, n_1, n_C	populations of the ground and excited states, and continuum, respectively
k_1, k_{-1}, k_2	rates of absorption, stimulated emission, and ionization, defined by $k_1 = \int \sigma_i(n) I_i(n) dn$ where $\sigma_i(n)$ is the cross section for the process, and $I_i(n)$ is the photon intensity addressing atoms in state i.
k_0 and k_p	the rates, respectively, for spontaneous emission and collisional or diffusive loss.
T	time period of exposure to the photon flux
V	volume over which ions are collected
w_0	beam radius ($1/e^2$ intensity) at the focus
w_z	beam radius at a distance z from the focus
Z_c	confocal distance, $Z_c = \pi w_0^2/\lambda$
v_{ave}	average atomic velocity

for pulse durations long compared to the transit time, t_t, which can be roughly calculated as t_t = beam diameter/atomic velocity.

Clearly, the greater the detail of the model in reflecting the experimental conditions, the greater the difficulty in solving the coupled equations. In the special case of a temporally-invariant laser pulse (square pulse), the rate equations may be explicitly solved to give (Zakheim and Johnson, 1980):

$$n_c = \frac{k_1 k_2}{\ell_2 \ell_3} \left\{ 1 + \frac{\ell_3}{\ell_2 - \ell_3} e^{-\ell_2 T} + \frac{\ell_2}{\ell_3 - \ell_2} e^{-\ell_3 T} \right\} \tag{2a}$$

with

$$\ell_2 = \frac{P + Q}{2}, \quad \ell_3 = \frac{P - Q}{2} \tag{2b}$$

and

$$P = k_1 + k_{-1} + k_2 + k_0 + k_p \tag{2c}$$

$$Q = \{P^2 - 4k_1(k_2 + k_p)\}^{1/2}$$

These equations may be simply solved and serve as useful, rapid approximations for both pulsed and cw zeroth-order ionization efficiency calculations.

Pulsed Lasers

We consider first a typical pulsed laser application. Here, the exposure duration T is defined by the pulse duration of the laser, which we will take to be 10 nsec, at a repetition rate of 20 Hz. In a time-of-flight application, the collection efficiency for ions is roughly uniform over a relatively large volume, which we will define as a cube 0.5 cm on a side. A uniform atom source within this volume is assumed, providing atoms on a continuous basis with an average velocity of 5×10^4 cm/sec.

Previous experience has shown that at typical laboratory laser powers, saturated ionization can be achieved with minimal focussing of the laser (spot sizes ≈ 0.2 cm). Thus, it is possible to ignore the variation of the laser spot size along the direction of laser propagation. For completeness, and in anticipation of the cw case, we will include this variation. If w_0 is the beam diameter at the focus of the laser (centered in the cell), the confocal distance is defined as

$$z_c = \pi * w_0^2/\lambda \tag{3}$$

and the beam radius as a function of distance along the z axis is given by

$$w_z = w_0 * \left\{1 + \frac{z^2}{z_c^2}\right\}^{1/2} \tag{4}$$

Transverse to this, the intensity will be described by a Gaussian distribution. Thus, at any point in space,

$$I(x,y,z) = I(0) * w_0^2/w_z^2 * \exp\{-2(x^2+y^2)/w_z^2\} \tag{5}$$

and noting that the total power in the beam P is given by

$$P = I_0 \int_0^\infty r \exp(-2r^2/w_0^2)\, dr \tag{6a}$$

$$= I_0\{w_0^2/4\} \tag{6b}$$

In other words, $I_0 = 4*P/w_0^2$. Note also that laser pulse energies are conveniently quoted in mJ, while the natural unit for I is photons/cm$^2-$sec. A convenient conversion is that a 1 mJ pulse (or 1 mW for 1 sec) contains $\approx (5 \times 10^{12})*\lambda$(nm) photons.

Figure 4 displays the results of these calculations for the center of the ionization cell; that is, where the intensity of the laser is at its maximum. As is evident from the figure, efficient ionization is possible for relatively modest laser powers. However, not all of the atoms receive exposure at the central intensity, thus the overall efficiency is less than the central value. Figure 5 displays the ionization efficiency as a function of radial distance from the center of the cell at the middle of the z-axis (solid line), as well as a normalized Gaussian shape (dashed line). The fact that the ionization efficiency is higher than the Gaussian at the center of the cell indicates that saturation (complete ionization) is occurring at that location.

Figure 6 displays the overall ionization efficiency for the entire cell as a function of spot size for a variety of laser pulse powers. A Monte Carlo approach was used to generate this data (Hammersley and Handscomb, 1964), basically by choosing locations in the ionization volume at random. Two vertical axes are shown. To the right, the ionization efficiency per laser pulse is given (i.e., the

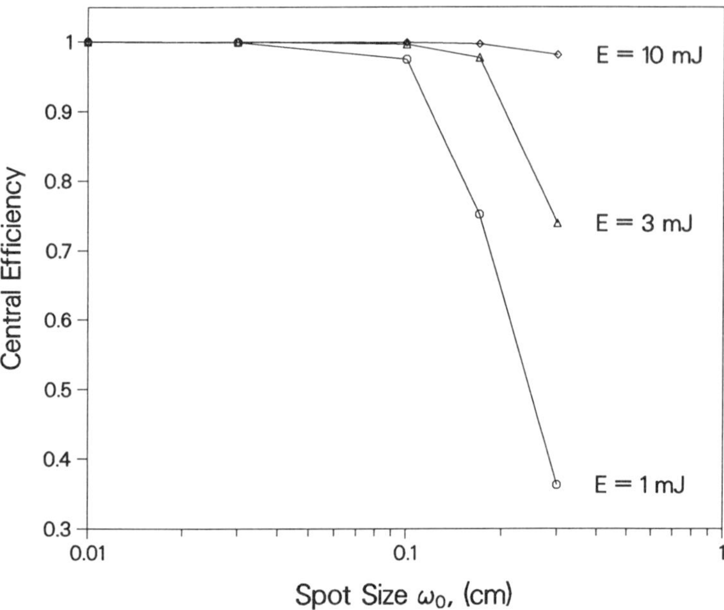

Figure 4. Ionization efficiency calculation for pulsed laser RIMS at the center of the ionization volume as a function of spot size at three different pulse energies. Note that saturated ionization occurs for pulse energies > 10 mJ.

efficiency with which the atoms in the cell at any given pulse of the laser are ionized). To the left, the overall efficiency is shown (i.e., the fraction of the atoms ionized for a macroscopic period of time). The difference between the two axes is a reflection of the lack of ionization during the interpulse period of the laser, and reflects the effective duty factor: [(repetition rate × cell width)/atomic velocity]. Note that as the cell width becomes smaller, the per pulse efficiency rises, while the overall efficiency remains constant. This ''dark'' interpulse period presents the greatest limitation to pulsed laser resonance ionization. Novel methods by which to recover and recycle the atoms lost during this period have been used (Hurst et al., 1984; Andreev et al., 1986; Rimke et al, 1987; Davis and Thonnard, 1989) and laser repetition rates of perhaps two orders of magnitude greater are available, but with a sacrifice in experimental simplicity.

The shape of the curves in Figure 6 simply reflects the focal volume in the ionization region. That is, for very small central focal spot sizes (<10 μm), the laser beam diameter expands at the ends of the region, leading to large total irradiated volumes. For moderate central spot sizes (≈10 μm), the beam is a constant diameter throughout the cell, leading to relatively small irradiated volumes. As the focal spot size is increased (>10 μm), the ionization volume again expands, leading to the production of ions over a large volume. Shifts

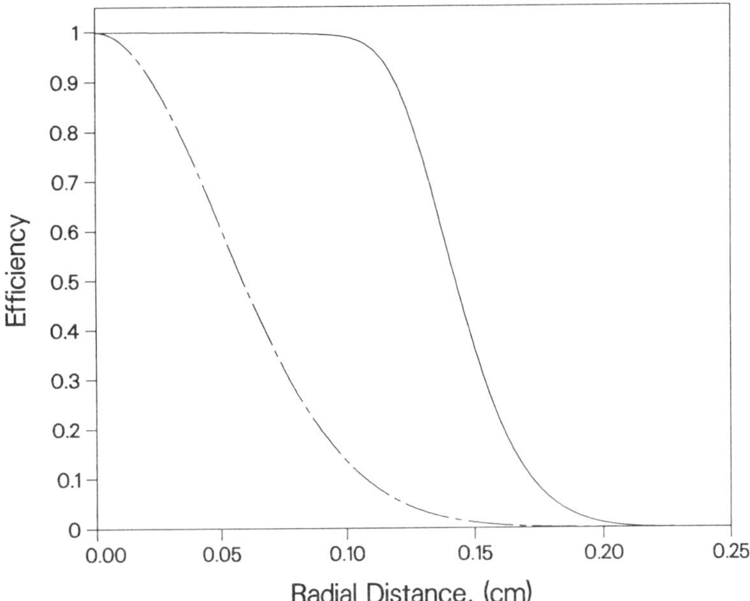

Figure 5. Ionization efficiency calculation for pulsed laser RIMS as a function of radial distance from the center of the cell at the middle of the z-axis (solid line). Also shown is a normalized Gaussian shape (dashed line).

between the curves at different laser pulse energies reflect the ionization efficiency underlying the geometric configuration. Higher laser powers obviously lead to greater efficiencies, as total saturation is not achieved. At the largest beam diameters, the effect of intermediate state fluorescence can also be observed in the tailing off of ionization efficiency at lower laser pulse energies.

Continuous Wave (cw) Lasers

For the case of cw laser resonance ionization, the exposure of the atom to the laser field is determined by both the spatial intensity distribution of the laser and the atomic velocity through the beam. It can thus be seen that a time-varying exposure will result for most realistic situations. This is contrary to the time-invariant laser pulse assumed in the explicit solution of the rate equations; however, for our first-order approximation, we choose to simplify reality. We assume that the intensity of the laser is invariant within a radius r_c, defined by:

$$r_c = w_z/2^{1/2} \tag{7}$$

In terms of the Gaussian distribution, this implies a beam of uniform intensity defined by the 1/e intensity radius of the true beam. In cases where deleterious

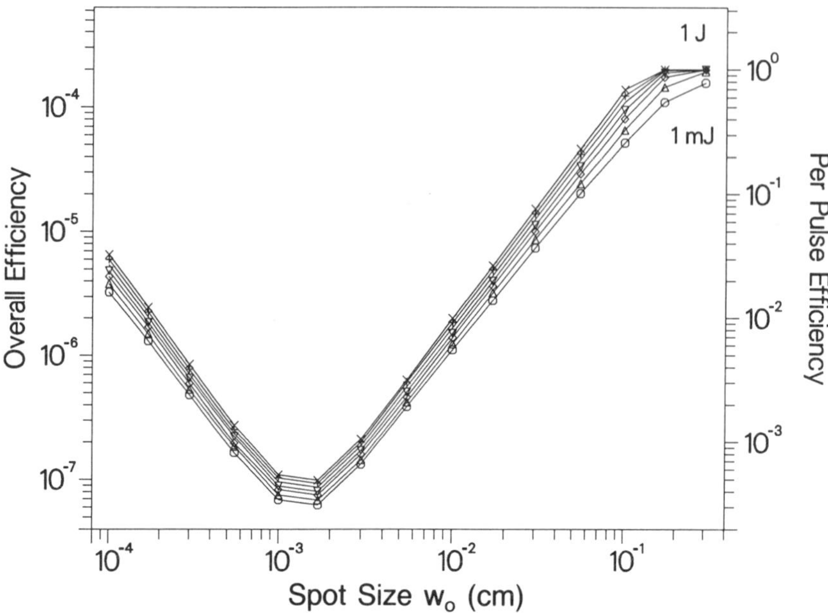

Figure 6. Ionization efficiency calculation for pulsed laser RIMS as a function of spot size for a variety of laser pulse powers. Left vertical axis displays the overall efficiency. Right vertical axis displays ionization efficiency per laser pulse.

optical pumping (transfer of atoms to nonaccessible states) may occur, this approximation may not be completely accurate. At any position z along the laser axis, the intensity is given by:

$$I(r,z) = I_0 * w_0^2/w_z^2 \text{ for } r < r_c \tag{8a}$$

with the power normalization again given by

$$I_0 = 4 * P/w_0^2 \tag{8b}$$

Atoms in this case are assumed to be traveling in the positive y direction, transverse to the laser axis. In this arrangement, the transit time of the atom through the beam may be calculated by geometrical arguments:

$$T = 2 * (r_c^2/y^2)^{1/2}/v \tag{9}$$

The experimental configuration for these calculations is applicable to a magnetic-sector mass spectrometer; specifically, the ionization volume was taken to be 0.1 cm transverse to the laser and 0.5 cm long, roughly corresponding to the size of a thermal ionization filament. Data points may be obtained by randomly selecting (y,z) points within the ionization volume (Hammersley and Handscomb, 1964).

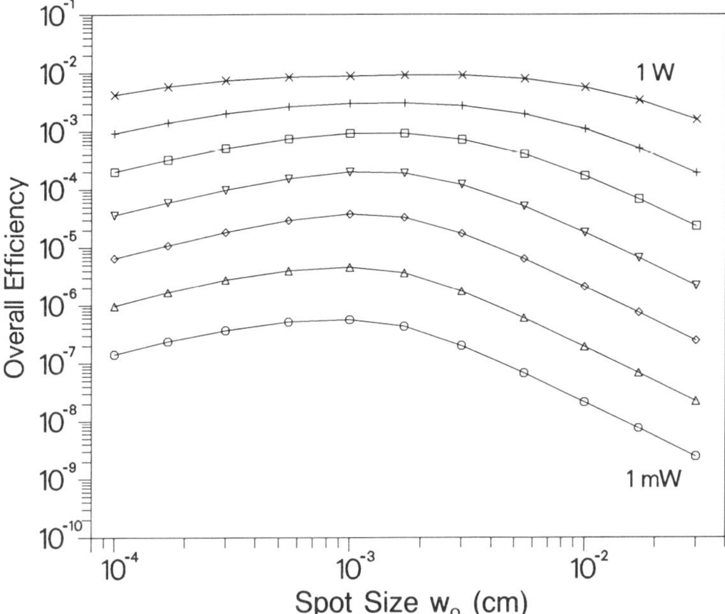

Figure 7. Ionization efficiency calculation for cw laser RIMS as a function of spot size for a variety of laser pulse powers. The left vertical axis displays the overall efficiency.

Figure 7 displays the results of cw laser calculations for various laser powers as a function of spot size. A single efficiency axis is shown in this figure, as this case exposes all atoms within the ionization volume and does not need to be corrected for the dark interpulse period of the pulsed laser case. As is evident from the figure, relatively small beam diameters are required for efficient ionization, in order to provide sufficient photon fluence. This is congruent with the small ion collection volume for typical magnetic instruments.

For cw ionization, at small central focal spot sizes (≈ 1 μm), efficient ionization only occurs at the center of the volume. As the laser beam expands toward the ends of the cell, the photon fluence drops off rapidly, and few ions are created. At moderate spot sizes (≈ 10 μm), ionization occurs along the total length of the volume. As the spot size increases further (100 μm), the ionization efficiency again drops, in spite of the increased volume exposed.

Both the pulsed and cw laser cases display similar overall ionization efficiencies. For the pulsed laser case, effective ionization occurs for the short pulse durations. For the cw case, a much larger fraction of the sample is exposed to a less effective laser fluence. The appropriate choice of mass spectrometer and laser is therefore likely to depend largely upon the specific equipment available to the experimentalist, although such factors as isotope shift, hyperfine structure, laser linewidth and mode structure, and power stability can all influence the final choice.

B. Mass Spectrometry

Mass spectrometry is the second half of the RIMS technique. Once ions have been created by the resonant laser(s), it is generally necessary to disperse the ions prior to detection in order to measure isotope ratios. (Isotopically-selective resonance ionization is typically insufficiently selective to remove the necessity for this step.) Various types of mass spectrometers are useful for this purpose, with the specific choice dependent upon factors such as the desired precision, accuracy, and dynamic range. In addition, the choice of a spectrometer can also reflect the choice of laser system and vice versa.

Spectrometers

The mass spectrometer of choice for much of the geologic community has traditionally been a magnetic sector instrument (a momentum dispersive instrument). This type of spectrometer has long been used to precisely measure isotope ratios (Perrin et al., 1985). This is due to several factors which include: good stability, high throughput efficiency, excellent abundance sensitivity, high dispersion, and low noise.

In this instrument, ions are accelerated to a fixed potential (~10 keV), extracted from the source region into a magnetic field in which the ions follow circular paths. The radius of the circular path is proportional to the mass of the ion; thus heavier ions are deflected to a smaller degree than lighter ions. Upon exit from the magnetic field, the ions encounter an exit slit placed in front of the detector. Modern systems have advanced to include pulse counting, moveable multiple Faraday cup detection, adjustable slits, and multiple electron multiplier systems. In addition, computing and modeling capabilities (Wollnik, 1987) have now advanced such that a complete system can be designed to include fringe fields, second, third and higher order effects, as well as complete ion optics systems. In general, magnetic sector mass spectrometers are continuous dynamic systems where the sample is generated in an ''always on'' mode.

Two detector configurations are common with magnetic sector instruments: (1) a single slit and detector; or (2) multiple slits with multiple detectors. In the single slit/detector form, only a single isotope may be measured at any particular moment. To measure isotope ratios, it is necessary to adjust either the accelerating potential or the magnetic field to pass successive isotopes to the detector. This is a disadvantage where the magnitude of the ion beam may be changing with time (i.e., a biased measurement may result), or where the amount of an analyte isotope is limited (i.e., valuable data can be lost during measurement of other isotopes). The multiplex advantage afforded by multiple detectors addresses these problems; however, interslit distances often require adjustment for the measurement of different elements, and space requirements can limit the type of detectors placed behind each slit.

Time-of-flight (TOF) mass spectrometers have also enjoyed much use in RIMS, particularly in combination with pulsed laser ionization. Among the advantages of time-of-flight (TOF) mass spectrometers are that they are relatively simple and robust. A TOF is velocity dispersive and must be operated in a pulsed mode to allow velocity discrimination (in time) of the different masses. In this method, ions are created at a well-defined time (the duration of the laser pulse), and are then accelerated out of the source region by potential fields. Both temporally- and spatially-varying and spatially-invariant voltages may be used to increase the resolution of the spectrometer (Wiley and McLaren, 1955). The transit time of an ion through a field-free region to the detector is inversely proportional to its mass; heavier ions travel more slowly and require a longer period to reach to detector. A reflectron geometry of the flight tube may also be employed to increase resolution; ions are reflected by electric fields at the end of one leg of the field-free region, to continue down a second leg (Watson, 1985). The advantage of this geometry is the correction for nonzero velocities of the original atoms. A time-of-flight instrument has the advantage that all ions created by a single laser pulse may be detected, allowing the measurement of an isotope ratio (in principle) with a single laser pulse. The effects upon isotope ratio of spatial and temporal variations in each individual laser pulse are in principle eliminated.

Quadrupole mass spectrometers (QMS) are basically mass selectors or filters, and are favored when high-abundance measurements are not required and rapid analysis is needed. The main advantages of a this type of mass spectrometer are (Roboz, 1968): (1) relative independence of energy distribution of the ion beam; (2) high transmission, particularly at low resolution; and (3) the ability to rapidly scan. Uses include routine noble gas measurements of major isotopes, organic sample analysis, and survey studies. In addition, when desired, the signal from a quadrupole MS can be detected in a gated mode as required for pulsed RIMS experiments.

Another type of mass spectrometer is an ion cyclotron resonance (ICR) mass spectrometer where the centrifugal force acting upon an ion in a magnetic field is balanced by the force from the field causing the ion to travel in a circle. Subjecting this ion simultaneously to an electric field perpendicular to the magnetic field causes the ion to proceed in a cycloidal path at right angles to both fields. The frequency of the ion motion is $u/r = eB/mc = w_c$ (Willard et al, 1974). If a rapidly alternating field is applied, as the frequency approaches the cyclotron frequency of a particular ion, the ion will absorb energy and be accelerated, causing the radius of its cycloidal radius to increase. Instead of collecting ions, the energy absorbed by the ion is detected by measuring the image currents induced on the cell. A mass spectrum is obtained by scanning the magnetic field at a fixed frequency, or the reverse. Because of long path lengths ($\sim$1000 times larger than in a sector instrument), ICR is useful for the study of ion molecule reactions at low pressures.

Advantages of ICR are speed, reasonable sensitivity, wide mass range, and high mass resolution. Disadvantages are poor dynamic range, low precision, poor abundance sensitivity, and lack of pulse counting ability. These features as well as physical constraints limit the use of this type of mass spectrometer for RIMS work. Nevertheless, laser microprobe and direct ionization applications of ICR (Brenna, 1988; Muller et al., 1989) and some isotopic analysis (Snyder, Lucatorto, Debenham and Geltman, 1985) have been demonstrated.

Detectors and Detection Electronics

Both electron multiplier and Faraday cup/electrometer detectors are used in conjunction with the mass spectrometers described above, depending upon the magnitude of the ion signal. For low ion count rates, an electron multiplier is applicable because of its low background and sensitivity to single ion counts. Disadvantages of this type of detector include limited dynamic range, $\leq 10^5$ (large-magnitude ion beams can destroy the dynode surfaces, especially in the later stages of the multiplier), finite detector dead time ($\approx$10 nsec), and pulse pileup (due to production of a large number of ions during a short laser pulse). Potential effects from these disadvantages include nonlinear response and inability to discriminate simultaneous ion signals.

High-precision instruments in our laboratory utilize 17-stage Cu-Be dynode multipliers because of their steady gain and low noise characteristics (typically $<$0.05 counts per minute), but are normally limited to count rates less than 250,000 ions per second. Continuous dynode electron multipliers (channeltrons) may also be used, with the advantage of a more robust nature but the disadvantage of nonconstant gain.

Also in development are Daly detectors (Daly, 1960; Agbaji, et al., 1985). In this detector, incoming ions are accelerated to a very high voltage to impinge upon a metal surface, where they scatter a large number of electrons. These electrons are then accelerated in the opposite direction, where they collide with a scintillator screen. A photomultiplier exterior to the spectrometer vacuum envelope then detects the scintillation burst. The advantage of this somewhat complicated chain of events is a detector of low noise, high dynamic range ($\geq 10^6$), and high linearity.

With either type of counting detector, single ions generate a signal which is amplified and discriminated. Pulses are then counted using a scalar or their time history recorded using timing electronics.

The Faraday cup detector is basically a current integrating device; all charge entering the cup should be trapped and read by a sensitive electrometer. Suppression electrodes may be added to the cup to prevent the loss of electrons scattered by ion impacts on the cup surface. This type of detector is especially suited to the measurement of relatively large ion currents, but noise limitations of the electrometer restrict the usefulness (sensitivity) below approximately 100,000 ions per second. In our laboratory, a moveable Faraday cup is coupled with the electron multiplier to provide both high sensitivity and high dynamic range.

Most of today's mass spectrometers are interfaced with some sort of computer control. Such a system generally will control the mass scanning of the mass spectrometer (e.g., change the RF-field, accelerating voltage, or the magnetic field) as well as collect the signal (either from a pulse counter, channeltron, or a Faraday cup) for processing. These data are then reduced by standard numeric techniques to analyze, calculate and interpret the information.

Sample Preparation

As in most mass spectrometry, sample preparation can be critical for RIMS. Although RIMS can potentially be performed without any sample preparation, some minimal amount is normally required. The main difference between RIMS and conventional MS is the degree of purification and separation required. Several methods can be used for preparation of samples for analysis. Briefly, samples are normally placed into solution which requires dissolution via microwave digestion, strong acids, or some more involved processes (especially for whole rock samples). Normally in geologic samples, some sort of gross separation is needed which separates glasses from magnetites. Then, as required, a separations chemist removes the bulk of unwanted materials and elements.

A more critical aspect of the RIMS process is that in order to maintain the selectivity, the species being evolved must be atomic in nature and in a neutral, low-lying electronic state. Importantly, this requires that speciation and complexation effects be minimized. Significant effort has been spent within the RIMS community to determine the optimum procedure to assure that this occurs. As with thermal ionization, most practices have developed by trial-and-error and ''black magic''. In fact, most of the techniques utilized incorporate methods initially developed for standard thermal ionization mass spectrometry. One favored method is to attempt to maintain a reducing atmosphere for the atom of interest, in order to inhibit the evaporation of oxides. The major techniques to achieve this have been to overcoat the stippled sample with either uranium (Cantanzaro et al., 1966) or graphite (Walker and Fasset, 1986). Additionally, resin beads from the separations process have also been used to provide a reducing environment (Downey et al., 1984b). Lastly, a successful way to achieve good neutral atomization has been to electroplate platinum over the sample (Rokop et al., 1982; Fearey et al., 1988).

Another important aspect of sample preparation for RIMS is the minimization of any thermal ionization. The easiest way to achieve this is to use filament materials which have a low work function. Finally, to attain maximum sample utilization, nonstandard filament configurations are often used. Examples include ultrasmall filaments (Fasset et al., 1984) boat shaped filaments to increase sample interaction and concentrate analyte within the laser volume, as well as enclosed source geometries (Andreev et al., 1986) which should limit sample loss.

C. Pulsed RIMS

Initial work in RIMS typically utilized pulsed lasers for excitation and ionization. The reasons for this are largely historical: the state-of-the-art in pulsed lasers was far advanced relative to that of cw lasers during the 1970s when resonance ionization was in its infancy. In addition, several of the initial applications of resonance ionization were directed toward the detection of metastable electronic states produced in a burst mode (Hurst et al., 1979). Hence, the duty cycle and high intensities of pulsed lasers were well-suited to these applications.

Pulsed lasers are still used in the majority of RIMS experiments. Motivations for their use include the wide wavelength coverage (near IR to XUV), relative ease of tuning, and high intensities possible. In addition, the relatively broad spectral bandwidth, and poorly formed mode structure tend to make pulsed laser ionization less susceptible to isotopic bias than with cw lasers. In general, the requirements on lasers for pulsed RIMS are relatively simple (Ruster, 1989): (1) tunability, to ensure that the laser can be tuned into resonance with an electronic resonance of the analyte of interest; (2) sufficient intensity to drive both the resonant (bound-bound) and bound-free transition(s); and (3) sufficient repetition rate to address a substantial fraction of the analyte atoms. The specific values required will depend on the particular application. Table 2 displays the characteristics of several commonly used laser systems for pulsed RIMS work. Laser linewidths for pulsed lasers are ultimately limited by the uncertainty principal, called the transform limit $t = 1/2\pi\Gamma_{fwhm}$. However, pulsed laser linewidths are typically 5 to 20 times the transform limit (0.02–2 cm^{-1}).

Depending on the pulse energy of the laser and the application, the pulsed laser beam is usually loosely focused to a spot size of 1 to 5 mm in the source region of the mass spectrometer. In most instances, this ensures that the beam cross-sectional area will be invariant over the dimensions of interest; therefore, several millimeters. This facilitates numerical modeling (Miller and Nogar, 1983a) of the ionization yield (see modeling section above), and minimizes the probability for accidental nonresonant ionization of interfering species in the "hot" spots that might occur in a tightly focused beam.

Because pulsed RIMS generates ions in a burst mode, many previous efforts have utilized mass spectrometers capable of multiplex detection, such as time-of-flight and Fourier transform ion cyclotron resonance instruments.

Table 2a. Pulsed Laser Systems Commonly Used for RIMS

Dye Laser Type	*Pulse Energy[a] (mJ)*	*Pulse-Width[b] (nsec)*	*Repetition Rate[c] (Hz)*	*Spectral[d] Bandwidth (cm^{-1})*
Flashlamp–pumped	10–1000	250–10^6	1–50	0.25–0.75
YAG laser–pumped	10–300	7–15	10–50	0.02–0.45
Excimer laser–pumped	5–100	10–25	10–500	0.02–0.45
Cu laser–pumped	1–5	10–25	6000	0.02–0.45

[a] Maximum pulse energy at the fundamental frequency—will vary with wavelength; will be reduced for frequency-doubled ouptut

[b] Nominal pulse-width—may vary with manufacturer

[c] Maximum repitition rate—dependent on manufacturer and model

[d] Assumes no use of etalon, bandwidth altered or frequency-doubled output

Table 2b. Cw Laser Systems Commonly Used for RIMS

Laser Type	*Power*	*Wavelength range (nm)*	*Spectral Bandwidth (cm^{-1})*
Kr, Ar ion laser[a]	5–25 (vis) 0.5–7 (uv)	407–799[b,c] 275–386[b,c]	0.01–1.0
Linear dye	5–25% conversion of pump laser	370–935[c,d]	0.1–5
Linear Ti:Sapphire		670–1100[c]	0.03–1.5
Ring dye	5–30% conversion of pump laser	370–935[c,d]	0.0001–1.5
Ring Ti:Sapphire		670–1100[c]	0.0001–1.5
Doubled Ring dye[e]	0.1–10 mW	270–395	≤ 0.0001
Doubled Ring Ti:Sapphire[e]	10–200 mW	350–460	≤ 0.0001
Nd: YAG[a]	5–100 W	1064	0.1–2
Diode[f]	1 mW–10 W	670–1550	≤ 0.0001–2
CO_2	1–100 W	9.2–10.8 μm[g]	0.01–1

[a] fixed frequency, not tunable

[b] either single line or multi-line outputs

[c] may require different mirror sets for entire range

[d] dye changes required for entire range

[e] tunable over limited range, must operate in single-frequency mode

[f] very limited tuning range; wavelength specified at purchase

[g] wavelength selected to single line ouputs

Sample preparation is crucial for effective application of any mass spectrometric technique, including RIMS. Early work was based on thermal sources using Langmuir evaporation, as this method was well-developed for thermal mass spectrometry (Rokop et al., 1982). It soon became apparent that the use of a continuously evaporated source with pulsed ionization resulted in substantial loss of sample.

Several alternatives have been explored to overcome this problem. The use of cw lasers (Miller and Nogar 1983a; 1983b), alternate source geometries (Andreev et al., 1986) and pulsed thermal sources (Fassett et al., 1984) have all been reported. More recently, a variety of sputter sources have been utilized, including particle bombardment (Becker and Gillen, 1984; Kimock et al., 1984; Pellin et al., 1984) and laser ablation or desorption (Beekman et al., 1980; Nogar et al., 1985; Kröenert et al., 1987; Nogar et al., 1990b). The variety of atomization sources commonly used for RIMS are listed and evaluated in Table 3.

The use of laser ablation coupled with RIMS detection of the sputtered neutrals has a number of interesting applications as well as contrasting advantages and disadvantages. Among the advantages of laser desorption is the ability of lasers to interrogate insulating materials (Estler and Nogar, 1988). The duty cycle argument mentioned above is also extremely important. In addition, laser desorption can allow high spatial resolution (≈ 1 μm) and the evaporation of difficult refractory or tightly bound samples. Lastly, this method can be used to study the fundamentals of laser-material interactions by diagnosing the evolution and speciation of evaporated material, as well as the kinetic and internal energy distributions of the evaporated atoms, molecules, and fragments (Estler et al., 1987). On the other hand, quantitative measurements by laser-desorption/RIMS have been relatively infrequent, in part because of the difficulty in removing reproducible amounts of material. In addition, the damage done to the material surface is in general greater than for sputter-initiated RIMS.

Table 3. Atomization Sources for RIMS Analyses

Source Type	*Advantages*	*Disadvantages*
Thermal Filament	Reproducibility, Stability, Literature, High duty cycle for cw lasers	Low effective duty cycle for pulsed lasers, Sample preparation
Laser Ablation	Pulsed Operation, Microprobe capability	Kinetic energy spread, Excited states & ions, Production of molecules & fragments
Particle Sputtering	Pulsed Operation, Microprobe capability, depth profiling	Kinetic energy spread, Excited states & ions, Matrix effects, Production of molecules & fragments
Glow Discharge	Pulsed Operation, Reproducibility, Stability	High Pressure, Sample preparation

It is informative to compare typical parameters for laser ablation (Nogar et al., 1985) and particle sputtering (Hand et al., 1989). In a laser ablation process, typical pulse energies are 10^{-1} J, delivered to a 1 mm^2 spot in a 10^{-8} sec pulse (Nogar et al., 1990b). In the near-IR (1 μm), this corresponds to a "particle" fluence of 5×10^{19} photons/cm^2. While it is difficult to estimate a general number of ejected particles for the laser ablation case, for Si an ablation yield of 5×10^{14} atoms per pulse has been reported for a deposited laser energy of 5×10^{-4} J (Mayo et al., 1982).

In the case of particle-induced sputtering, typical beam currents are ≈1 μA (Gruen et al., 1988). For a spot size of ≈1 mm^2, this corresponds to a flux of 6×10^{14} ions/cm^2—sec. In a 10 μsec, 1 keV ion pulse, this produces a fluence of 6×10^9 ions/cm^2, with the deposition of 10^{-6} J of energy. Assuming a sputtering yield coefficient of 2 (two-particles ejected for each incident ion), 1.2×10^8 atoms are ejected from the 1 mm^2 irradiated spot. Because of the dramatic differences in the mechanisms of sputtering and laser ablation, the velocity and spatial distributions also vary significantly (Nogar et al., 1990b).

For a two-dimensional approximation, the temporal evolution of a thermal plume generated by laser ablation from the surface can be calculated (Nogar et al., 1990a). Typical results are shown in Figure 8 for a vapor having an atomic

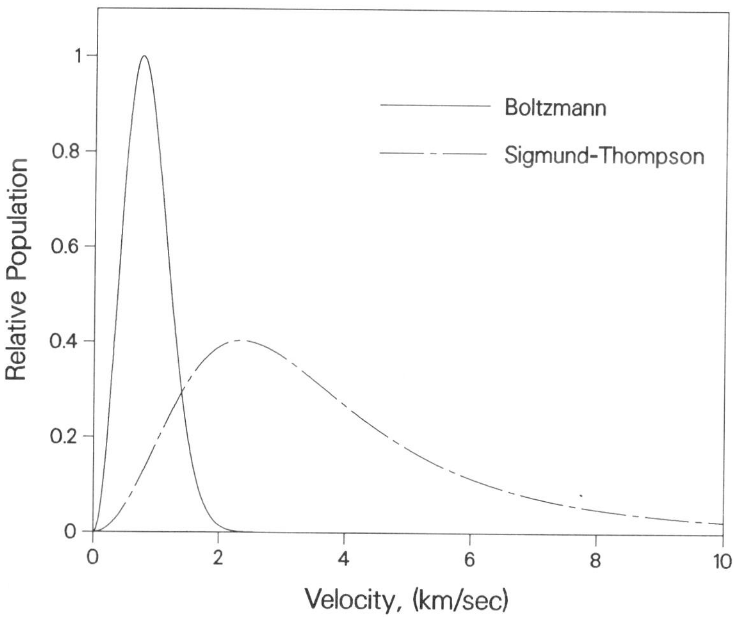

Figure 8. Velocity distributions for Maxwell-Boltzmann (m = 60, T = 2000 K) and Sigmund-Thompson (m = 60, E_b = 333 kJ/mole) descriptions of sputtered atoms.

mass of 60 and a thermal temperature of 2000 K. As the time after initiation increases, the plume moves away from the surface, and broadens spatially. Thus, in order to obtain the greatest sensitivity, the plume should be probed close to the surface and soon after initiation of the ablation event. On the other hand, if the goal is maximal information about the spatial and temporal distributions of the ejected material, the ionization laser should be delayed, and directed at a distance from the surface.

A glow discharge may also be used in conjunction with RIMS as an efficient source of gas phase atoms (Savickas et al., 1984). In this case, a low pressure gas (typically 0.1–1.0 Torr) supports a discharge in the source region. The analyte of interest is sputtered from a solid support, and extracted through an orifice into the ionization region or the mass spectrometer (Hess and Harrison, 1986). The sputter process is relatively non-selective, and thus exhibits reduced matrix effects. Differential pumping is required to maintain the ionization region at sufficiently low pressure so that ion separation and detection function properly. In conventional glow discharge mass spectrometry, the sputtered atoms are subsequently ionized (with ≈1% efficiency) in the same, or a secondary discharge. With resonant laser ionization, the efficiency can be much greater, approaching 100% for atoms present in the focal volume during the laser pulse. For RIMS applications, the glow discharge is normally run in a pulsed mode, and the ionization laser pulsed synchronously with the discharge. As in the discussions above, this serves to better match the duty cycles of the atom source, and the laser ionization efficiency.

D. Continuous Wave (cw) RIMS

The primary motivation for considering cw laser ionization for RIMS processes is that pulsed laser RIMS is often limited in both sensitivity and dynamic range (Miller and Nogar, 1983b). In principle, a cw RIMS process can greatly enhance the overall sample use by naturally matching the thermal evolution of the sample. In addition, pulse height analysis and pulse pileup difficulties are reduced. Further, the laser characteristics are in general more easily controlled and monitored. Lastly, using single-frequency cw lasers, isotopically selective analysis becomes viable.

The effective duty cycle is of particular importance when one attempts to make high dynamic range measurements. Effective duty cycles for pulsed laser are typically £ $\approx 10^{-3}$, while cw lasers are intrinsically 100% when utilizing thermal atomization sources. This allows high sensitivity and high dynamic range measurements. Unfortunately, cw lasers suffer from relatively low peak powers, so that purely resonant processes are usually required [i.e., no nonlinear processes such as two-photon excitation (Bonin and McIlrath, 1984; Melikechi and Allen, 1986].

Accuracy, precision, and reproducibility of isotope ratio measurements are particularly important within the isotope geochemistry arena. These aspects of RIMS have received much attention over the last few years. Characteristics that must be considered in assessing the accuracy of RIMS isotope ratio measurements are: laser mode structure (both spectral and temporal); laser linewidth effects; atomic hyperfine structure; and atomic selection rules. All of these effects can perturb (i.e., bias) the measurements. The precision and reproducibility of measurements reside primarily with laser stability (both in power and frequency) as well as frequency resettability.

Cw lasers offer the best method to control these parameters. The mode structure of cw lasers is relatively easy to monitor and control (Bomse and Keller, 1983; Connally and Morton, 1990). In addition, any changes in the mode characteristics of a cw laser are typically quite slow, whereas the mode structure of pulsed lasers can change from pulse to pulse (Nogar and Keller, 1985b).

It is also important to consider laser linewidths when making isotope ratio measurements because of hyperfine splittings and isotope shifts. Often, it is assumed that a "broad" linewidth uniformly excites all isotopes equally, but this has proven to not always be true (Miller et al., 1989). This may be due to mode structure, optical pumping effects, density matrix anomalies or selection rules (Cowan, 1981; Bonin and McIlrath, 1984; Melikechi and Allen, 1986). These effects can apply for either cw and pulsed lasers.

Cw lasers are essentially limited only by stability, and are readily available with ~1 MHz linewidths ($<0.0001\ cm^{-1}$). Such narrow linewidths allow for isotopically selective ionization (where ultrahigh dynamic range is needed) and detailed spectroscopic studies. Linewidths for cw lasers can also be made to operate with linewidths as large as 1 to 3 cm^{-1} when selectivity is not a criteria. Reproducibility is basically defined by power stability and frequency reproducibility. Continuous dye lasers normally exhibit a stability ~1% and can utilize "noise eaters" to further increase stability to 0.01%. Precision wavemeters or some other external standard are routinely used to limit frequency errors.

The spatial mode structure is easily controlled simply by varying the pump focus. Normal pure TEM_{00} mode is easily attained and maintained. The spectral mode structure requires more diagnostics but can be controlled in modern commercial lasers easily (Bomse and Keller, 1983; Connally and Morton, 1990).

The most common pump lasers are argon and krypton ion lasers (see Table 2b). Cw solid state lasers operate in the IR with only weak doubling and presently are not useful as pump lasers for widely tunable lasers.

Cw tunable lasers are available in basically two types, each with different gain mediums. The most common and generally most versatile are dye lasers. The other uses a solid state crystal as the gain medium, typified by Ti:Sapphire, F^+-center and Alexandrite. Tunable cw lasers also generally come in two configurations: linear and ring. The first is very easy to use but is limited in conversion efficiency due to saturation and hole-burning. Ring lasers do not have this problem but often have critical alignment requirements.

Intracavity excitation can be used in some instances to increase the effective power, (Downey et al., 1984a). For shorter wavelengths, frequency doubling is required with dramatic reduction in laser power. Powers above a milliwatt are often very difficult, although fractions of a milliwatt (focused) are often sufficient to saturate bound-bound transitions.

The future of RIMS will undoubtedly involve the use of semiconductor diode and solid-state cw and pulsed lasers. Presently, the use of the RIMS technique is hindered by the complexity of the tunable lasers as well as their cost. In contrast, diode and solid-state lasers are simple to operate and possess the properties of low cost, long life-time, (limited) tunability, and narrow bandwidth (~25 MHz for ''off-the-shelf'' items). Their implementation will significantly increase as the optical output of these devices continues to develop into the visible and near-UV. There are a few examples of diode laser use for ionization processes; most often they are used in only one step of the ionization process (due to their low photon energy output and limited spectral availability) (Lawrenz et al., 1987; Shaw et al., 1989).

It is sometimes possible to improve ionization efficiency by the use of an additional ionization laser. The purpose of this laser is to supply photons at some fixed frequency to efficiently promote the excited atoms to the continuum. The ionization process often has a very low cross section, and thus may have a low efficiency when driven with relatively low-power dye lasers. Some cw lasers available for this purpose include argon ion, krypton ion, CO_2, and Nd-YAG lasers.

III. APPLICATIONS

A. Thorium RIMS

Many nuclear clocks are available for geochronology; some of the most useful are based on the daughter products of uranium. The extensive applications of these nuclear clocks are collectively termed uranium-series disequilibrium (Friedlander et al., 1981; Ivanovich and Harmon, 1982) which is a well established valuable method in geochemistry and geochronology. $^{238}U/^{230}Th$ disequilibrium is particularly valuable due to its geochronologic time scale (10 to 350 ky) which accesses ages not readily obtainable by other methods. Important applications for the $^{238}U/^{230}Th$ decay-series are the dating of corals (Edwards et al., 1988) continental carbonate deposits (Li et al., 1989; Sturchio et al., 1989) ocean floor basalts (Goldstein et al., 1988a,b), and volcanic rocks (Condomines et al., 1988; Murrell et al., 1990; Volpe and Murrell, 1990).

Many potential applications of ^{238}U-^{230}Th disequilibrium have been severely restricted due to the relatively large sample size requirements (~10 grams) and the poor precision obtainable by alpha spectrometry. The relatively low concentration of U and Th, as well as the long half-life of ^{232}Th and ^{230}Th, contribute significantly to this limitation. Further, since large samples are typically needed, significant questions can arise concerning potential secondary alterations to the samples.

Mass spectrometers are now used for the longer-lived members of the U-decay series (Edwards et al., 1987, 1988; Goldstein et al., 1988a,b). The reasons for ion counting over alpha counting are obvious and straightforward: (1) smaller sample size and purer mineral separates (crucial for internal isochron determinations); (2) improved analytical precision; and (3) significantly shorter counting times. Using thermal ionization techniques, earlier workers (Goldstein et al., 1988a,b; Murrell et al., 1990) have been able to reduce sample size and improve precision both by about a factor of ten compared to alpha counting. They have successfully applied these techniques to measurements of ^{238}U-^{230}Th disequilibrium in bulk samples ($\sim$1 g) of ocean basalts (Goldstein et al., 1988a,b; Murrell et al., 1990); however, in order to apply the internal isochron technique (Ivanovich and Harmon, 1982; Faure, 1986; Condomines et al., 1988) to obtain crystallization ages, the use of mineral separates is required.

The uranium and thorium contents of many of the separated phases are quite low, and the thorium ionization efficiency currently available by thermal techniques [$\sim$0.04% efficiency (Goldstein, 1988a; Murrell et al., 1990)] is not always sufficient to make these measurements. Note that typical ^{230}Th/^{232}Th ratios in young volcanic rocks are $\sim 5 \times 10^{-6}$ which requires quite high signal levels in order to obtain good precision and accuracy for the minor isotope. The low thermal ionization efficiency of thorium is due to a combination of low vapor pressure and relatively high ionization potential.

RIMS has the potential to significantly increase ionization efficiency and promises to make the geochronology of the last 350 ky much more accessible. In fact, preliminary RIMS experimental results have demonstrated an approximately *40-fold increase* in ionization signal over TIMS, (Murrell et al., 1990; Fearey et al., 1990b). With further work, we anticipate another *10-50 fold increase* in relative ionization efficiency. This will bring the overall ionization efficiency into the several percent range, potentially approaching the tens of percent. Such a gain then allows for the internal isochron method (Ivanovich and Harman, 1982; Faure, 1986; Condomines et al., 1988) to be utilized on very small samples, as well as extend the age interval available for study with this technique.

Several factors must be examined when considering the utilization of a RIMS process, particularly to increase signal levels over those attainable by TIMS. We will use the RIMS ionization process of thorium as a working example. Important questions include:

1. At what isotopic ratio are the species of interest?
2. What are the precision requirements and desires?

3. What is the desired sample size?
4. What limitations are imposed by the element's ionization potential?
5. What is the best resonance ionization scheme?
6. What are the important spectroscopic factors (e.g., hyperfine structure, isotope shifts, potential optical pumping effects, etc.)?
7. What laser linewidth is most appropriate?
8. What laser system is best (i.e., pulsed or cw, follows from 1–7)?
9. What are the required mass spectrometer specifications?
10. How do all of these factors interrelate and effect each other?

The abundance of ^{230}Th relative to ^{232}Th is typically $\sim 5 \times 10^{-6}$ in young volcanic rocks with a required precision on the order of a 0.5 to 1.0%. Maximum sample size is on the order of 1 microgram with desired sample sizes of subnanogram. These limitations impose severe restrictions to the final RIMS process. The high dynamic range, precision, and sample size requirements imply that a highly efficient method be used maximizing sample interaction. Further, the effective stability of the ionization process must be high (i.e., very little noise can be allowed).

The thorium ionization potential is 6.08 eV or $\sim$49,000 cm^{-1}. If a simple $1+1$ process is used then photons of 407 nm (deep blue) or shorter are required. If a $2+1$ scheme is used, then photons of only 612 nm are needed. Obviously, more color ionization schemes are also possible but more complicated. With this information in hand, one can then examine atomic spectral data tables (Corliss and Bozman, 1962) for possible applicable transitions. Finally, important isotopic factors must be considered such as hyperfine structure and isotope shift. In a simple one-photon process, one determines whether an isotope has hyperfine structure by examining the nuclear spin (I) of the isotope and the J quantum number of the transition of interest (Cowan, 1981). The allowable transitions (Herzberg, 1944) then are $F'' \rightarrow F'$ with $\Delta F = 0$ and ± 1 (where $F = |J+I| \ldots |J-I|$, F'' is the ground state, while F' is the excited state. For higher order processes, see Bonin and McIlrath (1984) and Melikechi and Allen (1986). Since for ^{232}Th and ^{230}Th the spin is zero, neither will have any hyperfine structure. For estimating the isotope shift, it is useful to remember the general isotopic shift trend of the periodic table: the isotope shift is large at low masses (mass effects); reaches a minimum at mass number$\approx$100 (both mass and volume effects are small); and is again large at high masses (volume effects) (Bushaw, 1989). The isotope shift for thorium for the applicable transitions is $\sim$0.4 cm^{-1}, (Engleman and Palmer, 1983).

The last question effecting the optimal choice of RIMS process is the laser linewidth. Several aspects must be considered; these include the expected Doppler width (i.e., for a thermally evolved sample, typical linewidths are $\sim$0.05 cm^{-1}), the spectral width including hyperfine structure, and the isotope shift. The minimum desirable laser linewidth for isotopic ratio measurements would be equivalent to the Doppler width, while the maximum would overlap all hyperfine features and excite all isotopes equally regardless of the isotope shift. A linewidth narrower than the Doppler width reduces the number of atoms available to interact with the laser, while a greater linewidth will effectively decrease the number of resonant photons. However, resonant transitions are easily saturated, and linewidths larger than the Doppler linewidth or hyperfine envelope will still efficiently saturate the resonant transition.

Another aspect that will need to be examined in the final analysis of any isotopic ratio measurement is the potential for isotopic biases because of differing hyperfine structures between isotopes which can be nonequally excited by the laser and can bias the apparent ratio (Miller et al., 1989; Fairbank et al., 1989b).

For these experiments, we require a large dynamic range of measurement and the greatest possible precision. This suggests the use of a cw system to maximize duty cycle and eliminate pulse pile-up problems. A simple 1 + 1 resonance ionization scheme is therefore utilized. The choice of linewidth is less clear, but a standard standing-wave dye laser can easily be controlled to yield linewidths for <0.05 cm^{-1} to >3.0 cm^{-1}. Note that assuming a Doppler width of 0.05 cm^{-1} (matched with a Lorenztian laser linewidth), an isotope shift of 0.4 cm^{-1} and an excess of 10^6 of ^{232}Th would give a background of ~4000 counts per second (Bevington, 1969; Paisner and Solar, 1987) which can easily be discriminated against by most mass spectrometers.

In order to make best use of the high-duty cycle for ionization, and the relatively large ratios anticipated, a magnetic sector instrument with both a Faraday cup collector and an electron multiplier tube is used. With both types of collectors, a 10^6 measurement is not limited by the detection system. However, since a single sector will not have sufficient abundance sensitivity, a modest amount of isotopic selectivity is required which adds a degree of complexity to the experimental measurement system.

As mentioned above, we have succeeded in increasing the overall ionization efficiency via RIMS by a factor of ~40 times over TIMS (~10-fold with dye laser and ~4-fold with addition of Ar^+ laser) (Murrell et al., 1990; Fearey et al, 1990a). Our choices were as follows: (1) a simple 1 + 1 RIMS process, Figure 9; (2) one of 6 or 7 transitions between 404 and 372 nm. (Table 4 shows the most prominent observed RIMS thorium transitions) (Palmer and Engleman, 1983); (3) cw lasers with a nominal linewidth of ~1.3 cm^{-1}; (4) a magnetic single-sector mass spectrometer (with the capability to go to a two-stage after initial work); and (5) since there is no hyperfine structure for either ^{230}Th or ^{232}Th, only the isotope shift is important, therefore RIMS isotopic selectivity has the potential to allow $<5{:}10^8$ level measurements on only a single-sector MS (assuming the laser linewidth is narrowed to match the Doppler width).

Figure 10 shows a typical signal versus current plot for the strongest transition at 382.95 nm which attained ~19 volts of Th signal or $\geq 10^9$ counts per second. The overall ionization efficiency, based on extrapolation from TIMS work (Goldstein et al., 1988a; Murrell et al., 1990) is ~1%.

It is clear, therefore, that with careful consideration of the above 10 important

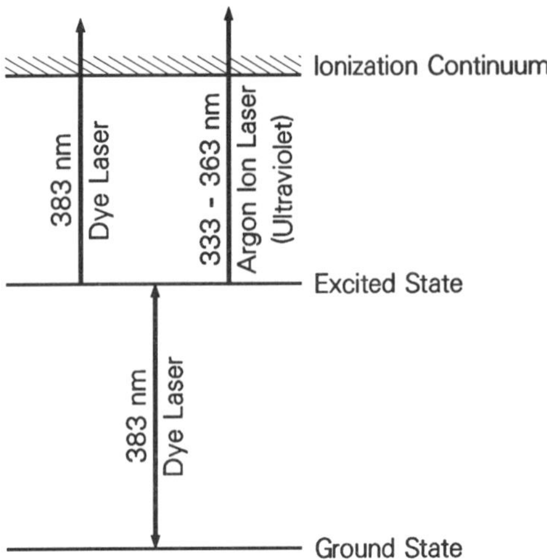

Figure 9. The applicable energy-level diagram for the cw-thorium RIMS process.

Table 4. Thorium Transitions Used in RIMS

Wavelength (nm, vac)	*Energy (cm^{-1})*	*Even*	*J''*	*Odd*	*J'*	*Rel. Int.*	*RIMS Int.*	*Comments*
403.71883	24769.7142	0	2	24769	3	0.71	—	dye edge
395.15131	25306.7615	0	2	25306	2	0.35	0.04	
392.62049	25469.8883	3687	3	29157	3	0.25	0.92	thermally populated
384.07845	26036.3476	0	2	26036	3	0.76	0.35	
383.18605	26096.9837	0	2	26096	3	0.37	0.17	
382.94709	26113.2683	0	2	26113	2	1.00	1.00	
380.41547	26287.0486	0	2	26287	1	0.73	0.21	
372.04927	26878.1606	0	2	26878	3	1.09	—	dye edge

questions, it is possible to reasonably choose the best approach to the RIMS process for a particular problem as well as the likelihood of success.

Lastly, we are pursuing a technique which is expected to raise the overall effective laser power 10- to 50-fold, which should raise our ionization efficiency significantly. In the final analysis, we anticipate utilizing isotopically selective ionization of ^{230}Th and ^{232}Th by locking the laser to an optogalvanic signal generated in isotopically doped hollow cathode lamps (Keller and Zalewski, 1980; Apostol et al., 1982; Keller et al., 1983). In addition, the dye laser will be stabilized by a "noise eater" (precision of $\leq$0.5% has been demonstrated for this system). By shifting laser frequencies and by using a multicollector mass spectrometer, we anticipate ^{230}Th/^{232}Th ratio measurements of subnanogram samples with a precision $\leq$0.1% and potential dynamic range of $\leq 1{:}10^9$. Such capa-

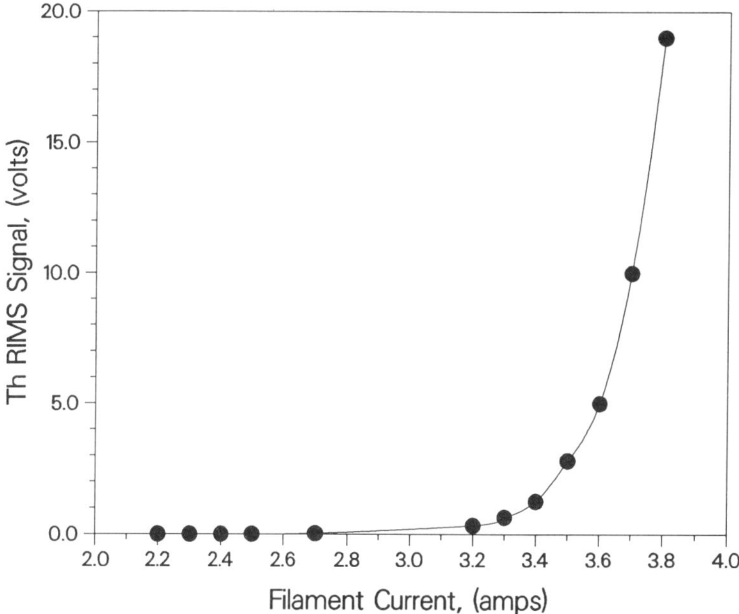

Figure 10. A thorium ion signal versus current for the strongest transition at 382.95 nm. The maximum signal corresponds to ~19 volts of Th signal or $\geq 10^9$ counts per second.

bilities will greatly enhance and facilitate the use of the internal isochron method (Condomines and Allégre, 1980; Ivanovich and Harmon, 1982; Faure, 1986) for U-Th disequilibrium.

B. Re/Os

Rhenium and osmium differ significantly from most other geological systems studies in isotope geochemistry; for example, ^{87}Rb-^{87}Sr, ^{147}Sm – ^{143}Nd, and the U-Th decay series. Unlike these other isotopic systems which are chemically stable in silicate materials, the Re and Os share the geochemical property of being "siderophile"; that is, they are incorporated into the metallic irons and iron-alloys, such as found in the Earth's core. This property allows the Re-Os system to be used for studying the geochemical fractionation, and hence dating of such things as rocks, meteorites, and ores (Riley and Delong, 1970; Luck et al., 1980; Luck and Allégre, 1984, Faure, 1986) which cannot be readily dated by other methods. When there is an absence of metallic iron, the usual condition in the Earth's crust and mantle (Faure, 1986) rhenium is enriched several hundred-fold relative to osmium during the melting process responsible for the production of crustal rocks for mantle rocks (Faure, 1986; Walker et al., 1987).

Thus, the Re-Os system has extreme sensitivity as a tracer of geochemical differentiation, and hence, a unique capability to document the evolution of the Earth's mantle and crust through time (Allégre and Luck, 1980; Walker et al., 1987).

Even though the Re-Os system appears to be an ideal tool for the study of various geochemical and geophysical processes, isotope geochemists have made only limited use of this tracer system. This is due to several factors: (1) the natural abundance of both elements in most common rocks is very low (ppb-ppt); (2) the ionization potentials of Re and Os are relatively high (7.88 and 8.7 eV, respectively); (3) both elements are very refractory (m.p. = ~2000 °C and ~3000 °C, respectively); and (4) the major problem of isobaric interference between ^{187}Re and ^{187}Os. These factors have basically precluded standard TIMS and limited the application of SIMS (Allégre and Luck, 1980) and ICP-MS (Lichte, 1986; Russ, 1987). RIMS offers a potential solution to this problem.

This has been demonstrated recently by both Walker and co-workers (1986, 1987) and, most recently, by Blum and co-workers (1990b). Unlike the U-Th series disequilibrium problem where the primary limitation is the Th ionization efficiency, both Re and Os must be ionized through a RIMS process. The ratios of interest are typically ~one-third for $^{187/186}$Os and ~two-thirds for $^{187/185}$Re. Precision requirements are 3 to 5%. Sample sizes are typically nanogram quantities from ~10 grams of rock. The combination of high IP and moderate isotope ratios allows pulsed-lasers to be used reasonably, so that a potentially large number of ionization schemes may be used.

Walker and Fassett (1986) have developed systematics that allow them to use a single laser dye with excellent stability and high doubling efficiency, for both Re and Os laser ionization. A single dye process allows the analysis of both elements with the same system simply by changing the wavelength slightly. Figure 11 shows the applicable energy-level diagrams for their RIMS processes. The resonant transition at 297.69 nm for Re is from the ground state, which is ~99% populated at their operating temperatures (~2000°C) (Walker and Fassett, 1986), whereas at the operating temperature of Os (~3000 °C) only 66% in the ground state and over 8% in the "optimum" resonant transition at 297.16 nm. Walker et al. (1989c) and Fassett have obtained an effective duty cycle of $\sim 10^{-4}$. The utility of this particular RIMS process has been demonstrated for several geochemical applications (Walker et al., 1987; Walker and Morgan, 1989a; Walker et al., 1989b,c).

Blum et al. (1990a,b) and co-workers chose a different tack of searching for the highest ionization efficiency RIMS process for each element to maximize overall signal. While this allows for the analysis of smaller samples, it is offset by the increased instrumental complexity. Figure 12 shows the energy level diagram used by Blum et al. (1990b) for Re and Os. For Re, a more efficient transition from the ground state is used (relative to that used by Walker and

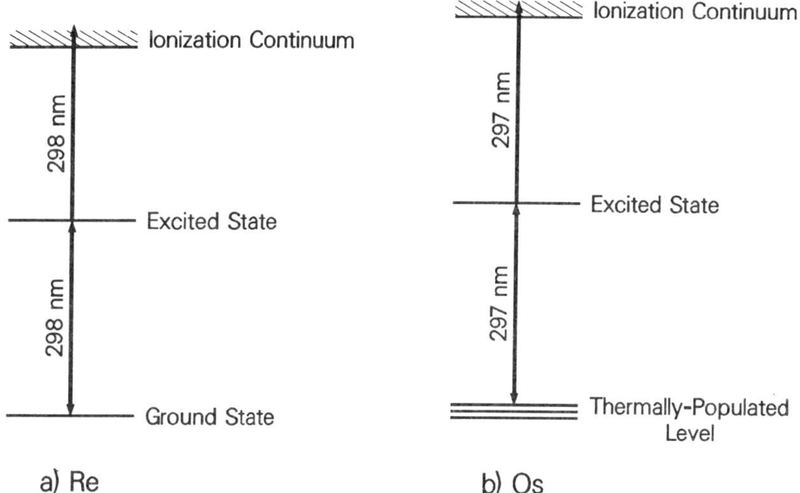

Figure 11. Energy levels and transitions used for Re and Os (Fassett and Walker, 1987a).

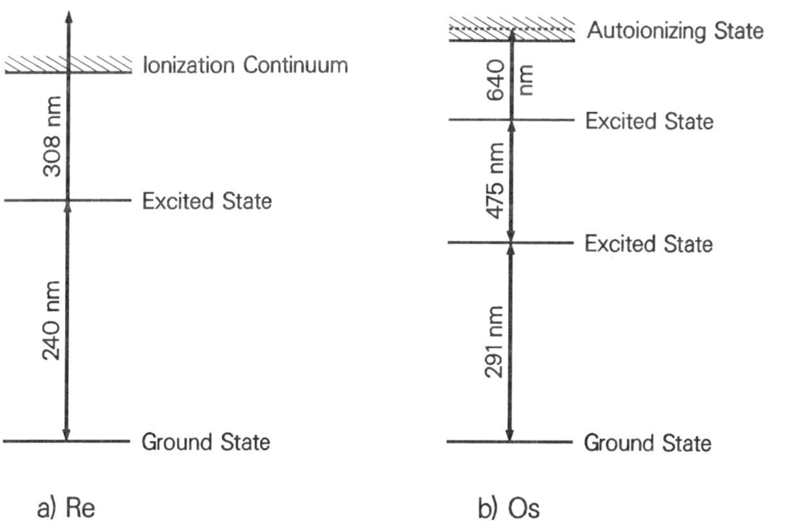

Figure 12. Energy levels and transitions used for Re and Os (Blum et al., 1990b).

Fassett, 1986). However, for Os, a significantly more complicated three-color (1 + 1 + 1) ionization RIMS process (2-resonant transitions and 1 auto-ionizing level) is used (Figure 12b). In addition, they have also utilized a pulsed Ar^+ ion beam sputter source (synchronized to the pulsed laser) to create the sputtered atoms for the RIMS process. Through the combination of these factors, an overall effective ionization yield of ~1% for osmium was achieved. This RIMS process has also been shown to be applicable to geochemical analysis (Blum et al., 1990a,b).

It is interesting to consider the potential in utilizing a cw system for the Os RIMS analysis. Frequency doubling would be required to attain the required photon energies which, in turn, requires single-frequency operation of the dye lasers. Although, a cw single-frequency dye laser would interact with relatively few atoms because of the large Doppler width, by using two counter-propagating beams in the first two steps of the 1 + 1 + 1 process, Doppler-free excitation of the entire velocity distribution can be achieved (Demtröder, 1981). A nonselective Ar^+ laser could be used to promote the excited atoms into the continuum. Several features of this approach are clear: (1) effectively 100% duty cycle; (2) very high efficiency (potentially approaching several tens of percent); and (3) tremendous isotopic selectivity. Although the experimental complexity would be increased, the potential gain may be worth the effort.

C. Technetium

Technetium is a rare element exhibiting no stable isotopes. Although it is produced in high yields from the fission of U and Pu, it is extremely scarce in nature (Curtis et al., 1987). Because of this, technetium is particularly useful as a hydrologic or geologic tracer (Stierman and Ruedisill, 1988). In addition, it has been proposed (Cowan and Haxton, 1982) that isotopes ^{97}Tc, ^{98}Tc, and ^{99}Tc in deep underground molybdenite ore are an integrated measure of the Earth's exposure to ^{8}B solar neutrinos, and can be used to measure the rate of solar neutrino production. Because most Tc isotopes of interest have very low specific activity, methods are needed for nonradiochemical determinations. Mass spectrometric measurements have been pursued (Anderson and Walker, 1980; Unger et al., 1987), but in some instances have been made difficult by isobaric interferences due to naturally occurring molybdenum and ruthenium. RIMS is thus a natural tool with which to address this problem.

Initial experiments utilized the frequency doubled output of a Nd^{+3}:YAG-pumped dye laser to generate mJ pulses near 300 nm for 1 + 1 ionization of Tc, with detection in a time-of-flight mass spectrometer (Nogar et al., 1983). A number of resonances were identified (Bozman et al., 1968) in this spectral region, and the ionization was demonstrated to be relatively free of interferences from molybdenum. However, interferences from unidentified hydrocarbons, along with a need to use cw lasers for ultimate accuracy, dictated that longer wavelength ionization processes be identified and tested. Several such processes, all using a visible photons and "1 + 1 + 1" ionization schemes, were demonstrated (Downey et al., 1984b). Using this apparatus, a detection efficiency of 10^{-5} was estimated, with a detection limit of 10^8–10^9 atoms in the sample. Here, detection efficiency is defined as the fraction of atoms in the sample which are detected as mass-resolved ions. It was pointed out in this work that a substantial fraction of technetium in the sample was apparently vaporizing in a molecular form that was not addressable via laser ionization.

Subsequent investigations utilized cw lasers with the "1 + 1 + 1" ionization scheme previously identified, and a magnetic mass spectrometer to measure the ionization efficiency and saturation parameters for these transitions (Miller et al., 1987). Saturation was demonstrated for bound-bound transitions, and previous estimates of the detection limit were confirmed.

More recently, laser desorption has been used with RIMS for the detection of Tc via a 1 + 1 + 1 ionization process utilizing UV and near IR photons (Trautmann et al., 1988). Using a Cu-pumped dye laser and a time-of-flight mass spectrometer, this group reported a detection efficiency $\approx 10^{-6}$, also with a detection limit of 10^8–10^9 atoms in the sample (Sattelberger et al., 1989).

D. Krypton

Isotope ratio measurements of krypton have important applications in geochemistry, particularly in connection with groundwater characterization. Such studies are relevant for the study of subsurface reservoirs and evaluating possible sites for radioactive waste disposal. Two isotopes are pertinent in this connection: ^{81}Kr and ^{85}Kr. ^{81}Kr is produced by cosmic-ray spallation reactions on stable krypton isotopes as well as by neutron capture on ^{80}Kr (Lehmann, 1989). With a decay half-life of 210,000 years, this isotope is applicable for dating groundwaters with ages of 50,000 to 1,000,000 years (Thonnard et al., 1987). ^{81}Kr may also be useful as a historical solar neutrino monitor (Kuzminov et al., 1988) through its production from ^{71}Br by ^{7}Be neutrinos; deep underground deposits of bromine salts exist at the Volgograd bishofite deposit. ^{85}Kr is produced through fission reactions, with the primary source being reactor fuel element reprocessing facilities. With this source and a half-life of 10.76 years, ^{85}Kr is most applicable to dating young waters (Lehmann and Loosli, 1989). Decay counting of the krypton isotopes is not feasible for many situations because of the low concentrations; ^{81}Kr also is hampered by its long half-life.

Resonance ionization has been extensively explored for the quantitation of the rare krypton isotopes (Kramer et al., 1983; Hurst et al., 1985). The first demonstration of measurements of ^{81}Kr in groundwater occurred in 1985 (Lehmann et al., 1985b). Since that time, efforts have continued to improve the speed, efficiency, sensitivity, and reproducibility of the measurement technique (Willis et al., 1989; Thonnard, 1987). We summarize here the work of these groups.

Water sample sizes are typically on the order of 50 liters. These samples are degassed, and then the krypton fraction is isolated from the other gases. Because a pulsed-laser/time-of-flight instrument will be used for the final measurement, several pre-enrichment steps must occur to reduce the background of other krypton isotopes to manageable levels (Thonnard et al., 1987; Willis et al., 1989). The principal reason for this is the extremely small amount of ^{81}Kr in the sample, relative to the other isotopes. This leads to the twin problems of inadequate spectrometer resolution and space charge effects in the ion source. The pre-enrichment is accomplished using a plasma ion source and velocity filter (Lehmann et al., 1987) followed by two passes through an electron impact source quadrupole mass spectrometer (Thonnard et al., 1987). At each step, the filtered ions are collected in aluminum or silicon targets for recycling, and spike isotopes are used to quantify the isotopic discrimination.

The result of the above separation and enrichment is a sample containing a few million total krypton atoms and a few thousand atoms of ^{81}Kr, embedded in silicon. This sample is introduced into the resonance ionization time-of-flight mass spectrometer and cooled to 20 to 60 K. Atoms are released from the sample by pulsed Nd:YAG laser annealing; those that are not ionized return to the cold surface for another pass. This "atom buncher" serves to conserve the scarce sample and significantly improve the overall effective ionization efficiency of the pulsed laser ionization scheme (Hurst et al., 1984). Ionization is accomplished through a three-color, three-photon process, driven by a single Nd:YAG laser and three dye lasers, using the wavelengths 116.5, 558.1, and 1064.0 nm (Kramer et al., 1983).

The overall accuracy of the resonance ionization technique for ^{81}Kr measurements of such groundwater samples is presently around 30%. A proposed improvement is the use of a single magnetic sector instrument to replace the dual quadrupole enrichment steps, using the greater abundance sensitivity of the magnetic instrument to permit decreased sample handling and increased sample throughput. Eventual accuracies of 10% are envisioned (Thonnard et al., 1987). Isotopic selectivity in the resonance ionization step has also been discussed (Payne, 1989).

Photon burst mass spectrometry for krypton analysis is a technique closely related to resonance ionization (Fairbank, 1987; Fairbank et al., 1989a). In this method, krypton is ionized in a plasma discharge and passed through a mass separator for initial enrichment of the desired isotope. At the exit of the separator, the ion beam is decelerated and encounters a charge exchange cell, through which metastable Kr atoms are formed. This metastable level is the lower level of an essentially closed two-level system. The metastable atoms are then exposed to a collinear laser beam within the confines of a high collection efficiency optics system. Atoms in resonance with the laser emit a burst of photons as they cycle between the two states. This burst can be distinguished from scattered laser light through statistical arguments. In addition to the normal isotope shifts and hyperfine structure differences, an artificial isotope shift between krypton isotopes is imposed by the differing velocities of the different masses. This allows substantial discrimination against undesired isotopes, and should permit an efficient, single-pass measurement system.

REFERENCES

Ackerhalt, J. R. and Shore, B. W. (1977). Rate equations versus Bloch equations in multiphoton ionization. *Phys. Rev.*, **A16**, 277–282.

Agbaji, E. B., Capey, W. D., Gilbert, J. R. and Al-Mashhadani, A. M. A. (1985). HV scans using a single-focusing mass spectrometer fitted with a modulated Daly metastable ion detector. *Int. J. Mass Spectrom. Ion Processes*, **64**, 49–54.

Agostini, P., Georges, A. T., Wheatley, S. E., Labropoulos, P. and Levenson, M. D. (1978). Saturation effects in resonant three-photon ionization of sodium with a non-monochromatic field. *J. Phys.*, **B11**, 1733–1747.

Allégre, C. J. and Luck, J. M. (1980). Osmium isotopes as petrogenetic and geologic tracers. *Earth Planet. Sci. Lett.*, **48**, 148–154.

Anderson, T. J. and Walker, R. L. (1980). Determination of picogram amounts of technetium-99 by resin bead mass spectrometric isotope dilution. *Anal. Chem.*, **52**, 709–713.

Andreev, S. V., Mishin, V. I. and Letokhov, V. S. (1986). High-efficiency laser resonance ionization of strontium atoms in a hot cavity. *Opt. Commun.*, **57**, 317–320.

Apostol, D., Blanaru, C., Ionescu, A., Popescu, G., Popescu, I. I. and Vasiliu, V. (1982). Frequency-locked helium-neon laser by optogalvanic effect. *Rev. Roum. Phys.*, **27**, 581–585.

Becker, C. H. and Gillen, K. T. (1984). Nonresonant multiphoton ionization as a sensitive detector of surface concentrations and evaporation rates. *Appl. Phys. Lett.*, **45**, 1063–1065.

Beekman, D. W., Callcott, T. A., Kramer, S. D., Arakawa, E. T., Hurst, G. S. and Nussbaum, E. (1980). Resonance ionization source for mass spectroscopy. *Int. J. Mass Spec. Ion Phys.*, **34**, 89–97.

Bevington, P. R. (1969). *Data Reduction and Error Analysis for the Physical Sciences*. McGraw-Hill, New York.

Blum, J. D., Pellin, M. J., Calaway, W. F., Young, C. E., Gruen, D. M., Hutcheon, I. D. and Wasserburg, G. J. (1990a). Resonance ionization mass spectrometry of sputtered osmium and rhenium atoms. *Anal. Chem.*, **62**, 209–214.

Blum, J. D., Pellin, M. J., Calaway, W. F., Young, C. E., Gruen, D. M., Hutcheon, I. D. and Wasserburg, G. J. (1990b). In situ measurement of osmium concentrations in iron meteorites by resonance ionization of sputtered atoms. *Geochim. Cosmochim. Acta*, **54**, 875–881.

Bomse, D. S. and Keller, R. A. (1983). Modified wide-band dye laser tunable over 5 cm-1 with 0.08 cm-1 bandwidth. *Anal. Chem.*, **55**, 977–978.

Bonin, K. D. and McIlrath, T. J. (1984). Two-photon electric-dipole selection rules. *J. Opt. Soc. Am.*, **B1**, 52–55.

Bozman, W. R., Corliss, C. H. and Tech, J. L. (1968). Energy levels and classified lines in the first spectrum of technetium (Tc I). *J. Res. Nat. Bur. Stand.*, **72**, 559–608.

Brenna, J. T. (1988). *Quantitative Elemental Isotope Ratio Determinations by Fourier-Transform Ion-Cyclotron-Resonance Mass Spectrometry*. Wiley, New York, pp. 327–330.

Bushaw, B. A. (1989). High-resolution laser-induced ionization spectroscopy. *Prog. Analyt. Spectrosc.*, **12**, 247–276.

Cantanzaro, E. J., Murphy, T. J., Garner, E. L. and Shields, W. R. (1966). Absolute isotopic abundance ratios and atomic weight of magnesium. *J. Res. Nat. Bur. Stds.*, **70A**, 453–458.

Condomines, M. and Allégre, C. J. (1980). Age and magmatic evolution of Stromboli volcano from thorium-230-uranium-238 disequilibrium data. *Nature*, **288**, 354–357.

Condomines, M., Hemond, C. and Allégre, C. J. (1988). Uranium-thorium-radium radioactive disequilibria and magmatic processes. *Earth Planet. Sci. Lett.*, **90**, 243–262.

Connally, W. J. and Morton, R. G. (1990). Modification of a krypton ion pumped dye laser for selected multimode operation. *Appl. Opt.*, **29**, 763–766.

Corliss, C. H. and Bozman, W. R. (1962). *Experimental Transition Probabilities for Spectral Lines of Seventy Elements*. Natl. Bur. Stand. Monograph, Washington, D.C.

Cowan, G. A. and Haxton, W. C. (1982). Solar neutrino production of technetium-97 and technetium-98. *Science*, **216**, 51–54.

Cowan, R. D. (1981). *The Theory of Atomic Structure and Spectra*. University of California Press, Berkeley.

Curtis, D. B., Cappis, J. H., Perrin, R. E. and Rokop, D. J. (1987). The geochemistry of natural technetium and plutonium. *Mat. Res. Soc. Symp.*, **84**, 735–736.

Daly, N. R. (1960). Scintillation type mass spectrometer ion detector. *Rev. Sci. Instr.*, **31**, 264–267.

Davis, W. A. and Thonnard, N. (1989). Improved Isotope Enrichment System for Processing Minute Noble Gas Samples. *Inst. Phys. Conf. Ser.* **94**, 233–236.

Demtröder, W. (1981). *Laser Spectroscopy: Basic Concepts and Instrumentation*. Springer-Verlag, Berlin.

Downey, S. W., Nogar, N. S. and Miller, C. M. (1984a). Resonance ionization mass spectrometry of uranium with intracavity laser ionization. *Anal. Chem.*, **56**, 827–828.

Downey, S. W., Nogar, N. S. and Miller, C. M. (1984b). Resonance ionization mass spectrometry of technetium. *Int. J. Mass Spec and Ion Proc.*, **61**, 337–345.

Edwards, R. L., Chen, J. H. and Wasserburg, G. J. (1987). Uranium-238/uranium-234/thorium-230/thorium-232 systematics and the precise measurement of time over the last 500,000 years. *Earth Planet. Sci. Lett.*, **81**, 175–192.

Edwards, R. L., Taylor, F. W. and Wasserburg, G. J. (1988). Dating earthquakes with high-precision thorium-230 ages of very young corals. *Earth Planet. Sci. Lett.*, **90**, 371–381.

Engleman, R. and Palmer, B. A. (1983). Precision isotope shifts for the heavy elements. II. neutral thorium. *J. Opt. Soc. Am.*, **73**, 694–701.

Estler, R. C., Apel, E. C. and Nogar, N. S. (1987). Laser mass-spectrometric studies of optical damage in calcium difluoride. J. *Opt. Soc. Am.*, **B4**, 281–286.

Estler, R. C. and Nogar, N. S. (1988). Chemical precursor to optical damage detected by laser ionization mass spectrometry. *Appl. Phys. Lett.*, **52**, 2205–2207.

Fairbank, W. M. (1987). Photon burst mass spectrometry. *Nucl. Inst. Meth. Phys. Res.*, **B29**, 407–414.

Fairbank, W. M., LaBelle, R. D., Keller, R. A., Miller, C. M., Poths, J. and Fearey, B. L. (1989a). Prospects for large dynamic range isotope analysis using photon burst mass spectrometry. *Inst. Phys. Conf. Ser.*, **94**, 53–56.

Fairbank, W. M., Spaar, M. T., Parks, J. E. and Hutchinson, J. M. R. (1989b). Anomalous odd-to-even mass isotope ratios in resonance ionization with broad-band lasers. *Phys. Rev.*, **A40**, 2195–2198.

Fassett, J. D., Moore, L. J., Shideler, R. W. and Travis, J. C. (1984). Pulsed thermal atom source for resonance ionization mass spectrometry. *Anal. Chem.*, **56**, 203–206.

Fassett, J. D. and Walker, R. J. (1987a). Ultratrace elemental and isotopic analysis of osmium and rhenium using resonance ionization mass spectrometry and thermal vaporization. *Inst. Phys. Conf. Ser.*, **84**, 115–120.

Faure, G. (1986) *Principals of Isotope Geology*. Wiley, New York.

Fearey, B. L., Johnson, S. G., Murrell, M. T. and Miller, C. M. (1991). Thorium RIMS for geochemical applications. *Inst. Phys. Conf. Ser.*, **114**, 311–314.

Fearey, B. L., Miller, C. M., Rowe, M. W., Anderson, J. E. and Nogar, N. S. (1988). Pulsed laser resonance ionization mass spectrometry for elementally selective detection of lead and bismuth mixtures. *Anal. Chem.*, **60**, 1786–1791.

Fearey, B. L., Parent, D. C., Keller, R. A. and Miller, C. M. (1990b). Doppler-free saturation spectroscopy of lutetium isotopes through resonance ionization mass spectrometry. *J. Opt. Soc. Amer.*, **B7**, 3–8.

Friedlander, G., Miller, J. M., Macias, E. S. and Kennedy, J. W. (1981) *Nuclear and Radiochemistry*. Wiley, New York.

Goldstein, S. J., Murrell, M. T. and Janecky, D. R. (1988a). Thorium and uranium isotopic systematics of basalts from the Juan de Fuca ridge (JDF) by mass spectrometry. *Trans. Am. Geophys. Union (EOS)*, **69**, 1502.

Goldstein, S. J., Murrell, M. T. and Janecky, D. R. (1988b). Thorium and uranium isotopic systematics of basalts from the Juan de Fuca and Gorda ridges by mass spectrometry. *Earth Planet. Sci. Lett.*, **96**, 134–136.

Gruen, D. M., Pellin, M. J., Calaway, W. F. and Young, C. E. (1988). *Laser Post-Ionization Secondary Neutral Mass Spectroscopy*. Wiley, New York, pp. 789–796.

Hammersley, J. M. and Handscomb, D. C. (1964) *Monte Carlo Methods*. Metheun, London.

Hand, O. W., Emary, W. B., Winger, B. E. and Cooks, R. G. (1989). Depth profiling of multilayered samples and comparisons of secondary ion and laser desorption mass spectra in the same instrument. *Int. J. Mass Spectrom. Ion Processes*, **90**, 97–118.

Herzberg, G. (1944). *Atomic Spectra and Atomic Structure*. Dover, New York.

Hess, K. R. and Harrison, W. W. (1986). Laser resonance ionization in a glow discharge. *Anal. Chem.*, **58**, 1696–1702.

Hurst, G. S., Payne, M. G., Kramer, S. D., Chen, C. H., Phillips, R. C., Allman, S. L., Alton, G. D., Dabbs, J. W. T., Willis, R. D. and Lehmann, B. E. (1985). Method for counting noble gas atoms with isotopic selectivity. *Rep. Prog. Phys.*, **48**, 1333–1370.

Hurst, G. S., Payne, M. G., Kramer, S. D. and Young, J. P. (1979). Resonance ionization spectroscopy and one-atom detection. *Rev. Mod. Phys.*, **51**, 767–819.

Hurst, G. S., Payne, M. G., Phillips, R. C., Dabbs, J. W. T. and Lehmann, B. E. (1984). Development of an atom buncher. *J. Appl. Phys.*, **55**, 1278–1284.

Ivanovich, M. and Harmon, R. S. (1982). *Uranium Series Disequilibrium* (with special reference to environmental aspects). Clarendon Press, Oxford.

Keller, R. A., Warner, B. E., Zalewski, E. F., Dyer, P., Engleman, R. and Palmer, B. A. (1983). The mechanism of the optogalvanic effect in a hollow-cathode discharge. *J. Phys., Colloq.*, **C7**, 23–33.

Keller, R. A. and Zalewski, E. F. (1980). Noise considerations, signal magnitudes, and detection limits in a hollow cathode discharge by optogalvanic spectroscopy. *Appl. Opt.*, **19**, 3301–3305.

Kimock, F. M., Baxter, J. P., Pappas, D. L., Korbin, P. H. and Winograd, N. (1984). Solids analysis using energetic ion bombardment and multiphoton resonance ionization with time-of-flight detection. *Anal. Chem.*, **56**, 2782–2791.

Kramer, S. D., Chen, C. H., Payne, M. G., Hurst, G. S. and Lehmann, B. E. (1983). Tunable VUV light generation for the low-level resonant ionization detection of krypton. *Appl. Opt.*, **22**, 3271–3275.

Kröenert, U., Becker, S., Hilberath, T., Kluge, H.-J. and Schulz, C. (1987). Resonance ionization mass spectroscopy with a pulsed thermal atomic beam. *Appl. Phys.*, **A44**, 339–345.

Kuzminov, V. V., Pomanskii, A. A. and Chikhladze, V. L. (1988). New possibilities for the geochemical bromine-81-krypton-81 solar neutrino experiment. *Nucl. Inst. Meth. Phys. Res.*, **A271**, 257–258.

Lawrenz, J., Obrebski, A. and Niemax, K. (1987). Measurement of isotope ratios by Doppler-free laser spectroscopy applying semiconductor diode lasers and thermionic diode detection. *Anal. Chem.*, **59**, 1232–1236.

Lehmann, B. E. and Loosli, H. H. (1989). Noble gas isotopes in hydrology. *Inst. Phys. Conf. Ser.*, **94**, 207–212.

Lehmann, B. E., Oeschger, H., Loosli, H. H., Hurst, G. S., Allman, S. L., Chen, C. H., Kramer, S. D., Payne, M. G., Phillips, R. C., Willis, R. D. and Thonnard, N. (1985b). Counting Kr-81 atoms for analysis of groundwater. *J. Geophys. Res.*, **90**, 11547–11551.

Lehmann, B. E., Rauber, D. F., Thonnard, N. and Willis, R. D. (1987). An isotope separator for small noble gas samples. *Nucl. Inst. Meth. Phys. Res.*, **B28**, 571–574.

Li, W. X., Lundberg, J., Dickin, A. P., Ford, D. C., Schwarcz, R., McNutt, R. and Williams, D. (1989). High-precision mass-spectrometric uranium-series dating of cave deposits and implications for paleoclimate studies. *Nature*, **339**, 534–536.

Lichte, F. E., Wilson, S. M., Brooks, R. R., Reeves, R. D., Holzbecher, J. and Ryan, D. E. (1986). New method for the measurement of osmium isotopes applied to a New Zealand Cretaceous/Tertiary boundary shale. *Nature*, **322**, 816–817.

Luck, J. M. and Allégre, C. J. (1984). Rhenium-187-osmium-187 investigation in sulfide from Cape Smith komatiite. *Earth Planet. Sci. Lett.*, **68**, 205–208.

Luck, J. M., Birck, J. L. and Allégre, C. J. (1980). Rhenium-187-osmium-187 systematics in meteorites: early chronology of the solar system and age of the galaxy. *Nature*, **283**, 256–259.

Mayo, S., Lucatorto, T. B. and Luther, G. G. (1982). Laser ablation and resonance ionization spectrometry for trace analysis of solids. *Anal. Chem.*, **54**, 553–556.

Melikechi, N. and Allen, L. (1986). Two-photon electric-dipole selection rules and nondegenerate real intermediate states. *J. Opt. Soc. Am.*, **B3**, 41–44.

Miller, C. M., Fearey, B. L., Palmer, B. A. and Nogar, N. S. (1989). High-fidelity in isotope ratio measurements by resonance ionization mass spectrometry. *Inst. Phys. Conf. Ser.*, **94**, 297–300.

Miller, C. M. and Nogar, N. S. (1983a). Calculation of ion yields in atomic multiphoton ionization spectroscopy. *Anal. Chem.*, **55**, 481–488.

Miller, C. M. and Nogar, N. S. (1983b). Continuous wave lasers for resonance ionization mass spectrometry. *Anal. Chem.*, **55**, 1606–1608.

Miller, C. M., Nogar, N. S., Apel, E. C. and Downey, S. W. (1987). Resonance ionization mass spectrometry at Los Alamos National Laboratory. *Inst. Phys. Conf. Ser.*, **84**, 109–114.

Moore, C. E. (1971). Atomic energy levels as derived from the analysis of optical spectra, Vols. I, II, and III. *Natl. Stand. Ref. Data Ser. NSRDS-NBS*-**35**, National Bureau of Standards, Washington, D. C.

Muller, J. F., Pelletier, M., Krier, G., Weil, D. and Campana, J. (1989). A new generation of microprobe: laser ionization and FT/ICR mass spectrometry. *Microbeam Anal.*, **24**, 311–316.

Murrell, M. T., Goldstein, S. J., Volpe, A. M., Fearey, B. L., Perrin, R. E. and Williams, R. W. (1990). Measurement of long-lived members of the uranium decay series by mass spectrometry. *Proceedings of the V.M. Goldschmidt Conference*, Baltimore, MD, p. 68.

Nogar, N. S., Downey, S. W. and Miller, C. M. (1985a). Multiple photon processes in tantalum resonance ionization mass spectrometry. *Anal. Chem.*, **57**, 1144–1147.

Nogar, N. S. and Estler, R. C. (1990a). Laser desorption/laser ablation with detection by resonance ionization mass spectrometry. In: *Lasers and Mass Spectrometry*. Oxford, New York, pp. 65–83.

Nogar, N. S., Estler, R. C., Fearey, B. L., Miller, C. M. and Downey, S. W. (1990b). Materials analysis by laser and ion beam sputtering with resonance ionization mass spectrometry. *Nucl. Inst. Meth. Phys. Res.*, **B44**, 459–464.

Nogar, N. S., Estler, R. C. and Miller, C. M. (1985). Pulsed laser desorption for resonance ionization mass spectroscopy. *Anal. Chem.*, **57**, 2441–2444.

Nogar, N. S. and Keller, R. A. (1985b). Effect of very weak laser sidebands on optical spectra involving easily saturable intermediate states. *Anal. Chem.*, **57**, 2992–2993.

Nogar, N. S., Sander, R. K., Downey, S. W. and Miller, C. M. (1983). Resonant multiphoton ionization for the detection of technetium. *Proc. SPIE*, **380**, 291–295.

Paisner, J. and Solarz, R. W. (1987). *Resonance Photoionization Spectroscopy*. Marcel Dekker, New York, pp. 175–260.

Palmer, B. A. and R. Engleman, J. (1983). Atlas of the thorium spectrum.

Payne, M. G. (1989). Method for the isotopically selective detection of inert gases. *Inst. Phys. Conf. Ser.*, **94**, 221–224.

Pellin, M. J., Young, C. E., Calaway, W. F. and Gruen, D. M. (1984). Trace surface analysis with pico-Coulomb ion fluences: direct detection of multiphoton ionized iron atoms from iron-doped silicon targets. *Surf. Sci.*, **144**, 619–637.

Perrin, R. E., Knobeloch, G. W., Armijo, V. M. and Efurd, D. W. (1985). Isotopic analysis of nanogram quantities of plutonium by using a SID ionization source. *Int. J. Mass Spectrom. Ion Processes*, **64**, 17–24.

Riley, G. H. and Delong, S. E. (1970). Osmium isotopes in geology. *Int. J. Mass Spec. Ion Phys.*, **4**, 297–304.

Rimke, H., Herrman, G., Muehleck, C., Sattleberger, P., Trautmann, N., Ames, F., Kluge, H.-J., Otten, E.-W., Rehklau, D. and Ruster, W. (1987). Detection of trace amounts of actinides by resonance ionization mass spectrometry. *Inorg. Chim. Acta*, **140**, 277–278.

Roboz, J. (1968). *Introduction to Mass Spectrometry, Instrumentation and Techniques*. Interscience, New York.

Rokop, D. J., Perrin, R. E., Knobeloch, G. W., Armijo, V. M. and Shields, W. R. (1982). Thermal ionization mass spectrometry of uranium with electrodeposition as a loading technique. *Anal. Chem.*, **54**, 957–960.

Russ, G. P., Bazan, J. M. and Date, A. R. (1987). Osmium isotopic ratio measurement by inductively coupled plasma source mass spectrometry. *Anal. Chem.*, **59**, 984–989.

Ruster, W., Ames, F., Kluge, H.-J., Otten, E.-W., Rehklau, D., Scheere, F., Herrmann, G., Muehleck, C., Riegel, J., Rimke, H., Sattelberger, P. and Trautmann, N. (1989). A resonance ionization mass spectrometer as an analytical instrument for trace analysis. *Nucl. Inst. Meth. Phys. Res.*, **A281**, 547–558.

Sattelberger, P., Mang, M., Herrmann, G., Riegel, J., Rimke, H., Trautmann, N., Ames, F. and Kluge, H.-J. (1989). Separation and detection of trace amounts of technetium. *Radiochim. Acta*, **48**, 165–169.

Savickas, P. J., Hess, K. R., Marcus, R. K. and Harrison, W. W. (1984). Glow discharge atomization source for resonance ionization mass spectrometry. *Anal. Chem.*, **56**, 817–819.

Shaw, R. W., Young, J. P. and Smith, D. H. (1989). Diode laser initiated resonance ionization mass spectrometry of lanthanum. *Anal. Chem.*, **61**, 695–697.

Shaw, R. W., Young, J. P., Smith, D. H., Bonanno, A. S. and Dale, J. M. (1990). Hyperfine structure of lanthanum at sub-Doppler resolution by diode-laser-initiated resonance-ionization mass spectroscopy. *Phys. Rev.*, **A41**, 2566–2573.

Smith, C. B. (1988). Mass spectrometry in geochemistry: an established tool in the Earth sciences. *Nucl. Inst. Meth. Phys. Res.*, **B35**, 364–369.

Snyder, J. J., Lucatorto, T. B., Debenham, P. H. and Geltman, S. (1985). Ultrasensitive laser isotope analysis in an ion-storage ring. *J. Opt. Soc. Am.*, **B2**, 1497–1502.

Stierman, D. J. and Ruedisill, L. C. (1988). Integrating geophysical and hydrogeological data: an efficient approach to remedial investigations of contaminated ground water. *ASTM Spec. Tech. Publ.*, 963, 43–57.

Sturchio, N. C., Bohlke, J. K. and Binz, C. M. (1989). Radium-thorium disequilibrium and zeolite-water ion exchange in a Yellowstone hydrothermal environment. *Geochim. Cosmochim. Acta*, **53**, 1025–1034.

Swain, S. J. (1980). Generalized multiphoton rate equations. *J. Phys.*, **B13**, 2375–2396.

Thonnard, N., Willis, R. D., Wright, M. C., Davis, W. A. and Lehmann, B. E. (1987). Resonance ionization spectroscopy and the detection of Kr-81. *Nucl. Inst. Meth. Phys. Res.*, **B29**, 398–406.

Trautmann, N., Herrmann, G., Muehleck, C., Peuser, P., Rimke, H., Sattelberger, P., Ames, F., Kluge, H.-J. and Otten, E. W. (1988). Laser resonance ionization mass spectrometry as a sensitive analytical method for actinides and technetium. *Isopenpraxis.*, **24**, 217–219.

Unger, S. E., McCormick, T. J., Treher, E. N. and Nunn, A. D. (1987). Comparison of desorption ionization methods for the analysis of neutral seven-coordinate technetium radiopharmaceuticals. *Anal. Chem.*, **59**, 1145–1149.

Volpe, A. M., Olivares, J. A. and Murrell, M. T. (1991). Determination of radium isotope ratios and abundances in geologic samples by thermal ionization mass spectrometry. *Anal. Chem.*, **63**, 913–916.

Walker, R. J., Carlson, R. W., Shirey, S. B. and Boyd, F. R. (1989b). Osmium, strontium, neodymium, and lead isotope systematics of southern African peridotite xenoliths: Implications for the chemical evolution of the subcontinental mantle. *Geochim. Cosmochim. Acta*, **53**, 1583–1595.

Walker, R. J. and Fassett, J. D. (1986). Isotopic measurement of subnanogram quantities of rhenium and osmium by resonance ionization mass spectrometry. *Anal. Chem.*, **58**, 2923–2927.

Walker, R. J., Fassett, J. D. and Travis, J. C. (1989c). The use of resonance ionization mass spectrometry for measuring the isotopic composition of rhenium and osmium extracted from silicate rocks. *Inst. Phys. Conf. Ser.*, **94**, 337–342.

Walker, R. J. and Morgan, J. W. (1989a). Rhenium-osmium isotopic systematics of carbonaceous chondrites. *Science*, **243**, 519–522.

Walker, R. J., Shirey, S. B. and Stecher, O. (1987). Comparative rhenium-osmium, samarium-neodymium and rubidium-strontium isotope and trace element systematics for archean komatiite flows from Munro township. Abitibi Belt, Ontario. *Earth Planet. Sci. Lett.*, **87**, 1–12.

Watson, J. T. (1985). *Introduction to Mass Spectrometry*. Raven Press, New York.

Wiley, W. C. and McLaren, I. H. (1955). Time-of-flight mass spectrometer with improved resolution. *Rev. Sci. Inst.*, **26**, 1150–1157.

Willard, H. H., Merritt, L. L. and Dean, J. A. (1974). *Instrumental Methods of Analysis*. Van Nostrand, New York.

Willis, R. D., Thonnard, N., Wright, M. C., Lehmann, B. E. and Rauber, D. (1989). Counting Kr-81 atoms in groundwater using RIS-TOF. *Inst. Phys. Conf. Ser.*, **94**, 213–216.

Wollnik, H. (1987). *Optics of charged particles*. Academic Press, New York.

Young, J. P., Shaw, R. W., Goeringer, D. E. and Smith, D. H. (1989b). Influence of laser characteristics on resonance ionization mass spectrometry. *Inst. Phys. Conf. Ser.*, **94**, 367–370.

Zakheim, D. S. and Johnson, P. M. (1980). Rate equation modeling of molecular multiphoton ionization dynamics. *Chem. Phys.*, **46**, 263–272.

Zenobi, R., Philippoz, J. M., Buseck, P. R. and Zare, R. N. (1989). Spatially resolved organic analysis of the Allende meteorite. *Science*, **246**, 1026–1029.

X-RAY FLUORESCENCE SPECTROMETRY

Gerald R. Lachance

I. Introduction 156
  A. Analytical Geochemistry Requirements 156
  B. XRF Spectrometry 157
II. X-Ray Physics 158
  A. Theory of XRF Emission 158
  B. Attenuation of X-Radiations 160
  C. Secondary Fluorescence (Enhancement) 161
III. Instrumentation 162
  A. Wavelength Dispersive Spectrometers 162
  B. Energy Dispersive Spectrometers 163
IV. Principles of Quantitative XRF Analysis 164
  A. Theoretical Expressions for Absorption Effects 165
  B. Theoretical Expressions for Enhancement Effects 168
  C. Theoretical Expressions for Total Matrix Effects 169
V. Compensation Methods 170
  A. Dilution Methods 170
  B. Standard Additions Method 172
  C. Compton Scatter Methods 173
VI. Mathematical Methods 175
  A. The Fundamental Parameters Approach 176
  B. Alpha Coefficient Methods 176

Advances in Analytical Geochemistry
Volume 1, pages 155–192

ISBN: 1-55938-332-1

C. Alpha Coefficients, Oxide Systems 180
D. Empirical Coefficient Methods 182
VII. Analytical Strategies 183
A. Specimen Preparation: Ideal or Compromise 184
B. Intensity Measurement Scheme 185
C. Data Reduction Process 188
D. Analytical Reproducibility, Detection Limits 190
References 192

I. INTRODUCTION

X-ray fluorescence spectrometry (XRF) continues to be a widely used technique for providing elemental composition of geological materials. Briefly stated, the analytical requirements for geochemical studies in particular can be summarized as "a very wide range of elements occurring in very wide ranges of concentration", which also aptly describes the analytical potential of the XRF technique. The primary goal of this chapter is to provide the chemist, geologist, or geochemist who is likely to submit samples for analysis with the basic principles and current methodology underlying quantitative XRF analysis.

A. Analytical Geochemistry Requirements

The requirements may vary somewhat on whether the submitter's main concern is with the academic aspects of a geological study or whether the main purpose is to conduct a geochemical exploration within a given time frame. In a recent paper, Harvey (1989) outlines several constraints which make the analysis of geochemical exploration samples different from the situations in the geological research laboratory or in a mining environment. These include:

1. A wide range of elements that will include some that frequently occur at very low levels of concentration in common rocks but may run up to levels several times their crustal abundance.
2. A full suite of major elements may not be required.
3. The sample types may be wide ranging; that is, soils, sediments, ores, concentrates, peats and other organic matter in addition to mineralized rocks.
4. The number of samples may run to several thousand.
5. Costs must be kept competitive and turn-around time minimized.

Harvey goes on to point out the obvious consequences of these factors:

1. The laboratory must be versatile, implying a well experienced staff and ideally one having a good working relationship with its clients.

2. The methods used should, as far as possible, be independent of having prior knowledge of the major constituents.
3. The need for a comprehensive quality control system.
4. Specimen preparation should be streamlined, preferably automated, yet remain sufficiently versatile.
5. The flow of samples through the various stages should be simple and efficient.

In short, one can safely say that over a period of a few years, the analytical laboratory servicing a geological clientele will likely be requested to provide quantitative composition data for many, if not most, of the elements within the range atomic number 3 to 92. While XRF spectrometry cannot offer a single process or recipe for achieving all these aims, it does offer the possibility of combining a number of instrumental components and customized spectrometers to meet most of the client's specific needs.

B. XRF Spectrometry

In essence, XRF analysis involves the irradiation of specimens with a source of primary radiations which, if of high enough energy, will in turn cause the emission of secondary X-radiations characteristic of the elements present in the specimen. The emitted spectrum is dispersed and the intensities of selected radiations are measured and ultimately converted to weight fractions.

The source of the primary beam may be that of a conventional X-ray tube or a micro X-ray tube or, simply the emission from a radioactive material. Combined with highly stabilized power supplies, X-ray tubes with their choice of target material, operating voltage, and current, provide a versatile and dependable energy source. Spectrometers must also provide for the sequential placement of specimens in the path of the incident radiation with a very high degree of reproducibility. Holders have been designed to accept solids, powders, and liquid specimens.

The emitted fluorescence radiations must be dispersed in order to resolve the spectrum in its individual characteristic lines. Two options are available. Dispersion may be as a function of wavelength or as a function of energy. In the case of the wavelength dispersive spectrometer (WDS), this is achieved by the use of analyzing crystals, whereas in the case of the energy dispersive spectrometer (EDS), a multichannel energy analyser is used. WDS spectrometers come in two categories; sequential and simultaneous. The sequential XRF spectrometers are single-channel instruments in the sense that a crystal-collimator-detector combination is used to measure emitted intensities "one wavelength at a time". This offers a high degree of versatility, but somewhat reduced productivity compared to the simultaneous models. Higher productivity is achieved in simultaneous models by selecting a specific crystal-collimator-detector-integrator combination for each wavelength under consideration, an option that evidently results in a costlier instrument. At the heart of the EDS spectrometer is a solid-state detector (such as Li drifted Si) combined with a multichannel analyzer. Spectra similar to WDS are generated but the characteristic radiations are dispersed as a function of the energy of the emitted photons. Because energy is inversely proportional to wavelength, WDS spectrometers offer better resolution for longer wavelengths (e.g., elements Na, Mg, Al, Si, . . .), while EDS instruments offer better resolution for radiations of shorter wavelengths (e.g., elements Rb, Sr, Y, Zr, . . .).

Lastly, but of major importance, has been the combination of X-ray spectrometers with dedicated computers, especially due to the fact that presently microcomputers only contribute a very small fraction of the total monetary outlay. Most manufacturers now offer software packages that not only control the spectrometer's operations for any one of a number of preselected analytical schemes, but provide for calibration, conversion of intensities to concentrations, and the printout of the analytical report. If attached to a carousel (hundred or more specimens) from which individual specimens can be sequentially loaded, XRF spectrometers can operate unattended on a continuous basis between successive reloads. The current status of the various components is visually displayed during the measurement sequences and any major drift flagged so that remedial action may be taken. Some manufacturers offer the possibility of carrying out diagnostics with their specialized equipment in their own application laboratories via modems.

II. X-RAY PHYSICS

Although there are no set lines of demarcation in defining the X-ray region in the electromagnetic spectrum, XRF spectrometry is mainly concerned with radiations within the range 0.2 to 20 Å. Recent advances in instrumentation are gradually extending this range to the 200 Å range; that is, down to atomic numbers 4 to 5. While it is important for XRF analysts to have a thorough knowledge of the properties of X-radiations, it is certainly desirable that the client be familiar at least with the basic principles of XRF emission and of the absorption of X-radiations by matter.

A. Theory of XRF Emission

X-ray spectra are the resultant of the interaction between photons and atoms. When a photon of sufficient energy reacts with an atom, one of the processes involves the transfer of the photon energy to one of the electrons of the atom which leads to its ejection from the atom. The ionized atom is then out of equilibrium and returns to its normal state via the transfer of electrons from outer shells. The shells are designated by the letters K, L, M, . . . , and in turn the energy levels of the subshells are designated as LI, LII, etc. The K shell has one energy level, the L shell has three, the M shell has five, and so forth. Table 1a lists the atomic energy levels for Ca and Fe that are involved in the generation of the K_α and K_β characteristic lines of these two elements. The transfers from the L shell to the K shell represents a loss in the potential energy of the atom which reappears as K_α photons whose energies are equal to the difference between the two levels involved. Similarly, transfers from the M shell to the K shell result in K_β photons, whereas transfers from the M and N shells to the L shell result in the emission of L_α and L_β photons, respectively. Thus,

$$E_{K_{\alpha 1}} = E_K - E_{LIII}$$

$$E_{K_{\alpha 2}} = E_K - E_{LII}$$

$$E_{K_{\beta 1}} = E_K - E_{MIII}$$

Table 1b lists the photon energies for the $K_{\alpha 1}$, $K_{\alpha 2}$, and $K_{\beta 1}$ radiations of Ca and Fe calculated from data in Table 1a. The corresponding wavelengths of a given photon energy is given by $\lambda = \frac{12.398}{E}$ where λ is the wavelength in Å and E is the energy in keV.

Table 1c lists the wavelengths generated by the data in Table 1b. The binding energies of the atomic shells are also referred to as critical energies (or critical wavelengths) to indicate the minimal energy (or maximum wavelength) that an X-ray photon must have in order to generate the K, L, M, . . . , series of characteristic lines. Comprehensive tabulations of the commonly used characteristic lines, in wavelengths and energies, can be found in most handbooks of physics, XRF textbooks, and other references.

Table 1. Selected Atomic Energy Levels, Photon Energies and Wavelength for Calcium and Iron

a) Atomic Energy Levels* (binding energies) in keV				
	E_K	E_{LIII}	E_{LII}	E_{MIII}
Ca	4.0381	0.3464	0.3500	0.0254
Fe	7.1112	0.7081	0.7211	0.0540
b) Photon Energies, in keV				
		$E_{K_{\alpha 1}}$	$E_{K_{\alpha 2}}$	$E_{K_{\beta 1}}$
Ca		3.6917	3.6881	4.0127
Fe		6.4031	6.3901	7.0572
c) Wavelengths, in Å				
	$K_{\alpha 1}$	$K_{\alpha 2}$	$K_{\beta 1}$	
Ca	3.358	3.362	3.090	
Fe	1.936	1.940	1.757	

* Data from Bearden and Burr (1967)

B. Attenuation of X-Radiations

When an X-ray beam passes through a thickness of material, its intensity is reduced due to absorption and scattering. The attenuation process can be expressed as

$$I_x = I_o e^{-\mu m} \qquad (1)$$

where I_o is the intensity of the incident monochromatic beam, I_x is the intensity after passing through a thickness of material, μ is mass attenuation coefficient or mass absorption coefficient (in cm^2/g units), and m is mass of material in a unit section of the beam. Although μ is actually the sum of two coefficients—that is, the mass photoelectric absorption coefficient and the mass scattering coefficient —the fact that the former is several times greater (often accounting for ~ 95% of the total), it is generally considered that experimental values of μ adequately reflect the properties of the photoelectric coefficient. The next step is an examination of the relation between mass absorption coefficients, wavelengths, and atomic numbers.

Because the absorption phenomenon is ever present and, proportionally, is the major contributor to the nonlinearity of the intensity-concentration relation in XRF, the knowledge of μ at any given wavelength for any given element is mandatory in order to theoretically quantify the absorption effect. For a given element, the value of μ increases rapidly with wavelength, but a plot of μ versus λ exhibits a number of sharp discontinuities. These are related to the critical energies (or their corresponding wavelengths) of the K, L, M, . . . , shells of the element in question. Tabulated values of μ are the result of critical studies of a very large number of experimental data. The values are then optimized by some form of fitting using various algorithms, for example

$$\mu = C\lambda^n \qquad (2)$$

where C and n are constants that depend on the element under consideration and change at each discontinuity, referred to as K, LI, LII, LIII, MI, absorption edges. Typical C and n values taken from Heinrich (1966) are listed in Table 2a and examples of μ values generated from these for various wavelengths are listed in Table 2b. The latter may then be used to calculate the total mass absorption coefficient for complex specimens (oxides, mixtures, minerals, etc.) at any given wavelength. For example, given that the weight per cent of Ca and O in CaO are 71.4 and 28.6, respectively, then μ_{CaO} for λ = 0.50 Å is

$$\mu_{CaO} = C_{Ca}\,\mu_{Ca} + C_O\,\mu_O = (0.714 \times 7.24) + (0.286 \times 0.54) = 5.32 \qquad (3)$$

A similar calculation carried out for Fe_2O_3 yields 10.34, thus for a mixture CaO 60% and Fe_2O_3 40%, the total mass absorption for λ = 0.50 Å is given by

$$\mu_{mix} = C_{CaO}\,\mu_{CaO,\lambda} + C_{Fe_2O_3}\,\mu_{Fe_2O_3,\lambda} = (0.60 \times 5.32) + (0.40 \times 10.34) = 7.33 \qquad (4)$$

C. Secondary Fluorescence (Enhancement)

The term secondary fluorescence designates that component of the emitted intensity of a characteristic line that is the resultant of excitation by one or more characteristic radiations of other elements in the specimen. It is best described by referring to specific elements; in the present case, the comparison of the process involved in the emission of Fe radiation in the Fe-Ca system (absorption is the only effect present) and in the Fe-Sr system (absorption and enhancement are both present).

Consider a specimen containing equal concentrations of Fe and Ca irradiated by a primary radiation, wavelength equal to 0.70 Å. The incident radiation will be absorbed by the factor

$$\mu_{s,\lambda} = C_{Fe}\ \mu_{Fe,\lambda} + C_{Ca}\ \mu_{Ca,\lambda} \quad (5)$$

Substitution of the appropriate data from Table 2 yields

$$\mu_{s,\lambda} = (0.50 \times 36.31) + (0.5 \times 18.21) = 27.36 \quad (6)$$

and the emitted Fe_{Ka} radiation will be absorbed by the factor

$$\mu_{s,\lambda_i} = C_{Fe}\ \mu_{Fe,\lambda_i} + C_{Ca}\ \mu_{Ca,\lambda_i} \quad (7)$$

where $\lambda_i = 1.937$Å.

Substitution of the appropriate values from Table 2 yields

$$\mu_{s,\lambda_i} = (0.50 \times 71.43) + (0.5 \times 296.20) = 183.82 \quad (8)$$

Ca_{K_α} and Ca_{K_β} radiations are also generated, but their wavelengths, 3.360 Å and 3.090 Å, are both *longer* (less energetic) than the K absorption edge of iron (critical excitation voltage), hence cannot eject electrons from the K shell of iron atoms. Therefore, the total matrix effect as far as the emission of Fe radiations is concerned can be expressed as

$$\mu_s = \mu_{s,\lambda} + \mu_{s,\lambda_i} = (27.36 + 183.82) = 211.08 \quad (9)$$

and is the resultant of absorption only.

A quite different situation exists if strontium is the element combined with iron. The absorption process is identical to that for calcium except that the values for μ_{Sr} would be different from those for μ_{Ca}. Just as for calcium, Sr_{K_α}, and Sr_{K_β}, radiations are now generated within the specimen, but contrary to calcium, their wavelengths, 0.877Å and 0.783Å, respectively, are both shorter (more energetic) than the iron K absorption edge. Therefore both strontium radiations can eject electrons from the K shell iron atom which leads to more Fe_{K_α} and Fe_{K_β} photons being emitted, hence the term enhancement. The theoretical expression for the enhancement process is somewhat more complicated than that for the absorption process (more parameters, some involving natural logarithms) but need not concern the client reader per se. The combination absorption effect and enhancement effect is generally termed matrix effect and is most pronounced in alloys.

Matrix effects are generally less of a problem in XRF analysis of geological materials due to the fact that oxygen constitutes a high proportion of most specimens and, even more so, if the specimens are in the form of fused discs. Nevertheless, the compensation or correction for matrix effects remains an important aspect of quantitative XRF analysis.

III. INSTRUMENTATION

Although radioactive sources and, more recently, synchrotron radiations have been used to generate characteristic radiations in specialized application, analytical laboratory XRF instruments must combine not only overall stability and reliability but also speed and convenience of operation. Modern XRF spectrometers have met these objectives by gradual improvements in the stability of X-ray tube sources, by operating under computer control (hence potentially on a continuous basis) and, and by providing for all input (specimen identification, instrumental operating conditions, etc.) and output information to be conducted from a dedicated computer terminal (analytical reports, estimates of precision, etc.). Initially restricted to using crystals for the dispersion of spectra, XRF spectrometers now fall in two general categories: those based on wavelength dispersion, and those based on energy dispersion.

A. Wavelength Dispersive Spectrometers

To meet the demands of laboratories in which the highest priority is that instrumentation be versatile in order to adapt to changing requirements and the demands of laboratories in which high productivity has greater priority, manufacturers have designed two types of instruments, namely, the sequential wavelength dispersive spectrometries (WDS) and the simultaneous WDS. The modern sequential type is basically a single channel instrument programmed to measure intensities automatically at any preset series of wavelengths. The analyst has the option of preselecting the X-ray tube potential, current, any one of up to six or eight analyzing crystals, one of two detectors, coarse or fine collimation, use of filters, duration of intensity measurement, and use of pulse discrimination for each individual intensity measurement. The simultaneous WDS, on the other hand, is basically a number of single channel instruments, each having its own collimator, crystal, detector, integrator etc., each selected for optimal conditions. Thus some 25 to 30 measurements can be carried out for a preselected integration time, simultaneously. The fact that each channel has its own collimator-crystal-detector-integrator arrangement, will necessarily translate into a costlier instrument which may in the long run be compensated by a lower overall cost per determination. The decision of whether the added cost of upgrading or incorporating new components (that manufacturers are continually producing in order to improve sensitivity, resolution, stability, etc.) will be beneficial is always a concern. Recently, synthetic multilayer analyzers (e.g., W/Si layers) with d-spacing of several tens of Å have been developed and shown to increase peak intensities by factors of 15 and 25 for magnesium and sodium, respectively. Detection limits of 150 ppm for carbon and boron have been reported. Software packages are available that require minimal input from the operator. From the analyst's point of view, there is no doubt that the ability to display calibration graphs on a monitor on command, identify outliers, etc., is a desirable convenience.

B. Energy Dispersive Spectrometers

The main advantages of energy dispersive spectrometers (EDS) stem from the simplicity of the instrumentation (lower cost) and the simultaneous integration and display of the *entire* spectrum emitted by the specimen. The latter enables the analyst to detect the presence of any unexpected elements and pass this information on to the submitter. Two limitations of EDS are due to their poorer resolution at decreasing energy of emitted characteristic radiations and the need to minimize coincidence losses (i.e., detector choking occurs more readily). The former is of some concern in a geological context as it affects the elements P, Si, Al, Mg, . . . , while the latter translates into operating the X-ray tube at lower power, hence necessitating longer counting times. Conversely, lower power operation results in instruments operated directly from 120-volt lines and that rely on air cooling for heat dissipation. On the other hand, the commonly used Si(Li) detector must be kept at very low temperatures; therefore, attached to a liquid nitrogen vacuum cryostat unit. Harding (1989) describes a recently introduced thermoelectrically cooled Si(Li) detector instrument and outlines its advantage over traditional cryogenic liquid cooling.

The detector output is fed to a multichannel analyzer-memory unit. Spectra can be examined in real time, or stored and recalled. Software provides for peak smoothing, deconvolution, expansion to facilitate viewing, superimposing spectra to facilitate comparison, etc. Sensitivity can be altered through the use of filters or by providing the option to use secondary radiations as the exciting source. For example, the radiations can be used to excite secondary radiations from any one of a number of targets (Sn, Mo, Ge, Fe, . . .). These radiations are then used to irradiate the specimen. The radiations emitted by these secondary targets will essentially be monoenergetic (K_α and K_β lines) and very effective for exciting elements having slightly lower critical excitation voltages. Thus one would select Mo radiations for added Sr and Rb sensitivity, Ge radiations for added Zn, Cu, . . . , intensities, Fe for Cr, V, etc. Margolin et al. (1985) assess the analytical capabilities of the MECA-10-44 EDS for solving geological problems when complete analysis is required in geochemical investigations. Optimal excitation conditions for 50 elements are tabulated.

IV. PRINCIPLES OF QUANTITATIVE XRF ANALYSIS

In common with all instrumental techniques, quantitative XRF analysis involves measuring the intensity of a characteristic signal emitted due to the presence of an element in specimens of unknown composition. The emitted signal is integrated over a period of time and its intensity expressed, in the case of XRF, in counts per second or kilocounts per second. Inherent in all experimental methods is the calibrating step; that is, carrying out a number of instrumental measurements on specimens of *known* composition in order to determine the calibration factor needed to convert intensity units to concentration units. A rough parallel may be made between the determination, let us say of iron, by a volumetric method and XRF methods. The chemist would proceed to dissolve a portion of a reference material of known iron concentration ($C_{Fe,s}$) and ultimately titrate (i.e., convert ferrous to ferric iron) using V_s ml of titrant. If an equal weight of an unknown sample is treated similarly, and V_u ml is the volume of titrant used in this case, then the concentration of iron in the unknown ($C_{Fe,u}$) is given by

$$C_{Fe,u} = V_u \left(\frac{C_{Fe,s}}{V_s} \right) \tag{9}$$

where $\frac{C_{Fe,s}}{V_s}$ is the calibration factor.

The XRF analyst can proceed in a similar fashion in that a measure of the intensity of a characteristic line of iron emitted by the unknown ($I_{Fe,u}$) and by a specimen of pure iron ($I_{(Fe)}$) are made. Assuming that the matrix effects in the unknown and in pure iron are identical, the conversion of intensity to concentration would then be expressed by

$$C_{Fe,u} = I_{Fe,u} \left(\frac{1.0}{I_{(Fe)}} \right) \tag{10}$$

However, XRF analysts must contend with the fact that the magnitude of matrix effects vary from specimen to specimen and with the fact that it is not always possible nor practical to measure intensities on specimens of pure analytes. Taking matrix effects into consideration results in the need for an additional term in the above intensity-concentration expression. Therefore XRF analysis must contend with

$$C_{i,u} = I_{i,u} \text{ (calibration factor)[matrix effect correction factor]}_u \tag{11}$$

and

$$C_{i,s} = I_{i,s} \text{ (calibration factor)[matrix effect correction factor]}_s \tag{12}$$

where $C_{i,u}$ and $C_{i,s}$ refer to the concentration of analyte i in a specimen of unknown composition and a standard, respectively; $I_{i,u}$ and $I_{i,s}$ refer to the net emitted intensities of element i in the specimen of unknown composition and the standard, respectively; and $[\ ]_u$ and $[\ ]_s$ are expressions that quantify the total matrix effect on the analyte in the specimen of unknown composition and the standard, respectively. Ratioing the two expression, gathering terms relating to the standard and solving for $C_{i,u}$ yields

$$C_{i,u} = I_{i,u}\left(\frac{C_{i,s}}{I_{i,s}[\ ]_s}\right)[\ ]_u \tag{13}$$

This is the more general expression indicating that calibration and analysis may be carried out using one or more complex reference standards provided that the value of $[\ ]_s$ and $[\ ]_u$ can be evaluated. If the evaluation of the matrix effect correction factor is based on theoretical principles, where the effect is defined as a sum of individual effects, the term numerical or mathematical methods is generally used when referring to this approach. If the evaluation of the matrix effect correction factor is based on experimental measurements on the specimens concerned, the approach is generally referred to as compensation or comparative methods. The basic principles underlying numerical methods will be examined next.

A. Theoretical Expressions for Absorption Effects

A complete derivation from first principles is beyond the scope of this presentation but as a result of contributions by a host of authors over the years, the absorption effect can be stated in fairly simple terms (Table 2). For a monochromatic incident beam of wavelength λ and intensity I,

$$P_{i,\lambda} = (\mu_{i,\lambda} I_{o,\lambda}) \frac{C_i}{\mu^*_{s,\lambda}} \tag{14}$$

where $P_{i,\lambda}$ is the emitted intensity due to primary emission; $\mu^*_{s,\lambda}$ is the total mass absorption coefficient of the specimen at that wavelength; and the asterisk is used to indicate that a geometry factor is included.

Depending on the angle of incidence of the primary beam and the angle of emergence of the characteristic radiation, their path lengths will be different and this must be taken into consideration. The total mass absorption coefficient is the sum of the mass absorption coefficient for the incident wavelength plus the mass absorption coefficient for the characteristic wavelength, the geometry of the instrument having been taken into consideration. Thus,

$$\mu^*_{s,\lambda} = \mu_{s,\lambda} \csc \Psi' + \mu_{s,\lambda i} \csc \Psi'' \tag{15}$$

where Ψ' and Ψ'' refer to the angle of incidence and the angle of emergence, respectively.

For the pure analyte, the above expression becomes

$$P_{(i),\lambda} = (\mu_{i,\lambda} I_{o,\lambda}) \frac{C_{(i)}}{\mu^*_{i,\lambda}} \tag{16}$$

Table 2. Theoretical Expressions for Absorption Effects

a **Coefficients for Calculating μ in Eqn (x.1)**

	λ < K edge			λ > K edge < L III edge		
Z	**C**	**n**	**K edge**	**C**	**n**	**L III edge**
8 (O)	3.8	2.82	23.23			
20 (Ca)	48.4	2.74	3.070	5.10	2.73	30.7
26 (Fe)	95.8	2.72	1.743	11.75	2.73	14.65
38 (Sr)	251.3	2.68	0.770	36.50	2.73	5.59

b **Mass Absorption Coefficients, μ**

λ	8 (O)	20 (Ca)	26 (Fe)	38 (Sr)
0.877 Sr_{K_α}	2.62	33.78	67.04	25.51
1.938 Fe_{K_α}	24.52	296.20	71.43	221.90
3.360 Ca_{K_α}	115.89	139.47	321.32	998.16
0.30	0.13	1.79	3.62	9.97
0.50	0.54	7.24	14.54	39.21
0.70	1.39	18.21	36.31	96.62
0.90	2.82	36.26	71.93	27.38
1.20	6.35	79.76	157.30	60.04
1.50	11.92	147.01	288.62	110.41
1.80	19.94	242.27	58.47	181.63
2.40	44.87	532.87	128.24	398.35
3.00	84.19	982.10	235.82	732.54
3.50	130.03	155.91	359.21	1115.83
5.00	355.53	412.82	951.10	2954.48

where $P_{(i),\lambda}$ is the emitted intensity for the pure analyte; $C_{(i)}$ is equal to 1.0, the weight fraction of i (pure analyte); and $\mu^*_{i,\lambda}$ is the total mass absorption coefficient of the analyte, at that wavelength, and by the same token,

$$\mu^*_{i,\lambda} = \mu_{i,\lambda} \csc \Psi' + \mu_{i,\lambda i} \csc \Psi'' \tag{17}$$

Ratioing the two equations (14 and 16) defining $P_{i,\lambda}$ and $P_{(i),\lambda}$ and, solving for $P_{i,\lambda}$ yields

$$P_{i,\lambda} = \frac{P_{(i),\lambda}\, \mu^*_{i,\lambda}\, C_i}{\mu^*_{s,\lambda}} \tag{18}$$

Observing that $\mu^*_{s,\lambda}$ may equally be defined as

$$\mu^*_{s,\lambda} = \sum_i C_i \mu^*_{i,\lambda} \tag{19}$$

Thus for a three-element specimen, the emitted intensity when absorption is the only matrix effect involved may be expressed as

$$P_{i,\lambda} = \frac{P_{(i),\lambda}\, \mu^*_{i,\lambda}\, C_i}{C_i\mu^*_{i,\lambda} + C_j\mu^*_{j,\lambda} + C_k\mu^*_{k,\lambda}} \tag{20}$$

In reference to the pure analyte, a three-element system may be visualized as a specimen of a pure element in which a portion C_i' has been replaced by an equal concentration of C_j and, another portion C_i'' has been replaced by an equal concentration of C_k leading to

$$P_{i,\lambda} = \frac{P_{(i),\lambda}\mu^*_{i,\lambda}\,(C_{(i)} - C_i' - C_i'')}{(C_{(i)} - C_i' - C_i'')\mu^*_{i,\lambda} + C_j\mu^*_{j,\lambda} + C_k\mu^*_{k,\lambda}} \tag{21}$$

Since $C_j = C_i'$ and $C_k = C_i''$, substitution and gathering terms gives

$$P_{i,\lambda} = \frac{P_{(i),\lambda}\ \mu^*_{i,\lambda}\ C_i}{C_{(i)}\mu^*_{i,\lambda} + C_j(\mu^*_{j,\lambda} - \mu^*_{i,\lambda}) + C_k(\mu^*_{k,\lambda} - \mu^*_{i,\lambda})} \tag{22}$$

Dividing numerator and denominator by $\mu^*_{i,\lambda}$ yields

$$P_{i,\lambda} = \frac{P_{(i),\lambda}\ C_i}{\left[1 + C_j\left(\dfrac{\mu^*_{j,\lambda} - \mu^*_{i,\lambda}}{\mu^*_{i,\lambda}}\right) + C_k\left(\dfrac{\mu^*_{k,\lambda} - \mu^*_{i,\lambda}}{\mu^*_{i,\lambda}}\right)\right]} \tag{23}$$

Solving for $P_{(i),\lambda}C_i$, expanding and defining

$$X_{ja,\lambda} = P_{i,\lambda}\left(\frac{\mu^*_{j,\lambda} - \mu^*_{i,\lambda}}{\mu^*_{i,\lambda}}\right) \tag{24}$$

yields

$$P_{(i),\lambda}\ C_i = P_{i,\lambda} + X_{ja,\lambda}\ C_j + X_{ka,\lambda}\ C_k \tag{25}$$

Solving for $P_{i,\lambda}$

$$P_{i,\lambda} = P_{(i),\lambda}\ C_i - X_{ja,\lambda}\ C_j - X_{ka,\lambda}\ C_k \tag{26}$$

which can be generalized to

$$P_{i,\lambda} = P_{(i),\lambda}\ C_i - \sum_j X_{ja,\lambda}\ C_j \tag{27}$$

for multielement systems.

While monochromatic beams are sometimes used as the excitation source, it is much more common to use the output of an X-ray tube directly in which case the exciting source is polychromatic. This being the case, Criss and Birks (1968) proposed that the primary spectral distribution be divided into a number of wavelength intervals with corresponding intensities. The emitted intensity is now the resultant of summing the values for each effective interval (i.e., between the minimum wavelength and the absorption edge related to the characteristic line used to measure the intensity). It therefore follows that the emitted intensity of an element in a multielement specimen irradiated by a polychromatic source when absorption is the only matrix effect involved can be expressed as

$$P_i = P_{(i)}\ C_i - \sum_j X_{ja}\ C_j \tag{28}$$

where

$$P_i = \sum_\lambda P_{i,\lambda}\ \Delta\lambda \tag{29}$$

$$P_{(i)} = \sum_{\lambda} P_{(i),\lambda}\, \Delta\lambda \tag{30}$$

$$X_{ja} = \sum_{\lambda} X_{ja,\lambda}\, \Delta\lambda \tag{31}$$

Thus, the expressions for absorption effects, whether the incident beam is monochromatic or polychromatic are similar and express the fact that the emitted intensity due to primary emission is directly proportional to the product $P_{(i)}\, C_i$ *minus* the absorption effect of each element that constitutes the matrix.

B. Theoretical Expressions for Enhancement Effects

While the derivation of expressions for enhancement from first principles is even more complex than those for absorption, enhancement effects can also be expressed in simple terms. For a binary system, Tertian and Claisse (1982) derived an expression of the form

$$P_{i,\lambda} + S_{i,\lambda} = P_{i,\lambda}\,(1 + e_{ij,\lambda}\, C_j) \tag{32}$$

for a monochromatic incident beam, wavelength, λ, where $P_{i,\lambda} + S_{i,\lambda}$ is the emitted intensity due to primary and secondary emission; and $e_{ij,\lambda}$ is the influence coefficient quantifying the enhancement of analyte i by matrix element j. Expanding the above expression for a three component system gives

$$P_{i,\lambda} + S_{i,\lambda} = P_{i,\lambda} + P_{i,\lambda}\, e_{ij,\lambda}\, C_j + P_{i,\lambda}\, e_{ik,\lambda}\, C_k \tag{33}$$

defining

$$X_{je,\lambda} = P_{i,\lambda}\, e_{ij,\lambda} \text{ and } X_{ke,\lambda} = P_{i,\lambda}\, e_{ik,\lambda} \tag{34}$$

yields

$$P_{i,\lambda} + S_{i,\lambda} = P_{i,\lambda} + X_{je,\lambda}\, C_j + X_{ke,\lambda}\, C_k \tag{35}$$

which can be generalized as

$$P_{i,\lambda} + S_{i,\lambda} = P_{i,\lambda} + \sum_{j} X_{je,\lambda}\, C_j \tag{36}$$

For a polychromatic incident beam, a similar summation process is used, i.e., the primary spectral distribution is divided into a number of $\Delta\lambda$ intervals and the total enhancement effect is the sum of each contribution

$$P_i + S_i = P_i + \sum_{j} X_{je}\, C_j \tag{37}$$

where

$$P_i + S_i = \sum_{\lambda} (P_{i,\lambda} + S_{i,\lambda})\, \Delta\lambda \tag{38}$$

is the total emitted intensity for a polychromatic incident beam, and

$$X_{je} = \sum_{\lambda} X_{je,\lambda}\, \Delta\lambda \tag{39}$$

is the prorated summation of each monochromatic enhancement term, P_i has been previously defined. As for absorption, the above expressions for enhancement effects, whether the incident beam is monochromatic or polychromatic, are similar and express the fact that the emitted intensity due to secondary emission is equal to the emitted intensity due to the primary emission *plus* the enhancement effect of each element whose characteristic lines have enough energy to eject electrons from atomic shells of the analyte.

C. Theoretical Expressions for Total Matrix Effects

Although the absorption and enhancement effects are expressed simply enough in Equations (28) and (37), the fact that appears P_i in both these expressions poses a problem in practice. This stems from the fact that only one element (the one having the shortest absorption edge) will involve primary emission, the more common situation being that the emitted experimental intensity, $I_{i,}$ is the resultant of both primary and secondary emissions. The link between experimental and theoretically calculated intensities is given by

$$I_i = g_i(P_i + S_i) \text{ or, } I_{(i)} = g_i\, P_{(i)} \tag{40}$$

where g_i is an instrumental proportionality constant which cancels out when intensities are expressed in relative terms, for example,

$$R_i = \frac{I_i}{I_{(i)}} \quad \frac{g_i(P_i + S_i)}{g_i P_{(i)}} = \frac{P_i + S_i}{P_{(i)}} \tag{41}$$

What is required in order to relate C_i directly in terms of emitted intensities; that is, I_i and $P_i + S_i$ are expressions that do not involve $P_{i.}$ Substitution of Equation (28) in Equation (37) leads to

$$P_i + S_i = P_{(i)}C_i - \sum_j X_{ja}C_j + \sum_j X_{je}C_j \tag{42}$$

which can be interpreted as: the emitted intensity due to primary and secondary fluorescence is *directly proportional* to $P_{(i)}$ C_i (i.e., the intensity that would be emitted if matrix effects were nil), *minus* the sum of all the individual absorption effects *plus* the sum of all the individual enhancements effects. For analytical purposes, the above expression is transformed by solving for $P_{(i)}$ C_i, namely

$$P_{(i)}\, C_i = P_i + S_i + \sum_j X_{ja}C_j - \sum_j X_{je}C_j \tag{43}$$

which can then be interpreted as: the intensity corrected for absorption and enhancement effects is equal to the emitted intensity due to the primary and secondary fluorescence *plus* the sum of all the individual absorption effects *minus* the sum of all the individual enhancement effects. Taking C_j in factor yields

$$P_{(i)}\, C_i = P_i + S_i + \sum_j (X_{ja} - X_{je})C_j; \tag{44}$$

defining $X_{jn} = X_{ja} - X_{je}$; and, taking $P_i + S_i$ in factor yields

$$P_{(i)}C_i = (P_i + S_i)\left[1 + \sum_j \frac{X_{jn}}{P_i + S_i} C_j\right]; \tag{45}$$

solving for C_i leads to

$$C_i = \frac{P_i + S_i}{P_{(i)}}\left[1 + \sum_j \frac{X_{jn}}{P_i + S_i} C_j\right] \tag{46}$$

and

$$C_i = R_i\left[1 + \sum_j n_{ij}C_j\right] \tag{47}$$

where R_i is the relative intensity to the pure analyte; and n_{ij} is the fundamental influence coefficient quantifying the net matrix effect of matrix element j on analyte i.

V. COMPENSATION METHODS

The common feature in the following methods for converting intensities to concentration is that the factor for the correction due to matrix effects is evaluated "globally". The term globally is used to indicate that the process involves the direct or indirect computation of a single value quantifying the total matrix effect. Contrary to theoretical expressions, it is not necessary to know the spectral distribution of the primary beam, the angles of incidence and emergence, mass absorption coefficients, etc. The choice of a compensation method will vary depending on the analytical context. One procedure is based on the fact that if a number of specimens of varying composition (hence significant variation in the magnitude of the matrix effect) are diluted equally using a common diluent, the resulting specimens will show much less variation in the magnitude of the matrix effect. Another procedure involves adding a known concentration of a judiciously chosen element to a portion of the sample and measuring its intensity which will be a function of the specimen's total matrix effect. A variant of the above is to add a known concentration of the analyte to a portion of the sample and measuring intensities "before and after addition". A procedure that is specific to XRF involves the use of Compton scatter which can serve as an indicator of absorption effects. The fact that a complete analysis of the specimen is not required is a definite advantage when only a few elements are to be determined. Dilution, standard addition, and Compton scatter, especially the latter, are widely used in quantitative XRF analysis of geochemical samples and will be discussed in some detail.

A. Dilution Methods

The principle underlying dilution methods is more easily described by reference to numerical examples using Equation (47). The fact that a theoretical expression is used in no way alters the premise that the practical application of dilution methods does not require knowledge of primary spectral distribution, instrument geometry, etc. The demonstration is based on comparing the difference in the value of the matrix effect correction factor in two specimens wherein the matrix effect is quite different, and observing how this difference decreases as a result of dilution.

Consider specimen A where $C_i = 0.2$, $C_j = 0.8$, and $n_{ij} = 3.0$, and specimen B where $C_i = 0.2$, $C_k = 0.8$, and $n_{ik} = -0.5$. For the undiluted specimens, substitution of these values in the [ ] term in Equation (47) gives

$$[\ ]_A = 1 + (3.0)0.8 = 3.4$$

$$[\ ]_B = 1 + (-0.5)0.8 = 0.6$$

Consider the case when both specimens are diluted by a factor of 10, using diluent (d), with $n_{id} = 1.0$; C_i now equals 0.02, C_j and $C_k = 0.08$, and the diluent $C_d = 0.90$ in both specimens. The equation for the diluted specimens is

$$C_i = R_i\,[1 + n_{ij}C_j + n_{id}C_d] \tag{48}$$

Substitution of the above values gives

$$[\ ]_A = 1 + (3.0)0.08 + (1.0)0.90 = 2.14$$

$$[\ ]_B = 1 + (-0.5)0.08 + (1.0)0.90 = 1.86$$

Now consider the case if both samples are diluted by a factor of 100 using the same diluent. For these *highly diluted* specimens, substitution of the appropriate values in Equation (48) gives

$$[\ ] = 1 + (3.0)0.008 + (1.0)0.99 = 2.014$$

$$[\ ] = 1 + (-0.5)0.008 + (1.0)0.99 = 1.986$$

If either one of the original samples had been used as the reference and the other as the unknown, the error due for not compensating for the matrix effect is given by the ratio $[\ ]_A/[\ ]_B$; that is, the concentrations would either be 5.67 times too high or 5.67 times too low. In the case of the medium dilution, the ratio now indicates that using either specimens as reference would lead to results that are 1.15 times too high or too low. For the highly diluted specimen, the ratio now shows that the results would now be either 1.014 times too high or 1.014 times too low (i.e., 20 ± 0.28 %).

There are, however, a number of points that must be taken into consideration. Dilution by a factor of 100 will drastically reduce the emitted intensity, hence this technique is only applicable for the determination of major or minor constituents. The loss of intensity can be made up in part by longer counting times, which in turn leads to much lower productivity. Another point to be considered is inherent to the dilution procedure itself. If a portion of the crushed rock is diluted by mixing, say with finely ground Al_2O_3, particle size effects and mineralogical effects are still present and are not compensated for. It is a well established fact that emitted intensity not only varies with particle size but that the variation is different depending on the wavelength of the characteristic line being used. The mineralogical effect is twofold. First, there is the effect due to the composition of the mineral; oxygen (cuprite) is a very weak absorber of Cu_{K_α} characteristic radiations, sulphur (covellite) is a weak absorber of Cu, iron (chalcopyrite) is a strong absorber, while arsenic (enargite) enhances Cu radiations. Second, elements in minerals with basal cleavage will emit quite different intensities should the plates be horizontal or perpendicular at the specimen surface. It is therefore preferable to dilute the sample, say with a borate salt, which can then be heated to its melting point in order to dissolve the sample. The melt is then cast as an amorphous glass disk, or cooled, crushed, and pelletized. A fusion procedure is necessarily more involved (more time consuming, specialized equipment) but it does deal effectively with particle size and mineralogical effects and compensates in part for matrix effects.

B. Standard Additions Method

The addition or spiking method is a subset of the internal standard method which consists in adding a known concentration of an element to the original sample and measuring its intensity. The added element must be chosen such that it will undergo a similar matrix effect as the analyte. In contrast to the dilution approach, the addition method is generally limited to the determination of constituents in the trace concentration range.

The problems associated with choosing an element that will undergo similar matrix effects as the analyte has led to the standard addition method being the more commonly used in XRF analysis. The method consists of adding to a portion of the original sample a known amount of the analyte itself, say ΔC_i. After an appropriate mixing step (fusion preferred), the fluorescence intensity is then measured on a specimen of the original sample, $I_{i,1}$, and on the specimen with the addition, $I_{i,2}$. The difference in intensity, ΔI_i, due to the ΔC_i addition is given by

$$\Delta I_i = I_{i,2} - I_{i,1} \tag{49}$$

In order to be applicable, the addition must be kept relatively small. Referring back to the general concentration—intensity relation, the relation can be expressed as

$$C_i = I_i \text{ (calibration factor)[matrix effect original sample]}_1 \tag{50}$$

and

$$\Delta C_i = \Delta I_i \text{ (calibration factor)[matrix effect sample + addition]}_2 \tag{51}$$

The two expressions are ratioed; the object being that both the calibration factor *and* the matrix effect cancel out, thus the concentrations are directly proportional to the intensities; that is

$$\frac{C_i}{\Delta C_i} = \frac{I_{i,1}}{\Delta I_i} \qquad (52)$$

and

$$C_i = I_{i,1} \left(\frac{\Delta C_i}{\Delta I_i} \right) \qquad (53)$$

Generally the analyst also aims at making an addition such that the concentration added, ΔC_i, is roughly equal to the estimated concentration C_i. A typical procedure would be to add 0.03 g of a mixture containing 10% analyte to a 3.0-g portion of the original sample. This implies an estimated value of 0.1% for C_i in the original sample. The specimen to which the addition has been made then consists of 99% of the original sample; hence the ratio $[\ ]_1/[\ ]_2$ is essentially equal to unity. For a sample estimated to contain 0.2% C_i, one would use similar proportions but add a mixture containing 20% C_i. Care must be taken however to remain within the linear intensity-concentration region; that is, generally 1 to 2% range in the original sample.

C. Compton Scatter Methods

Although absorption at any wavelength is the sum of a scattering component and a photoelectric absorption component, the following treatment is limited to how incoherent or Compton scatter may be used to compensate for *absorption* effects in XRF analysis. Readers may refer to a recent paper by Willis (1989) for an in-depth treatment of trace element analysis of geological materials using Compton scatter.

Compton scattering is the process by which the scattered photons lose some of their energy with a resultant increase in wavelength. The increase, $\Delta\lambda$, is given by

$$\Delta\lambda = 0.0243\ (1 - \cos\emptyset) \qquad (54)$$

where $\emptyset$ is the scatter angle; that is, the sum of the incident and take off angles. For example, with a Mo target X-ray tube, a fraction of the Mo_{K_α} line is scattered and two peaks are observed: a coherent peak at λ_{MoK_α}, and an incoherent Compton peak at $\lambda_{MoK_\alpha} + \Delta\lambda$. If the sum of the incident and take-off angles = 90° (60° + 30°, 45° + 45° are common), then the wavelength displacement is 0.024 Å; that is, the Mo_{K_α} peak at 0.71, and the Mo_{K_α} Compton peak at 0.734 Å. The property underlying the wide use of Compton scatter intensities for the correction of *absorption* effects is that the Compton peak intensity is inversely proportional to mass absorption coefficient.

The fact that a knowledge of the intensity of the Compton peak, I_c, measured on a suite of specimens may be used to assess the relative absorption effect within the suite is of capital importance when one considers that knowledge of the relative absorption can be used to correct for absorption effects for a wide range of elements. The fact that the mass absorption of a specimen changes drastically when crossing a major or minor element absorption edge means that a mass absorption coefficient determined using a Compton peak can only be applied to another wavelength *provided* that no major or minor absorption edges fall *between* the Compton peak and the analyte wavelength. For geological materials, it is generally the absorption edge of iron that sets the demarcation line. The K absorption edge of iron is equal to 1.743 Å, therefore elements whose K or L analyte spectral lines are <1.743 Å will not be subject to enhancement effects by major constituents and thus be amenable to be determined using Compton scatter for the correction due to absorption. For K lines, this means elements Z = 28 (Ni) through Z = 53 (I); and, for L lines, Z = 73 (Ta) through Z = 92 (U). A method for estimating mass absorption coefficients on both sides of absorption edges has been proposed, but because the assumptions made may not always be valid, the validity of the correction may suffer in consequence.

The following is a typical example of an application of Compton scatter to compensate for the variation in the absorption effect in a suite of rocks submitted for the determination of Rb and Sr in the context of a geochronology project. For space considerations, the numerical data is limited to three samples. An estimate of the accuracy of the results can be made, the samples having been subsequently analyzed for Rb and Sr using an isotope dilution technique. The samples were transferred as received (−200 mesh) to specimen holders (bottom mylar film) and irradiated with Mo radiations (Rh tube target, Mo secondary target) and Rb_{K_α}, Sr_{K_α}, and I_c measured using energy dispersion. Table 3 lists the net intensities and the computed concentrations with and without the correction based on the intensity of the Compton peak.

Given that I_i is proportional to concentration and inversely proportional to the total absorption when enhancement is not present, the expression for applying absorption effect corrections using Compton scatter is derived from

$$I_i = g_i \frac{C_i}{\mu_s} \text{ and } I_i^* = g_i \frac{C_i^*}{\mu_s^*} \tag{55}$$

where I_i, C_i, and μ_s refer to a specimen of unknown composition, and, I_i^*, C_i^*, and μ_s^* refer to a standard. Since I_c is inversely proportional to μ_s, substitution in the above leads to

$$I_i = g_i' C_i I_c \text{ and } I_i^* = g_i' C_i^* I_c^* \tag{56}$$

Ratioing and solving for C_i yields

$$C_i = I_i \left(\frac{C_i^*}{I_i^*}\right) \left[\frac{I_c^*}{I_c}\right] \tag{57}$$

where the ( ) brackets represent the usual calibration factor and the [ ] brackets represent the absorption effect correction factor. Conversely, C_i may be defined as

Table 3. Application of Compton Scatter for the Correction of Absorption

Analytical context						E_{keV}
Primary source:		Mo_{K_α} 0.711;		λ_c = 0.734;		$Mo_{K_{\alpha,\chi}}$ = 16.87
						Rb_{K_α} = 13.37
Counting time:		400 seconds;				Sr_{K_α} = 14.14
Reference Standard		642		I_c^* = 309.0		
Rb = 307.6 ppm,		Sr = 455.0 ppm,		$I_{Rb_{K_\alpha}}$ = 55.7,		$I_{Sr_{K_\alpha}}$ = 99.4
Calibration factors						
$Rb = \frac{307.6}{55.7} = 5.522$,			Sr = $\frac{455.0}{99.4} = 4.577$			
Unknowns						
Ident	$I_{Rb_{K\alpha}}$	$C_{Rb}{}^a$	I_c	**factor**	$C_{Rb}{}^b$	$C_{Rb}{}^c$
885	110.7	611.0	455.7	0.678	414.0	415.4
716	1.7	9.4	246.7	1.253	11.8	10.4
617	40.4	223.0	365.1	0.846	189.0	189.6
	$I_{Sr_{K\alpha}}$	$C_{Sr}{}^a$	I_c	**factor**	$C_{Sr}{}^b$	$C_{Sr}{}^c$
885	18.0	82.4	455.7	0.678	55.9	57.4
716	76.2	349.0	246.7	1.253	437.0	432.2
617	16.1	73.7	365.1	0.846	62.3	64.8

[a] uncorrected
[b] corrected
[c] by isotope dilution

$$C_i = \frac{I_i}{I_c}\left(\frac{C_i^* I_c^*}{I_i^*}\right) \tag{58}$$

where $\frac{C_i^* I_c^*}{I_i^*}$ is a constant. (59)

VI. MATHEMATICAL METHODS

Mathematical methods and numerical methods are the terms generally used to designate those procedures where the matrix effect correction factor is the result of the summation of the individual matrix effect for each element present. Some of these procedures are based exclusively on the time-honored equations developed many years ago using fundamental parameters available in the literature. Others are based on algorithms that approximate theory very closely, while a third category is based on statistical treatment of experimental data and are termed empirical. Due to space considerations only the essence or concepts underlying the more commonly used methods will be considered. The benefits of incorporating theoretical concepts in a correction procedure is that it extends the range with which specimens may differ from those used for the calibration, in contrast to the empirical approach where it is recognized that standards and unknowns must be very similar.

A. The Fundamental Parameters Approach

The breakthrough in the theoretical approaches for the correction of matrix effects was the development of a workable model (Criss and Birks, 1968), which, in addition to being an analytical method in its own right, has been the basis of most subsequent theoretical intensity-concentration expressions. In essence, the expression proposed by Criss and Birks may be envisaged as

$$P_1 + S_i = Q_i\, C_i \sum_{\lambda} \frac{\text{(theoretically calculated enhancement term)}}{\text{(theoretically calculated absorption term)}}\, \Delta\lambda \qquad (60)$$

where Q_i is a proportionality constant. As shown in Section IV-A and Section IV-B, the enhancement term is the resultant of the summation of each effective $\Delta\lambda$ enhancement effect, while the absorption term is the resultant of the summation of each effective $\Delta\lambda$ absorption effect. Given the integrated intensities of the primary incident beam as a function of $\lambda + \Delta\lambda$ and reliable fundamental parameters (mass absorption coefficients, fluorescence yields), the Criss and Birks' equation makes it possible to compute emitted *relative* X-ray fluorescence intensities for any specimens of known composition. The fact that analysis involves handling specimens of unknown composition is not a major drawback. The weight fraction in unknowns are found by an iterative process that involves making successively better estimates of mass fractions until the corresponding calculated relative intensities agree with the measured intensities.

In theory, a fundamental parameters method is only limited by the uncertainties in the values of the parameters used. As far as XRF analysis is concerned, this is not a major problem as many laboratories have devoted themselves to measuring the parameters accurately. The fact that the accuracy with which fundamental parameters are known decreases for the light elements, and that O, Na, Mg, Al, and Si constitute the bulk of most samples submitted for geochemical analysis, may be compensated for by calibrating with natural rock standards. A more serious drawback of the fundamental parameters approach for the analysis of geological samples is that summation over all effective must be carried out for each element for each iteration step. This may take an excessively long time in practice, hence the tendency to resort to influence coefficient methods (next section). Most manufacturers of X-ray spectrometers, WDS, and EDS offer software that includes the fundamental parameters method as an option for the correction due to matrix effects.

B. Alpha Coefficient Methods

The aim of influence coefficient methods is to bypass the need for carrying out the summations over all effective wavelengths for every analyte in each specimen analyzed. This translates into deriving expressions that condense in a single coefficient the quantification of the matrix effect of element j on analyte i, of element k on analyte i, etc. The simplest expression, Equation (47), has been termed the canonical form (Tertian, 1986), and is widely used in the analysis of geological materials in the form of fused discs. In order to clearly indicate that an assumption is made in such cases, the symbol α_{ij} is used instead of n_{ij}, thus in the expression

$$C_i = R_i\left[1 + \sum_j n_{ij}C_j\right] \tag{61}$$

it is implied that the influence coefficients are calculated explicitly from fundamental expressions involving fundamental and instrumental parameters. Their calculation involves summation over the effective wavelength range of the polychromatic primary beam. Extensive studies have shown that coefficients vary as a function of specimen composition, hence are unique to each specimen. The expression

$$C_i = R_i\left[1 + \sum_j \alpha_{ij}C_j\right] \tag{62}$$

on the other hand implies that the calculation of the influence coefficients ininvolved making one or more simplifying assumptions. The goal is to generate α_{ij} values from n_{ij} arrays (thereby retaining a link to the theory of XRF emission, absorption, and enhancement) that are then treated as constants. This is feasible in practice provided it can be established under what conditions will this assumption not introduce an unacceptable degree of error in the correction for matrix effects. With both the α_{ij} and n_{ij} expressions, analyte concentrations are computed by iteration, but the α_{ij} approach bypasses the somewhat lengthy summation step over the effective wavelength range of the incident beam required by the n_{ij} approach. It is therefore appropriate to examine why alpha coefficients can be treated as constants for the analysis of fused disc specimens and, also examine a number of adaptations that have been introduced by analysts when dealing with geological samples.

From a practical point of view, using fused discs specimens may be regarded as a medium dilution method. It is a middle-course approach in the sense that it aims at eliminating the problems associated with powdered specimens while minimizing the loss of intensity associated with dilution. Ideally, the end product of the fusion process is homogeneous amorphous glass discs with mirror-like surfaces involving dilution factors ranging from 5 to 10. Unfortunately, there still remains a need for matrix effect corrections which although certainly not as large, are still significant and must be taken into consideration in order to obtain accurate results.

Recalling Equation (48), let C_i, C_j, . . . , represent the weight fractions of the oxides in a geological material which is diluted by the addition of a fluxing agent. The composition of the fused disc specimen can now be represented by

$$C_i' + C_j' + \ldots + C_f = 1.0 \tag{63}$$

where C_i', C_j', . . ., now refers to weight fraction oxides in the fused disc specimen, and C_f is the weight fraction of the fluxing material or, in terms of weight fractions in the original material

$$C_i(1 - C_f) + C_j(1 - C_f) + C_f = 1.0 \tag{64}$$

and, for the pure oxide fused disc specimen

$$C_{(i)}(1 - C_f) + C_f = 1.0 \tag{65}$$

If I_i' and $I_{(i)}'$ represent the intensities for the geological material and pure oxide fused disc specimens, respectively, then

$$C_i' = R_i' \,[1 + n_{ij}C_j' + \ldots + n_{if}C_f] \tag{66}$$

where

$$C_i' = C_i\,(1 - C_f) \tag{67}$$

$$C_j' = C_j\,(1 - C_f) \tag{68}$$

$$R_i' = I_i'/I_{(i)}' \tag{69}$$

The fact that C_f will be fixed and generally greater than 80%, and that the original sample is approximately one-half oxygen, results in specimens that are 90%+ identical. This implies that the balance will be made up of one or two constituents in the 0 to 4% range, and all the remaining ones definitely less than 1%. Under these conditions, after theoretical verification that the arrays of coefficients n_{ij}, n_{ik}, . . . , vary but always within well defined narrow limits, *the assumption is made* that n_{ij}, n_{ik}, . . . n_{if}, can be treated as constants and not jeopardize the validity of the matrix effect correction, hence

$$C_i' = R_i'\left[1 + \sum_j \alpha_{ij}C_j' + \alpha_{if}C_f\right] \tag{70}$$

where the α_{ij}, . . . , α_{if} coefficients are now equated to: (1) an average value of n_{ij}, . . . , n_{if} arrays; (2) the values of n_{ij}, n_{ik}, . . . , n_{if} computed for binary systems wherein C_i is equal to the middle of the concentration range expected in the unknowns; and (3) the values of n_{ij}, . . . , n_{if} computed using the C_i, C_j, . . . , concentrations that represent a typical specimen, real or hypothetical. The usual iterative process is applied, and finally the concentrations in the original samples are given by

$$C_i = \frac{C_i'}{1 - C_f}\text{ ; and } C_j = \frac{C_j'}{1 - C_f} \tag{71}$$

A variant introduced by some analysts is based on the fact that if the term $\alpha_{if}C_f$ is considered to be constant, then it can be factored; that is, taken out of the [ ] brackets. The canonical form is retained

$$C_i' = R_i'\left[1 + \sum_j \alpha_{ij}'C_j'\right] \tag{72}$$

but the α'_{ij}, α'_{ik}, . . . , coefficients are now equated to

$$\alpha'_{ij} = \frac{\alpha_{ij}}{1 + \alpha_{if}C_f}\text{; and } \alpha'_{ik} = \frac{\alpha_{ik}}{1 + \alpha_{if}C_f} \tag{73}$$

and are generally referred to as "modified alphas"; that is, basic α_{ij}, α_{ik}, . . . , coefficients that have been modified as a result of eliminating the $\alpha_{if}C_f$ term within the [ ] brackets.

Another variant is sometimes used by analysis who wish to visualize the individual corrections in terms of the concentrations in the original samples rather than concentrations in the fused disc. Thus, instead of

$$C'_i = R'_i\left[1 + \sum_j \alpha'_{ij}C'_j\right] \tag{74}$$

the relation is stated in terms of the concentrations in the samples; therefore

$$C_i(1 - C_f) = R'_i\left[1 + \sum_j \alpha'_{ij}\{(1 - C_f)C_j\}\right] \tag{75}$$

Regrouping terms within the [ ] bracket and taking into account that

$$R_i = \frac{R'_i}{1 - C_f} \text{ (i.e., dividing both sides by } 1-C_f\text{) yields} \tag{76}$$

$$C_i = R_i\left[1 + \sum_j \{\alpha_{ij}(1 - C_f)\}C_j\right] \tag{77}$$

and again reverting to the canonical form

$$C_i = R_i\left[1 + \sum_j \alpha''_{ij}C_j\right] \tag{78}$$

where the α''_{ij} coefficient is now equated to the product α'_{ij} $(1-C_f)$; that is, a doubly modified basic coefficient.

Another variant is sometimes introduced in order to compensate for the loss on ignition (LOI) that may occur during fusion due to the presence of volatiles (CO_2, H_2O, . . .) in the samples. Some analysts opt for igniting all samples *prior* to fusion, proceed with the fusions using the ignited residues, and subsequently recalculating to the original samples. In other cases, if the determination of CO_2 and H_2O are requested and determined by non-XRF techniques, their concentrations can be used to correct for the LOI by adding a term in the expression used to calculate the matrix effect correction factor, thus

$$C_i = R_i\left[1 + \sum_j \alpha_{ij}C_j + \alpha_{iLOI}C_{LOI}\right] \tag{79}$$

where

$$C_{LOI} = C_{CO_2} + C_{H_2O} \tag{80}$$

A third alternative is based on the following algebraic transformation. Considering that

$$C_{LOI} = 1 - C_i - C_j - \ldots \tag{81}$$

substitution of $1 - C_i - C_j - \ldots$ for C_{LOI} in Equation (79) leads to

$$C_i = R_i\left[1 + \sum_j \alpha_{ij}C_j + (1 - C_i - C_j - \ldots)\,\alpha_{iLOI}\right] \quad (82)$$

which after expansion gives

$$C_i = R_i\left[1 + \alpha_{ij}C_j + \ldots + \alpha_{iLOI} - \alpha_{iLOI}C_i - \alpha_{iLOI}C_j - \ldots\right] \quad (83)$$

followed by regrouping gives

$$C_i = R_i\left[1 + \alpha_{iLOI} + (-\alpha_{iLOI})C_i + \sum_j (\alpha_{ij} - \alpha_{iLOI})C_j\right] \quad (84)$$

Taking $1 + \alpha_{iLOI}$ in factor, yields

$$C_i = R_i\,(1 + \alpha_{iLoI})\left[1 + \left(\frac{-\alpha_{iLOI}}{1 + \alpha_{iLOI}}\right)C_i + \sum_j\left(\frac{\alpha_{ij} - \alpha_{iLOI}}{1 + \alpha_{iLOI}}\right)C_j\right] \quad (85)$$

which once again tends to be stated in the canonical form

$$C_i = R_i''\left[1 + \alpha_{ii}'C_i + \sum_j \alpha_{ji}'C_j\right] \quad (86)$$

where

$$R_i'' = R_i\,(1 + \alpha_{iLOI}) \quad (87)$$

$$\alpha_{ii}' = \frac{-\alpha_{iLOI}}{1 + \alpha_{iLOI}} \quad (88)$$

$$\alpha_{ij}' = \frac{\alpha_{ij} - \alpha_{iLOI}}{1 + \alpha_{iLOI}} \quad (89)$$

the latter being referred to as ''modified hybrid alphas'', hybrid to designate that the coefficient is the resultant of combining two alpha coefficients. In effect, the correction for the LOI has been disseminated among all the other constituents which necessitated the introduction of the term $\alpha_{ii}'C_i$ to the matrix effect correction factor.

C. Alpha Coefficients, Oxide Systems

The conditions under which the alpha coefficients in the above expressions can be calculated from theory and treated as constants does not apply to unfused geological materials; that is, oxide specimens in the form of pressed pellets. In the following examination of matrix effects in oxide specimens, it is assumed that the binder is added in such a small quantity (a common practice) that it does not act as a diluent, and the particle and mineralogical effects will be disregarded. Recalling the concept of alpha coefficients and the fact that under some circumstances analysts may be compelled to work directly with nondiluted oxide specimens, we are then dealing with the case where the matrix effect of element j on analyte i can no longer be equated to a constant. However, the canonical form can be adapted to retain both a degree of the theoretical exactness of the fundamental influence coefficient approach and its inherent flexibility. It is based on concepts proposed by Claisse and Quintin (1967) as a result of a theoretical and experimental examination of absorption effects in XRF when the incident beam is polychromatic. A detailed examination of large number of binary n_{ij} coefficients shows that they vary within well-defined limits. While the plot of n_{ij} versus C_j for elemental systems is best approximated by an hyperbolic function, in the case of oxides systems, the curves are not as pronounced, and a linear fit is generally considered acceptable. Analysts may proceed by applying a linear least squares fit process, or use data for the two binaries ($C_i = 0.8$, $C_j = 0.2$; $C_i = 0.2$, $C_j = 0.8$) to establish the intercept, α_j, and the slope, α_{jj} of a linear relation α_{ij} versus C_j. Alternatively, the two coefficients can be calculated by solving the equations

$$n_{ij,1} = \alpha_j + \alpha_{jj}C_{j,1} \tag{90}$$

$$n_{ij,2} = \alpha_j + \alpha_{jj}C_{j,2} \tag{91}$$

which by Cramer's Rule yields

$$\alpha_j = \frac{n_{ij,1}C_{j,2} - n_{ij,2}C_{j,1}}{C_{j,2} - C_{j,1}} \tag{92}$$

and

$$\alpha_{jj} = \frac{n_{ij,2} - n_{ij,1}}{C_{j,2} - C_{j,1}} \tag{93}$$

For a binary system, Claisse and Quintin proposed the expressions

$$C_i = R_i[1 + \alpha_j C_j + \alpha_{jj} C_j^2] \tag{94}$$

which can be transformed to the canonical form

$$C_i = R_i[1 + (\alpha_j + \alpha_{jj}C_j)C_j] \tag{95}$$

Based on a judicious development using a different formalism, it was shown (Tertian, 1975) that for multicomponent systems, it is by far preferable to modify the Claisse-Quintin model; thus

$$C_i = \left[1 + \sum_j (\alpha_j + \alpha_{jj}\, C_M)C_j\right] \tag{96}$$

where

$$C_M = C_j + C_k + \ldots = \sum_j C_j \tag{97}$$

However, it is observed that even if C_M is substituted for C_j in the Claisse–Quintin expressions, the matrix effect correction term is still not adequate due to what is referred to as the third element effect. Claisse and Quintin had shown that the latter could be compensated for by introducing the additional term $\alpha_{ijk}C_jC_k$, thus yielding for a three-component system

$$C_i = R_i\,[1 + (\alpha_j + \alpha_{jj}C_M)C_j + (\alpha_k + \alpha_{kk}C_M)C_k + \alpha_{ijk}C_jC_k] \qquad (98)$$

or, in the canonical form

$$C_i = R_k\left[1 + \sum_j \left((\alpha_j + \alpha_{jj}C_M) + \alpha_{ijk}C_k\right)C_j\right] \qquad (99)$$

The coefficient α_{ijk} may be found by solving

$$\alpha_{ijk} = \frac{(C_i/R_i) - 1 - (\alpha_j + \alpha_{jj}C_M)C_j - (\alpha_k + \alpha_{kk}C_M)C_k}{C_jC_k} \qquad (100)$$

While the coefficient α_{ijk} is not a constant, experience has shown that in the vast majority of cases, an average value can be calculated and treated as a constant. For a multicomponent system, the canonical form of the modified Claisse–Quintin model is

$$C_i = R_i\left[1 + \sum_j \left(\alpha_j + \alpha_{jj}\,C_M + \sum_k \alpha_{ijk}C_k\right)C_j\right] \qquad (101)$$

and defining

$$\alpha_{ij} = \alpha_j + \alpha_{jj}C_M + \sum_k \alpha_{ijk}C_k \qquad (102)$$

yields

$$C_i = R_i\left[1 + \sum_j \alpha_{ij}Cj\right] \qquad (103)$$

This results in a model which allows for adjusting the alpha coefficients as a function of specimen composition during the iteration process.

D. Empirical Coefficient Methods

Although rejected by the theoretical school, empirical coefficient methods still find application in the XRF analysis of geochemical samples. All are based on using experimental data to generate the influence coefficients by regression processes, and are generally used when dealing with pressed pellet specimens. In one model, the individual correction terms are based on the concentration of the constituents, while in the other, the correction terms are based on the intensities of the constituents concerned. In order to facilitate recognition of the models and their coefficients, the symbols K_{ij}, K_{ik}, . . . , will be used to designate the influence coefficients in the concentration model, and k_{ij}, k_{ik}, . . . , to designate the coefficients in the intensity model.

The expression used to generate empirical influence coefficients in the concentration model is the canonical form transformed to

$$\frac{C_i}{R_i} - 1 = C_{j,1}K_{ij} + C_{k,1}K_{ik} + \ldots \qquad (104)$$

Algebraically, this can then be visualized as solving a set of linear equations; for example, for a three-component system

$$\frac{C_{i,1}}{R_{i,1}} - 1 = C_{j,1}K_{ij} + C_{k,1}K_{ik} \quad (105)$$

$$\frac{C_{i,2}}{R_{i,2}} - 1 = C_{j,2}K_{ij} + C_{k,II}K_{ik} \quad (106)$$

where the subscripts 1 and 2 refer to reference standards 1 and 2, respectively.

Unfortunately, experience has shown that the numbers generated by this approach have little or no physical meaning in that their algebraic signs are sometimes opposite to what is predicted by theory. It is acknowledged that the values merely constitute a "best set of numbers" that are only applicable to specimens of the same type as those used to generate the experimental data.

The intensity model (Lucas-Tooth and Price, 1961) is based on two sweeping assumptions: (1) enhancement is regarded as negative absorption and assumed to obey the same laws; and (2) absorption of element i by j is linearly proportional to C_j. Thus in terms of intensity, the actual departure from the perfect i-j binary calibration curves will be proportional to I_j which led the authors to propose the relation

$$C_i = k + I_i\left(k_o + k_{ii}I_i + \sum_j k_{ij}I_j\right) \quad (107)$$

where k, k_o, k_{ii}, and K_{ij} are constants.

The influence coefficients are generated by regression of experimental data. The advantages of this approach are that it is applicable in cases where only one or two elements are to be determined, and that the standards available need only be certified for the elements that are to be determined. In some manufacturers' software, the intensity model is simply the canonical form wherein the concentrations have been replaced by the corresponding emitted intensities; that is

$$C_i = k + k_iI_i\left[1 + \sum_j k_{ij}I_j\right] \quad (108)$$

It is generally stipulated that intensity models are only applicable if the standards used to generate the coefficients and the unknowns are *very similar*.

VII. ANALYTICAL STRATEGIES

The term strategy is used in the sense of "planning by XRF analysts to establish one or more analytical schemes in order to be in the most advantageous position to meet the requirements of a given analytical context"; in the present case, samples submitted in the course of geochemical explorations. Ultimately, analytical strategies are the end product of a tripartite contract. One party is the client, whose main concerns are turnaround time and the quality of the compositional data reported; another is management, which in addition to an overall interest must be concerned with costs. The third party is of course the analyst, who must make the judicial procedural choices and compromises in order to reach agreement on what will be the XRF laboratory's analytical responsibilities. The analyst's procedural options may be grouped in three categories: (1) those pertaining to specimen preparation; (2) those pertaining to intensity measurements; (3) those pertaining to the data reduction process, which involves a calibration step and some form of correction for matrix effects.

The three categories are linked in that specimen preparation affects the intensity measurement scheme which in turn affects the data conversion method selected. The number of experiments that have to be carried out will vary depending on the expertise already in place, thus the following should be considered as guidelines only.

A. Specimen Preparation: Ideal or Compromise

While it is fairly simple to describe the ideal specimen for XRF analysis—homogeneous, flat, and infinitely thick in respect to X-rays—preparation of ideal specimens in practice is a different matter. Harvey (1989) compares the problems associated with the fusion approach and the pressed powder method and concludes that the latter is the preferable choice for *routine* work. There is no doubt that the fusion approach can produce a disc which satisfies the criteria for an ideal specimen but a number of disadvantages are singled out:

1. Preparation of fused discs is relatively slow.
2. Requires specialized equipment (cost).
3. Use of a single flux may not always be possible.
4. Fusion usually carried out in platinum-gold crucibles (cost).
5. Damage to the crucibles due to the presence of sulfides, As, Pb, Zn, Sb, ...
6. Inherent dilution factor.

The advantages inherent to using pressed pellets are:

1. The possibility of using so little binding material (a few drops of a mixture of polyvinylpyrrolidine and methyl cellulose dissolved in ethanol and water) that when used with standards prepared in the same way, the dilution factor can be ignored.
2. No weighing is involved.
3. No significant increase in background intensities.
4. Less costly equipment needed.

while on the other hand

1. The grain size distribution and homogeneity of the powder is of great importance to successful analysis.

2. The grinding process is more critical; fine grinding in agate receptacles is preferable but costly.

An evaluation of the precision of specimen preparation procedures is primordial in assessing their feasibility. When provided by the client with: (1) a list of key elements, desirable elements, and their likely concentration range; (2) an estimate of the precision, accuracy and detection limits required, desirable, etc.; and (3) a suite of typical samples, the analyst is then in a position to assess the precision associated with the various steps of the preparation procedure. This may involve:

1. Splitting the samples (portion A and B) and treat each portion separately for crushing and grinding.
2. Preparation of duplicate specimens on each portion. It is preferable to prepare duplicate specimens on many samples rather than prepare many replicates on a few samples.
3. Measurement of peak and background intensities (nominal counting times) for the elements of interest.

Ratioing net intensities obtained on duplicate specimens of the same portion provides an estimate of the adequacy of the pelletizing step, while comparison of intensities of specimens from portions A and B provide a check of the adequacy of sample size, and of the crushing and grinding steps. Whatever the specimen preparation method ultimately chosen, it must be ascertained that it more than meets the required precision criteria. Willis (1989) points out that in the case of low ash coals, infinite thickness for Mo_{K_α} radiations is of the order of 5 cm (10 cm for Rh_{K_α} radiations), and that lack of infinite width yields nonlinear calibration curves when using $Mo_{K\alpha}$ Compton peak intensities. The next step involves the preparation of a few specimens of known composition. These need not be of the certified type since any in-house standards are acceptable; the object being to assess the sensitivity and precision of each element as a function of instrumental operating conditions.

B. Intensity Measurement Scheme

Given reliable specimens, the need for a carefully arranged and systematic plan for generating the necessary intensity data that is conducive to yielding high-quality analytical results is no less important. Correct operating conditions linked to an adequate quality control of the validity of the intensities ascribed to the elements being determined are a prerequisite of any analytical strategy. The word valid in reference to intensities relates to the need to generate *net* intensities as part of the intensity-to-concentration conversion process.

Generally one aims for one set of operating conditions that will avoid spectral overlap and maximize count rates, but it may be necessary to tailor conditions to individual sample batches. This is a particular problem in geochemical exploration if elements normally treated as occurring in low trace levels appear as significant peaks. There is no substitute in these situations to having prior knowledge (from the client) or having to repeat the analyses. Unlike many techniques, the quality of an X-ray measurement may be improved by counting for a longer time which in turn will improve detection limits. Thus it is possible to show graphically the variation in the theoretical detection of an element as a function of total count time. Harvey (1989) tabulates a set of operating conditions that may serve as a guide for the determination of 50 minor and trace elements. A list of 20 selected lines of trace elements and the potential error due to overlapping peaks is also tabulated along with estimated (2σ) detection limits for given total count times (peak plus background). Willis (1989) cautions analysts about a recently recognized source of error: the occurrence of Compton scatter of *specimen* spectral lines that have a measurable tailing associated with them.

The determination of valid background intensity at peak positions is crucial when elements at the trace level are being determined. Unfortunately, the background is almost always sloping and nonlinear, hence intensities have to be measured at a number of off-peak positions. Willis (1989) comments on nine possible ways in which the background at a peak position can be estimated. These range from the simple case when the background can be considered as horizontal to fitting a second order polynomial curve to background intensities measured at three or more positions. Willis stresses that background intensities at peak positions should *always* be interpolated and *never* extrapolated and, that *under no circumstances*, should third or higher order polynomial equations be used.

Consideration also has to be given to the development of an appropriate intensity measurement scheme that will serve for quality control. Harvey (1989) points out that quality control of analytical data in XRF spectrometry is potentially very rigorous due to the nondestructive nature of the technique, its high precision, and the degree of automation that can be achieved. Thus the problem at hand can be summarized as to what reference materials will be chosen to monitor the operation of the spectrometer in an assembly-line type operation. The following may be regarded as areas which require monitoring in order to ascertain the quality of XRF analyses: (1) the status of the spectrometer; (2) instrumental drift, both short and long term; (3) sensitivity; and (4) accuracy. Not only must the analyst select the appropriate reference materials but also fix the sequence and frequency with which they will be inserted in the intensity measurement scheme. On the one hand will be the desire to monitor all areas closely, while on the other hand the need to maintain an acceptable level of productivity. Thus care must be taken not to overmonitor nor have monitors serve more than one purpose if the primary goal is jeopardized.

Checking the status of the spectrometer can be achieved by using an ultrastable reference and measuring the necessary intensities on *extended count times* to verify that each crystal, detector, collimator, filter, etc. are functioning within specifications. This would normally be done immediately after each periodic instrument servicing, or change in one of the components, by comparison with intensities obtained when the spectrometer was originally installed. This test should also be carried out as soon as other less rigorous tests indicate a potential problem, or at some preset cycle (weekly, monthly, . . .) so that partial failures in any one component can be detected and remedied as soon as possible. If the intensities are within the range expected from counting statistics, the instrument is then considered to be ''operational''.

Although current spectrometers are very reliable and have a high degree of stability, it cannot be expected that repeated intensity measurements will be exactly reproduced week after week. Instrumental drift occurs for many reasons and can be checked by examining plots of successive counting monitor (CM) intensities for each element as a function of time; that is, its position in the counting sequence. Unless the instability is very short term (minutes) and random, CM intensities can be used directly to compensate for drift, which, in essence, means taking the necessary steps to ascertain that measurements made at the time of analysis are fully compatible with measurements made at the time of calibration. Aside from remaining stable after multiple use, the CM must generate an adequate count rate for each element. Since this is very unlikely to be the case with a naturally occurring material, CMs in a geochemical context are usually spiked materials; for example, a granite or limestone to which has been added an equivalent of 1000 pm or so of elements occurring at the trace level. The life span of CMs can be so-to-speak extended by preparing a large quantity of spiked material, preparing a number of replicates, measuring intensities on the whole suite in one run, and then selecting the four or five that are most similar which are then used interchangeably.

Except for the case when the standard addition method is used, at least one (preferably more) specimen must be a material of known concentration necessary to determine the factor to convert intensities to concentrations. This calibration step is of such importance in XRF spectrometry that it must be undertaken with great deliberation. The choice of what materials will be used as standards will depend on what approach will be used subsequently to compensate for matrix effects. Coupled with this is the availability of reliable standards. Although there are excellent summaries of the wide selection of geological reference standards (Abbey, 1983; Flanagan, 1986; special issues of *Geostandards Newsletter*), Harvey (1989) points out that there is a marked lack of reference standards that cater to the wider range of elements and concentration ranges encountered in geochemical explorations. The exception is the six United States Geological Survey's GXR samples which, unfortunately, still show large variances in the recommended values for many elements. Harvey considers this a serious problem, encourages organizations preparing standards to remember the exploration field, and refers to three main approaches which can be taken to cover greater ranges:

1. Spiking a suitable matrix (to be used with caution).
2. Mix proportions of reliable geochemical reference materials.
3. The preferred approach is to create secondary in-house standards. Suitable mineralized samples are analyzed carefully, preferably by more than one method, and different laboratories if possible.

Despite having met the criteria for a sound calibration, analysts would be foolhardy to proceed directly to analysis of unknowns. Confirmation of how well the overall analytical scheme performs must be established. Once again the analyst is faced with the task of selecting what materials will be used for this purpose. The problem is linked to the lack of available standards since a valid assessment of accuracy should imply that the standards used *did not serve in the calibration process*. Unfortunately, there are a number of instances in the XRF literature where accuracy data are based on reprocessing the intensity data used for the calibration as "unknowns".

Analysis must also consider to what degree there is a need for verification of consistency. The term is used in this instance to refer to the need to check that the overall analytical scheme does not deteriorate over both short- and long-term periods. The key word here is overall. The ideal verification for consistency would require the frequent periodic preparation of new specimens of a number of geochemical reference standards that did not serve for calibration. Their insertion in the intensity measurement sequence would provide the necessary data to monitor both precision and accuracy. The lack of standards precludes this approach. Thus another approach must be considered to assess the degree of errors that may creep in as a result of the myriad of variables that are part and parcel of day-to-day laboratory operations. Consider that over more or less extended periods of time, detector windows and detector gas cylinders have to be replaced, reagents from different stocks may be used, staff transmutations, etc. may all contribute to a slight deterioration of an initially sound analytical scheme. A second best approach to check for overall consistency is to settle for monitoring overall precision only. One possibility is the allowance for the random insertion of a number of *blind duplicates* in the counting sequence. By consensus, these can originate equally within the laboratory and the submitter. When identified at the end of processing, blind duplicates within batches provide for checking that the procedural steps are being followed, while blind duplicates between batches throughout the exploration provide for checking long-term consistency.

C. Data Reduction Process

It is taken for granted that computer facilities are available for processing the mass of data generated by a fully automated spectrometer. The software for processing is generally supplied by manufacturers of XRF instruments and may not contain features desirable in a geochemical context; thus consideration may have to be given to the development of in-house software if maximum versatility is to be achieved. While the specimen preparation and intensity measurement steps certainly have a bearing on the quality of the concentrations reported, it would be most unfortunate if their quality was compromised by inadequate processing of valid intensity data. Given high-quality intensity data, the analyst must ensure that both the calibration and matrix effect corrections are effectual. The dual role of reference standards in the data reduction process will now be examined. In the first case, the standards serve to establish the instrument's sensitivity while in the second, they serve to estimate the accuracy of the compositional data reported.

The term calibration has been used to describe three different processes in XRF spectrometry. In some instances calibration refers to the periodic measurement of analyte intensities in order to generate intensity-to-concentration conversion factors. It can be envisaged as establishing the instrument's sensitivity which is assumed to remain constant for a given time interval. This is the type of calibration that analysts have in mind when they state that *only one* standard is required. It is advisable that instrumental calibration be carried out at regular and frequent intervals either by using a standard reference material in which case the conversion factors can be updated directly, or a CM may be used to normalize emitted intensities to be compatible with those used to generate the initial conversion factors. In other instances, the term calibration refers to the process whereby analysts combine in one operation the establishing of the instrument's sensitivity *and* the adequacy of the correction for matrix effects. A direct plot of C_i versus I_i for a number of standards will generally show a wide degree of scatter. The analyst may then calculate matrix effect correction terms by any one theoretical approach and plot the products intensity times matrix effect correction term versus concentration. If the correction displaces the original values vertically so as to produce a linear $I_{i,s}[\ ]_s$ versus C_i relation, extrapolation to C_i = 1.0 provides an average value for $I_{(i)}$, hence the calibration factor, $1/I_{(i)}$. In the third instance, calibration refers to combining the calibration step with the generation of all coefficients required by the influence coefficient model chosen for the correction of matrix effects. This approach is empirical with the process involving multiple regression of experimental data, hence a fairly large number of standards are necessary; for example, Ma and Li (1989) use 50 in an application of XRF for the analysis of geochemical prospecting samples in China. In some instances, the intensities are regressed versus concentrations (concentration models), or intensities are regressed directly versus all the intensities measured for each specimen (intensity models).

The second role of reference standards is that of providing ongoing checks of the accuracy of an analytical scheme. An example of a data reduction process involving accuracy checks is given (Harvey, 1989) and may be summarized as follows: The philosophy adopted is to group samples in batches and process whole batches as a unit rather than process data into concentrations during analysis. This is deemed preferable in that all quality control checks may be assessed for the batch as a whole and compensated in the same manner. Processing is carried out, for convenience, in two parts. During the first part, two passes are made; one to evaluate the status of the spectrometer and to check for within batch drift. If no problems are noted during the first pass, elemental abundances are computed; otherwise, any problems are written to the terminal and a decision taken whether the complete batch is to be reanalyzed. During the second processing phase, the values obtained for the standards are compared with their recommended values. A summary of all quality control information is tagged to appropriate files and the analytical report generated. The process is simple and carried out interactively, thus allowing new methods to be tested and for the interrogation of data at every step if a problem arises. The automatic updating of laboratory files provides immediate information on the progress of any batch, and a full record and related quality assessments of completed batches.

D. Analytical Reproducibility, Detection Limits

An estimate of the analytical reproducibility and of the detection limit is often quoted by authors describing the performance characteristics of methods of analysis for trace analytes. Both the analyst and the geochemist need to know whether the laboratory variance is suitably low for subsequent geochemical interpretation of the compositional data. This requires a separation of the variations in reported compositions into that due to sampling and laboratory causes, and the residual; that is, the geochemical variation, which is of real interest to the client. It has been pointed out that many methods used for the estimation of analytical precision are usually unsatisfactory in some respects (Thompson and Howarth, 1978). Among these are methods that take no account of change in absolute and relative reproducibility over the concentration range of the analyte; data recording practices that can lead to grossly incorrect estimates of variance, and lack of clarity and consistency in the definition of the terms "precision" and "detection limit". To deal more adequately with the widely recognized fact that precision varies with analyte concentration, these authors showed how duplicate analytical results can be used to give realistic estimates of precision in a given analytical context. For trace elements, a simple linear function often models adequately this relation in practice, the variation is expressed as

$$\sigma_c = \sigma_o + kC_i \tag{109}$$

where σ_o is the standard deviation of the field blank (background) and k is a constant. Howarth and Thompson (1978) examined the proposed method and examples of its use. More recently, the Analytical Methods Committee of the Royal Society of Chemistry published a report from its Statistical Subcommittee (M. Thompson, chairman; see Thompson, 1987) on the concept of "detection limit''. The report seeks to clarify the concept of detection limit in a manner consistent with the IUPAC definition. The detection limit is given by

$$C_l = ks_B/S \tag{110}$$

where k is a numerical constant (a value of k = 3 is ''strongly recommended'' by IUPAC); s_B is the standard deviation of the measured background; and S is the sensitivity of the method.

The subject is of key importance to the geochemist because of the apparent simplicity of the detection limit concept and the degree of confusion that surrounds its estimation; for example, whether by scientists occupied with the development of XRF spectrometers and those concerned with their use. Two cases will be examined; namely precision and detection limits relative to instrumentation; and relative to analytical data in a given analytical context.

Scientists who develop XRF instrumentation often compute detection limits based on the standard deviation of contiguous background measurements. In relation to practical analysis, these ''instrumental detection limits'' are unrealistically low, cannot be matched in practical circumstances, and are therefore misleading. They do not reflect all of the many inherent sources of variation in the procedural steps of an analytical scheme. Referring to the precision of an analytical method can be misleading if all the sources of error are not taken into account. Thompson and Howarth (1976) demonstrated how the use of many duplicate specimens, prepared and measured randomly over a significant time span, can be used to provide much more realistic estimates of precision. A problem that is seldom taken into consideration in providing valid estimates of the detection limit of an analyte is due to interference effects from spectral overlap. Their magnitude can be estimated and subtracted but there is always an uncertainty in the correction term. In general, the greater the interference effect, the greater will be the additional error, and in some circumstances it may become the dominant factor in the total correction terms. Under these conditions, the normal detection limit is too low. Thus the reproducibility and detection limit of an analyte in a situation where a spectral line of element j is not resolved from the characteristic line used to measure I_i may involve expressions of the type

$$(\sigma_o + k_1C_i + k_2C_j) \tag{111}$$

where k, k_1, and k_2 are constants. Should the term k_2C_j be dominant, one should then follow the recommendation (Harvey, 1989) that it is preferable to count longer on a clean peak than introduce a correction for overlapping peaks if the choice exists.

In summary, it is recognized that the X-ray spectrometers are valuable tools for providing compositional data required in geochemical explorations. A good working relationship between analysts and clients is a prerequisite and should lead to a clear understanding of the XRF laboratory's potential towards providing all or part of the required data. It is in the geochemist's best interest to be well informed so as to neither underestimate nor overestimate that potential.

REFERENCES

Abbey S. (1983). Studies in standard samples of silicate rocks and minerals, 1969–1982. *Geol. Surv. Can. Paper*, **83–15**.

Bearden J. A. and Burr A. F. (1967). Reevaluation of X-ray atomic energy levels. *Rev. Mod. Phys.*, **39**, 125–142.

Claisse F. and Quintin M. (1967). Generalization of the Lachance-Traill method for the correction of matrix effects in X-ray fluorescence analysis. *Canadian Spectros.*, **12**, 129–133.

Criss J. W. and Birks L. S. (1968). Calculation methods for fluorescent X-ray spectrometry, Empirical coefficients vs fundamental parameters. *Anal. Chem.*, **40**, 1080–1086.

Flanagan F. J. (1986). Reference samples in geology and geochemistry. *Geostds. Newslett.*, **10**, 191–264.

Harding A. R. (1989). Process control applications of The Peltier cooled Si(Li) detector based EDXRF spectrometer. *Adv. in X-ray Analysis* (Barrett, C. S., Gilfrich, J.V., Jenkins, R., Huang, T.C. and Predecki, P.K., Eds.). Plenum, New York, pp. 31–37.

Harvey P. K. (1989). Automated X-ray fluorescence in geochemical explorations. In: *X-ray Fluorescence Analysis in the Geological Sciences, Advances in Methodology*. (Ahmedali, S. T., Ed.) Geological Association of Canada, Department of Earth Sciences, Memorial University of Newfoundland, St. John's Newfoundland, Canada, pp. 221–258.

Heinrich K. F. J. (1966). X-ray absorption uncertainty. In: *The Electron Microprobe*. McKinley, T.D., Heinrich, K.F.J. and Wittry, D.B., Eds.). John Wiley, New York, pp. 296–377.

Howarth R. J. and Thompson M. (1976). Duplicate analysis in geochemical practice. Part 2. Examination of proposed method and examples of its use. *Analyst*, **101**, 699–709.

Lucas-Tooth H. J. and Price B. J. (1961). A mathematical method for the investigation of inter-element effects in x-ray fluorescent analyses. *Metallurgia*, **64**, 149–152.

Ma G. and Li G (1989). Application of X-ray fluorescence spectrometry to the analysis of geochemical prospecting in China. *X-ray Spectrom.*, **18**, 199–205.

Margolin E. M., Pronin Y. I., Choporov D. Y., Shil'nikov A. M., Komyak N. I., Goganov D. A., Serebyakov A. S. and Fed'kov E. A. (1985). Some experience in using the MECA-10-44 (XR—500) X-ray fluorescence analyser for solving geological problems. *X-ray Spectrom.*, **14**, 56–61.

Tertian R. (1975). A self-consistent calibration method for industrial x-ray spectrometric analyses. *X-ray Spectrom.*, **4**, 52–61.

Tertian R. and Claisse F. (1982). *Principles of Quantitative X-ray Fluorescence Analysis*. Heyden, London.

Tertian R. (1986). Mathematical matrix correction procedures for X-ray fluorescence analysis: A critical survey. *X-ray Spectrom*, **15**, 177–190.

Thompson M. and Howarth R. J. (1976). Duplicate analysis in geochemical practice. Part 1. Theoretical approach and estimation of analytical reproducibility. *Analyst*, **101**, 690–698.

Thompson M. and Howarth R. J. (1978). A new approach to the estimation of analytical precision. *J. Geochem. Explor.*, **9**, 23–30.

Thompson M. (1987). Recommendations for the definition, estimation and use of the detection limit. (Analytical Methods Committee; Statistical Sub-Committee, M. Thompson, chairman), *Analyst*, **112**, 199–204.

Willis J. P. (1989). Compton scatter and matrix correction for trace element analysis of geological materials. In: *X-Ray Fluorescence Analysis in the Geological Sciences, Advances in Methodology* (Ahmedali, S. T., Ed.). Geological Association of Canada, Department of Earth Sciences, Memorial University of Newfoundland, St John's, Newfoundland, Canada, pp. 91–140.

NEUTRON ACTIVATION ANALYSIS

Gregory W. Kallemeyn

I. Introduction ... 193
II. Theory ... 195
A. Neutron Irradiation of Samples ... 195
B. Element Concentration from Induced Radioactivity ... 197
C. Sensitivity ... 198
III. INAA ... 199
A. Instrumentation ... 199
B. Experimental Procedure ... 203
C. Specialized Procedures ... 205
IV. RNAA ... 205
V. Applications ... 208
References ... 209

I. INTRODUCTION

Neutron activation analysis is a powerful analytical tool in the fields of geochemistry and cosmochemistry. Concentrations of many different elements can be determined by this technique, with typical sensitivities in the nanogram to microgram per gram range (see Figure 1). It is usually referred to as a "trace element" method, but it is even useful for determining concentrations of major and minor elements.

Advances in Analytical Geochemistry
Volume 1, pages 193–209

ISBN: 1-55938-332-1

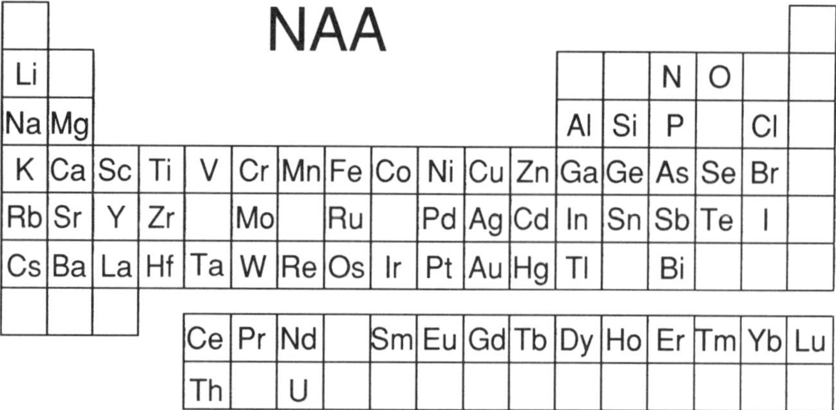

Figure 1. Elements whose concentrations can be determined by NAA.

There are two general methods involved in this analytical technique: (1) radiochemical neutron activation analysis (RNAA); and (2) instrumental neutron activation analysis (INAA). The greatest elemental concentration sensitivity is provided by the RNAA method, though the method is also destructive to the sample. Analysis by RNAA often requires considerable manpower and laboratory time because of the chemical separations involved. Thus, the number of elements that can be determined in any given sample is limited. The INAA method is the simplest as it does not require any chemical processing, and it also provides concentration data for the maximum number of elements. It has the added advantage in that the method is effectively nondestructive.

Unlike some other analytical methods, neutron activation analysis does not require large sample sizes. Multielement analysis can easily be performed on a sample <1 mg in size. Neutron activation analysis also does not require complex mathematical computations in determining element concentrations as does another trace-element technique, X-ray fluorescence analysis, since it does not suffer from matrix effects in typical geochemical samples.

All of these advantages make neutron activation analysis an especially good choice for the study of rare and valuable samples of lunar rocks and meteorites. The same advantages hold true for geological samples such as mineral separates, where it may be difficult or impossible to produce large amounts of samples.

The advent of modern, high-resolution, high-efficiency γ-ray detectors has lead to the preferential use of the INAA method in typical cosmochemical and geochemical neutron activation analysis studies. The RNAA method is now usually applied when determining extremely low concentrations of few elements is of the utmost importance. Often it may follow INAA analysis of the same sample. The bulk of this chapter will therefore concentrate on use of the INAA method.

II. THEORY

The basic principal underlying neutron activation analysis is the measurement of radioactivity induced in a sample following its irradiation by neutrons. The type of nuclear reaction that occurs is dependent upon the energy of the neutrons used for irradiating the sample. The different neutron energies are generally classified as in Table 1. The important types of neutrons when considering activation analysis are thermal, epithermal, and fast neutrons. The aspects of each will be discussed later.

A. Neutron Irradiation of Samples

Samples are usually irradiated in a nuclear reactor where there is a range of neutron energies, from thermal to fast. An irradiation position with the most homogeneous flux of neutrons is desired, and is usually also a position which maximizes the ratio of thermal to fast neutrons.

When a sample is placed in a flux of thermal-epithermal neutrons, an isotope of an element, X, may absorb a neutron and transform into a new species, X′, having the same atomic number, Z, but an atomic mass, A, that is one unit larger. The reaction, called an (n,γ) reaction, can be written as follows:

$$ {}^{A}_{Z}X(n,\gamma){}^{A+1}_{Z}X' \tag{1}$$

where n denotes the absorbed neutron and γ denotes a prompt γ-ray emitted as a result of the reaction.

The likelihood of this reaction occurring depends on several factors: (1) the average neutron capture cross section; (2) the number of target nuclides (related to the amount of an element in the sample and the isotopic abundance of the target nuclide); and (3) the intensity of the neutron flux. Some nuclides undergo resonance neutron capture in the epithermal energy range. Such a resonance cross section can be used to increase the detectability of an element, as will be discussed in a later section.

The new species, or nuclide, may be radioactive. Such a radionuclide would be more neutron-rich than the target-mulched, and therefore, would usually decay by β^- particle emission (plus an antineutrino):

$$ {}^{A}_{Z}X' \rightarrow {}_{Z+1}^{A}Y + \beta^- + \bar{\upsilon} \tag{2}$$

If the β^- decay leads to an excited daughter nuclide, Y, then one or more γ-rays are emitted. A common radionuclide produced in geological samples by neutron activation is ^{46}Sc which emits two γ-rays with energies 889 keV and 1120 keV (see Figure 2). It is the counting of these γ-rays that is the basis for most neutron activation analyses. Although it has the potential for great sensitivity, the β^- particle is rarely counted since it is not suitable for INAA applications.

Table 1. Concentrations of 40 Elements in Lunar Meteorite Allan Hills A81005 as Determined by INAA

		A81005,11
Na	μg/g	2.21
Mg	μg/g	48.8
Al	μg/g	139
K	μg/g	68
Ca	μg/g	106
Sc	μg/g	9.0
Ti	μg/g	1.77
V	μg/g	24.
Cr	μg/g	922
Mn	μg/g	584
Fe	μg/g	42.3
Co	μg/g	21.7
Ni	μg/g	182
Zn	μg/g	5.4
Ga	μg/g	2.7
Br	μg/g	90
Rb	μg/g	0.7
Sr	μg/g	141
Zr	μg/g	25
Cs	μg/g	25
Ba	μg/g	22
La	μg/g	1.71
Ce	μg/g	4.10
Nd	μg/g	2.90
Sm	μg/g	0.794
Eu	μg/g	0.66
Gd	μg/g	0.96
Tb	μg/g	0.170
Dy	μg/g	1.15
Ho	μg/g	0.25
Er	μg/g	0.72
Tm	μg/g	0.11
Yb	μg/g	0.69
Lu	μg/g	0.106
Hf	μg/g	0.61
Ta	μg/g	70
Ir	μg/g	6.0
Au	μg/g	1.9
Th	μg/g	0.26
U	μg/g	63

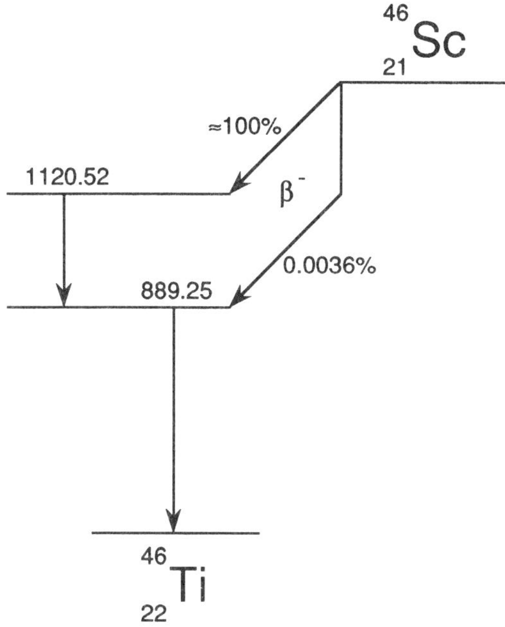

Figure 2. Energy decay diagram for ^{46}Sc.

B. Element Concentration from Induced Radioactivity

The amount of radionuclide produced by neutron activation in a given sample is proportional to the amount of the target nuclide in that sample. It then follows that the concentration of an element in a sample is proportional to the amount of the target nuclide, assuming a constant isotopic composition for the element. This is true for most elements; the exceptions being a few light elements, the rare gases, and Pb.

The disintegration rate or activity of a radionuclide can be calculated from the radioactive decay law:

$$A = dN_{X'}/dt = -\lambda N_{X'} \tag{3}$$

where $N_{X'}$ is the number of radioactive nuclei of X′ and λ is the decay constant given by $\ln 2/T_{1/2}$, where $T_{1/2}$ is the half-life of the radionuclide. During neutron irradiation some of the product radionuclides will decay. Therefore, the activity for a product radionuclide, X′, produced during an irradiation for time, t, at a neutron flux, ϕ, can be calculated as follows:

$$A = dN_{X'}/dt = \sigma\phi N_X(1 - e^{-\lambda t}) \tag{4}$$

where σ is the neutron capture cross section, ϕ is the neutron flux, and N_X is the number of target nuclei given by:

$$N_x = mN\theta/M \tag{5}$$

where m is the mass of the element, N is Avogadro's number, θ is the fractional isotopic abundance of X in the element, and M is the atomic weight of the element. The previous equations are sometimes useful for estimating the sensitivity for determining an element under given experimental conditions.

It follows from the previous discussion [Eqs. (3)–(5)] that if one irradiates a known amount of an element (as a standard) along with the unknown samples, then the ratio of the amount of a neutron-activated nuclide in the sample to that in the standard is equal to the ratio of the concentration of the element in the sample to the concentration in the standard, which in turn relate the mass of the element in the sample and standard. This relationship is shown mathematically by:

$$N[sample]/N[std] = conc[sample]/conc[std] = mass[sample]/mass[std] \tag{6}$$

C. Sensitivity

The sensitivity of neutron activation (i.e., the measure of how small an amount of an element can be detected in a sample) varies considerably among the elements that can be determined by the technique. The three basic factors affecting sensitivity are: (1) the fractional isotopic abundance of the target nuclide (θ); (2) the neutron capture cross section (σ); and (3) the half-life of the activated radionuclide. It is obvious from Eqs. (3) and (4) that large values of θ and σ will increase the radionuclide activity, and, in turn, improve sensitivity. Although less obvious, as the half-life of a radionuclide increases, the activity decreases (other factors being equal). But there is a practical limit to how short the half-life can be. If it is too short, the radionuclide may decay significantly before an NAA experiment can be initiated or completed.

Other factors can also play a role in determining the "detectability" of an element by neutron activation analysis. These factors, however, are more "site-dependent". Activity can be increased by irradiating samples at a higher flux or extending the period of irradiation. This depends on the maximum flux that can be generated by the nuclear reactor or other type of neutron generator that is available to the experimenter. The length of the irradiation is usually determined by the cost and the practical considerations surrounding transport of samples to and from the irradiation facility.

Sensitivity also depends on the decay characteristics of the radionuclide, such as the γ-ray energies and their absolute intensities (i.e., how many decays produce how many γ-rays). The importance of these characteristics in turn depend on the γ-ray detection systems used in the laboratory. A γ-ray from one radionuclide can also interfere with one of another radionuclide. Again, the effect this has on detectability of an element is dependent on the counting system used. More will be said about these "site-dependent" effects in a later section.

III. INAA

As noted earlier there are two general techniques of neutron activation analysis: instrumental (INAA) and radiochemical (RNAA). The INAA technique provides the maximum amount of elemental concentration data, and is the most suitable of the two methods for geochemical studies of a large number of elements and/or samples. Therefore, emphasis will be placed on this technique. However, the RNAA technique is still necessary for specialized experiments where concentrations of certain elements must be determined at levels too low for the INAA technique.

The INAA method is an essentially nondestructive method. The sample being studied is not processed chemically, but rather the γ-rays emitted by all decaying radionuclides are counted by the γ-ray spectroscopy system. The sample is left in a somewhat radioactive state, but for typical irradiations only low levels are present a few weeks after irradiation. Sometimes following an INAA study, a sample is analyzed for undetermined major and minor elements by other techniques such as XRF or fused-bead x-ray analysis.

A. Instrumentation

Instrumental neutron activation analysis became a viable method with the advent of commercially available high-resolution, lithium-drifted Ge γ-ray [Ge(Li)] detectors in the early 1970s. Earlier attempts at an instrumental-only method were with NaI(T1) γ-ray detectors which have relatively poor energy resolution (see Figure 3). The resulting γ-ray spectrum of an irradiated sample would usually be a jumble of overlapping γ-ray peaks, making quantitative analysis very difficult or impossible. For this reason, radiochemical separation of elements was usually necessary in order to provide useful results.

The Ge(Li) detector provided the necessary energy resolution to separate the multitude of γ-ray energy peaks in a spectrum. Early commercially available Ge(Li) detectors, though, had low efficiencies in detecting γ-ray [5–10% relative to the standard NaI(T1) detector]. Development of large, more efficient (15–20%) Ge(Li) detectors increased the element concentration sensitivity of INAA to a respectable level relative to the RNAA method. During the 1980s a new type of γ-ray detector began to replace the Ge(Li) detector. This was the high-purity Ge(HPGe) detector. This type of detector can be built in larger, more efficient models than Ge(Li) detectors. Typical relative efficiencies of 30 to 40% are obtainable, while still maintaining high resolution. Recently, efficiencies exceeding 100% relative to a NaI(T1) detector have been obtained.

As can be ascertained from the previous discussion, the detector is the heart of

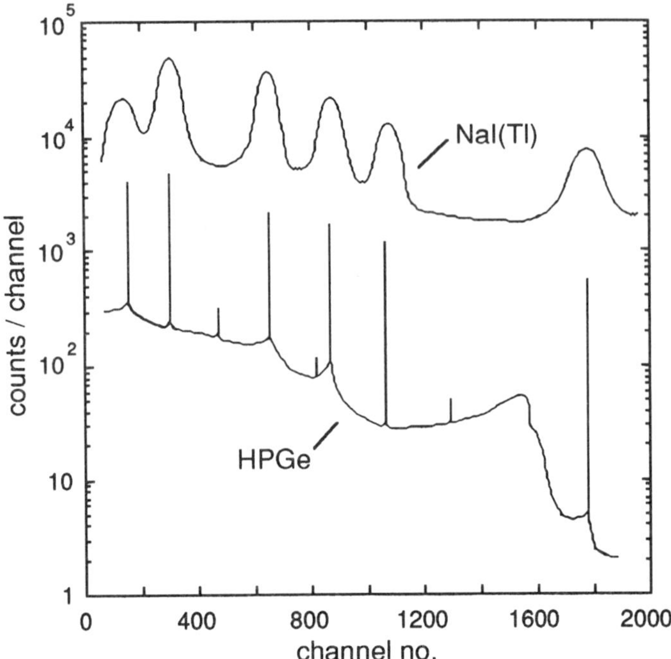

Figure 3. Comparison of resolution of γ-ray spectra for a NaI(T1) detector (*upper spectrum*) and a HPGe detector (*lower spectrum*).

the γ-ray spectroscopy system used for INAA. The typical modern γ-ray detector (in particular, the HPGe detector for the following discussions) consists of a cylindrical-shaped crystal of high-purity Ge mounted in an evacuated container usually made from A1 (see Figure 4). A preamplier is mounted directly on the detector, and the crystal assembly and amplifier FET are cooled to liquid nitrogen temperature to minimize thermal noise in the detection system (further explanation is beyond the scope of the present discussion). The HPGe detector is simply a semiconductor diode. Incident γ-rays produce electron-hole pairs which migrate to the oppositely charged ends of the detector crystal. The resulting charge is converted to a voltage pulse by a charge-sensitive preamplifier, the amplitude of which is proportional to the incident energy. There are three basic detector characteristics that provide a measure of how well it will perform for INAA experiments: (1) efficiency in detecting γ-rays, (2) ability to resolve closely-spaced γ-ray peaks in a spectrum; and (3) the peak-to-compton ratio.

The efficiency of γ-ray detectors was discussed in general terms above. Basically, the larger the detector, the greater the relative efficiency for detecting γ-rays. A typical HPGe detector with 30 to 35% relative efficiency will have dimensions of ~6 cm diameter and ~5 cm length. The detector efficiency is actually dependent on the γ-ray energy as seen in Figure 5.

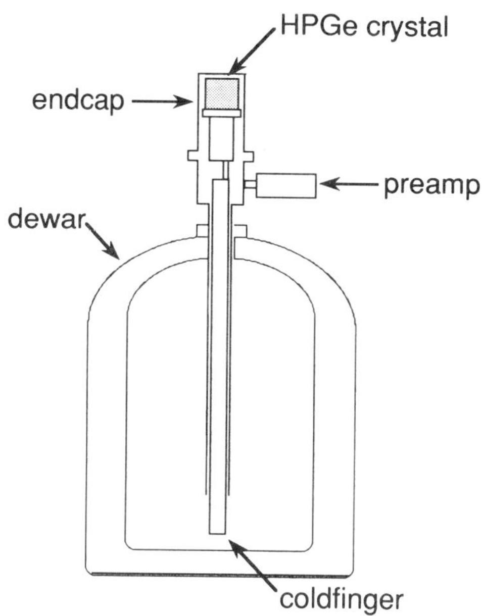

Figure 4. Schematic cross section of a typical HPGe coaxial detector.

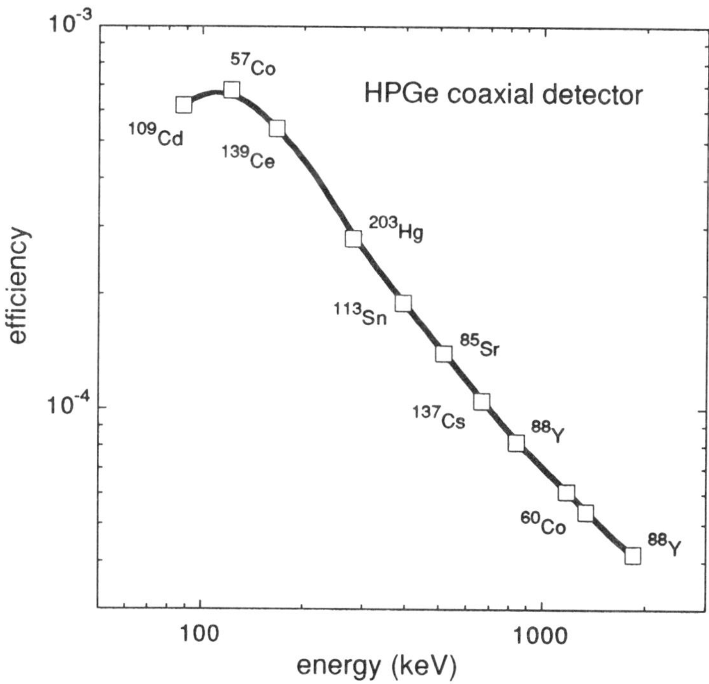

Figure 5. Detector efficiency vs. energy curve for a typical HPGe coaxial detector.

The relative efficiency as discussed above is that which is measured at particular γ-ray energy, usually the 1333 keV peak from the decay of ^{60}Co. Note that two detectors with the same relative efficiency at 1333 keV may have different efficiencies at energies around ~150 keV. This is because the geometry of the detector crystal also plays a role. Given equal efficiencies at 1333 keV, the detector with the greatest diameter-to-length ratio (which gives the greatest cross-sectional area) is preferred since it will give the best efficiency response at lower energies.

The resolution of a detector is a measure of the "energy width" of a γ-ray peak in a spectrum. This determines how close in energy two γ-rays can be and still be resolved in the γ-ray spectrum of a sample. It is usually measured as the full width at half-maximum (in keV) of the 1333-keV γ-ray peak. High resolution in a detector is essential since all of the activated radionuclides in the sample are counted simultaneously. A good detector for INAA has a resolution of 1.6 to 2.0 keV @ 1333 keV. Unfortunately, it is easiest to obtain high resolution in small-volume (and thus low-efficiency) detectors. A premium quality Ge crystal is necessary to provide the highest resolution in large-volume detectors. Large detectors also suffer from "ballistic effects" which limit the maximum sample count rate in order to avoid significant loss of resolution.

When a γ-ray interacts with a HPGe detector it scatters about the crystal depositing its energy with the production of electron-hole pairs, a process known as Compton scattering. Sometimes the scattered photon escapes, therefore depositing less than the original energy in the detector. The detector "sees" a γ-ray with a lower energy. The Compton interactions cause the formation of a Compton plateau at an energy lower than the true γ-ray energy peak (see the spectrum in Figure 3).

The peak-to-Compton ratio of a HPGe detector is a measure of its ability to "suppress" the Compton plateau. The higher the peak-to-Compton ratio, the easier it is to distinguish low-energy γ-rays in the presence of a very active high-energy γ-ray. It is intuitively obvious that large-volume detectors have the highest peak-to-Compton ratios (the quality of the detector crystals being equal).

The effects of Compton plateaus can be minimized with a special type of detector, the HPGe planar detector. This detector is made from a small, wafer-shaped crystal, usually ≤7 mm in thickness. The principle here is that high-energy γ-rays will usually pass through the detector without scattering. Therefore, fewer Compton events will be recorded in the spectrum. The small detector has a similar efficiency for detecting low-energy γ-rays as larger volume detectors. Typically, this detector is useful for γ-ray energies <200 keV. The small-sized detector has the advantage of greater resolution.

The basic γ-ray spectroscopy system for INAA consists of the HPGe detector described above with high-voltage power supply, an amplifier, an analog-to-digital convertor (ADC), and a multichannel analyzer (MCA) (Figure 6). A typical series of events that take place when a γ-ray is "counted". The incident

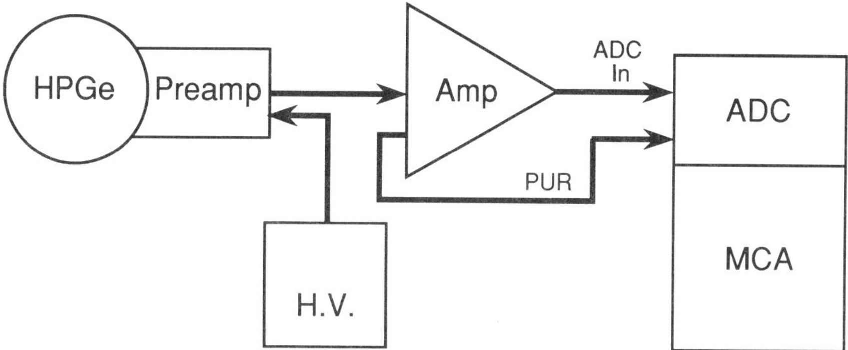

Figure 6. Schematic diagram of a typical γ-ray spectroscopy setup.

γ-ray deposits its energy in the detector crystal and the charge produced is converted to a voltage pulse by the preamplifier. This signal pulse is then amplified and reshaped such that the height of the new pulse is now proportional to the original γ-ray energy. This new pulse is digitized by an analog-to-digital convertor and counted by the MCA.

The MCA has a histogram memory broken into small increments or channels, each of which represents a certain discrete energy level. The pulse height of incoming pulse is used to determine which channel to place the count. As the "counts" build-up in the channels a γ-ray spectrum is formed.

B. Experimental Procedure

How samples are prepared for INAA usually depends on the type of samples. For geological samples where abundant mass is available, several grams are usually ground in a ceramic shatterbox or agate mortar and pestle. The actual amount to be ground up depends on how coarse-grained the sample is, and therefore how much is needed to produce a representative powder. Grinding tools made from tungsten carbide should not be used as the samples may become contaminated with tungsten. The activity of the radionuclide produced from tungsten can overwhelm the γ-ray spectrum. The powder is homogenized and a 150 to 300 mg aliquot is taken for analysis and placed in polyethylene irradiation vials.

If small mineral separates are to be analyzed, they may be placed in vials without any processing other than cleaning the surface. When extraterrestrial samples are to be analyzed, the amount of sample available is usually small. Gentle crushing, enough to efficiently pack the sample in the irradiation vial, is preferred to grinding to a fine powder. The less handling there is of a sample the fewer problems there are with contamination. However, if the sample is to be analyzed later by other methods, such as fused-bead X-ray analysis or isotopic studies, it may be necessary to fully powder the sample.

The following is the typical INAA procedure used a the UCLA neutron activation analysis laboratory (as described in Kallemeyn et al., 1989). In this laboratory a variety of samples are analyzed: geological samples, lunar samples, and meteorites.

Twelve to 16 samples are analyzed in each INAA experimental run. Samples are packaged in linear polyethylene vials. Sample masses are usually $\leq$300 mg. Geological samples are typically 150 to 250 mg, lunar samples 10 to 250 mg, and meteorite samples 1 to 300 mg. Along with these samples, an aliquot of a suitable USGS standard rock powder or Smithsonian meteorite powder is included in each run as a control. This allows monitoring of results from one run to the next.

Standards are prepared consisting of mixed-element solutions pipetted into polyethylene vials onto a matrix of high-purity MgO powder. The latter helps to minimize the problem of neutron self-absorption. Elements in the mixed standards are grouped as follows: La, Pr, Sm, Eu, Dy, Ho, and Er; La, Nd, Sm, Yb, and Lu; Ce, Gd, Tb, Eu, and Tm; Sc, Cr, Fe, Co, Ni, Ir, and Au; As, Se, and Zn; Br and Sb; Na and Mn; K and Ga; Ru, Re, and Os; Ba, Cs, Rb, and Sr; Zr, Hf, and Ta; U and Th; and Mg, Al, V, Ca, and Mn. Primary standard $CaCo_3$ powder is used for Ca. The mixed solutions are prepared from stock solutions prepared from high-purity compounds of the elements of interest. The rare earth elements (REE) are grouped into three standards according to the half-lives of the radionuclides produced. The groupings of the REE are as designed to minimize spectral interferences in the standards.

Samples and standards are irradiated for two minutes in a TRIGA reactor at a neutron flux of $\sim 2 \times 10^{12}$ using a pneumatic fast transfer system. They are counted after a 2 minute decay for Al, Mg, V, Ca, and Mn. All of these elements, except Mn, produce radionuclides with half-lives $<$10 minutes. The short half-lives necessitate the use of the transfer system and the counting of samples onsite at the nuclear reactor. The counting data are stored on magnetic tape for later computer processing.

For the main analysis, the samples and standards are irradiated for 4 hours at a neutron flux of $\sim 1.8 \times 10^{12}$ n sec^{-1} cm^{-2}. After a 30 minute decay in the reactor, the samples are removed and transported to the laboratory for counting. Elements determined in this experiment have half-lives $\geq$2.5 hours and include: Na, K, Ca, Sc, Cr, Mn, Fe, Co, Ni, Zn, Ga, As, Rb, Sr, Zr, Se, Cs, Ba, Sb, REE, Hf, Ta, Os, Ir, Au, Th, and U.

Samples and standards are usually counted four times on a 35% HPGe detector over a period of 4 to 6 weeks. The first count generally begins about 5 hours after the end of the irradiation. The second count begins the day after the irradiation, the third about 5 to 7 days later, and the fourth about 3 to 5 weeks later (depending on the type of sample; meteorites earlier, geological samples later).

A single count is also taken on a HPGe planar detector. As noted earlier, the planar detector minimizes the effect of Compton scattering and provides higher resolution. Several rare earth elements have γ-ray energies in the range 80–208, making them ideal for this detector. The planar is also used to determine Re (137 keV) in meteorite samples since the high resolution of the detector permits separating it from the 136-keV Se peak.

Spectral data collected by the MCA during each count is transferred to a computer for analysis. Data analysis is performed using the SPECTRA γ-ray analysis program (Grossman and Baedecker, 1986). During data analysis, the area of each peak of interest in the spectrum is determined. This area is proportional to the amount of the target nuclide in the sample. By comparing standards and samples as discussed in the previous section on theory (See Eq. 6), the concentration of an element can easily be calculated.

The biggest problem during data analysis is the accurate determination of peak areas. This is especially true for very small γ-ray peaks, and for overlapping γ-ray peaks. The SPECTRA program allows for interactive analysis with each spectrum during processing using a graphics terminal. Peak areas can thus be determined with greater accuracy and precision.

C. Specialized Procedures

In order to improve the detectability of an element producing a weakly active radionuclide, the Compton plateaus in a γ-ray spectrum can be suppressed electronically. An annulus of NaI(T1) or plastic scintillator is placed around the HPGe detector (Figure 7). Scattered photons which escape the HPGe detector are detected in the annulus. In this anti-coincidence system, if the detector and annulus detect nearly simultaneous γ-ray events, the event is not counted by the system. Of course, if the radionuclide produces coincident γ-rays these can be used to advantage to increase the detectability of an element. Such a setup is known as a γ-γ coincidence counter (Figure 8). Two detectors, looking at the sample at 180° to each other, are set to look only for the two γ-rays of interest. If both detectors see a count in coincidence, that count is recorded. This method virtually eliminates counts from other radionuclides in the sample. A coincidence counter setup at the Lawrence Berkeley Laboratory can determine Ir in geologic samples in concentrations down to 30 pg/g instrumentally.

IV. RNAA

Sometimes the concentration of an element of interest is too low in a sample to accurately determine, or even detect. Often this is because one or more very active radionuclides in the samples are masking a much weaker γ-ray peak of the element of interest. Radiochemical methods can then be used to separate the radionuclide of interest from the masking activities. Figure 9 shows some of the elements that are determined by RNAA in geological samples. As noted earlier, this method is totally destructive to the sample.

Sample preparation is somewhat different. Because the samples are often irradiated for a longer period and at a higher flux than for INAA (in order to further increase sensitivity), the samples are usually sealed in high-purity quartz vials. Standard solutions are pipetted either into quartz vials, or more likely, onto Al foils and dried. Sample sizes are about the same as for INAA, ranging from 10 to 300 mg.

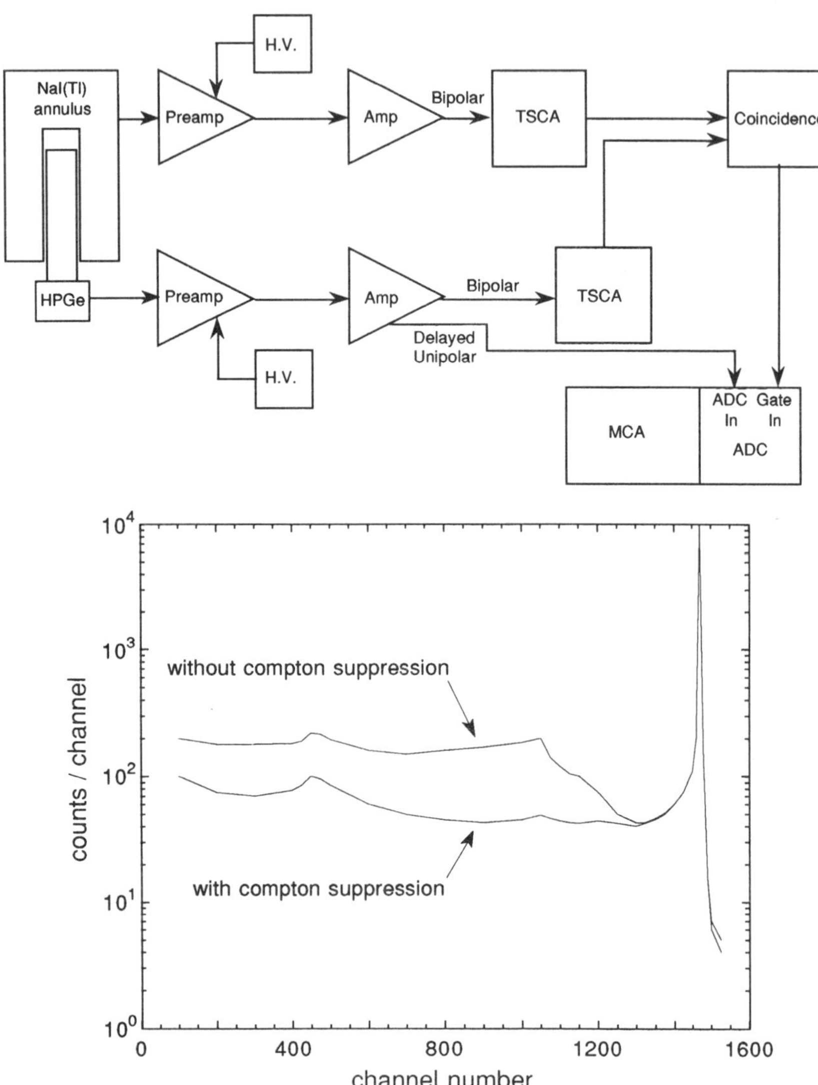

Figure 7. Schematic diagram (*top*) of a compton-suppression γ-ray spectroscopy system; γ-ray spectrum (*bottom*) showing the effect of compton suppression on the background.

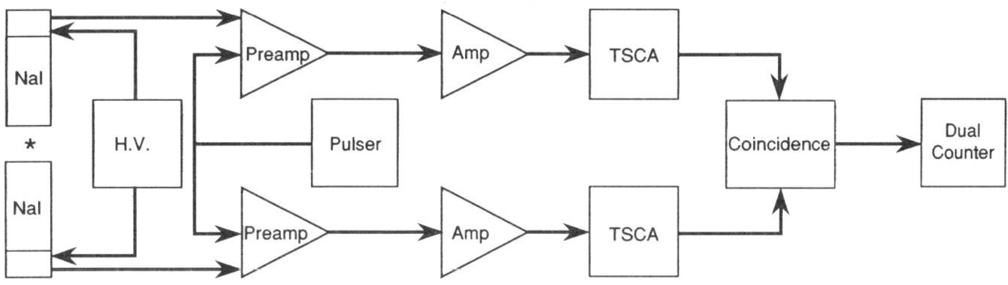

Figure 8. Schematic diagram of a γ-γ coincidence spectroscopy system.

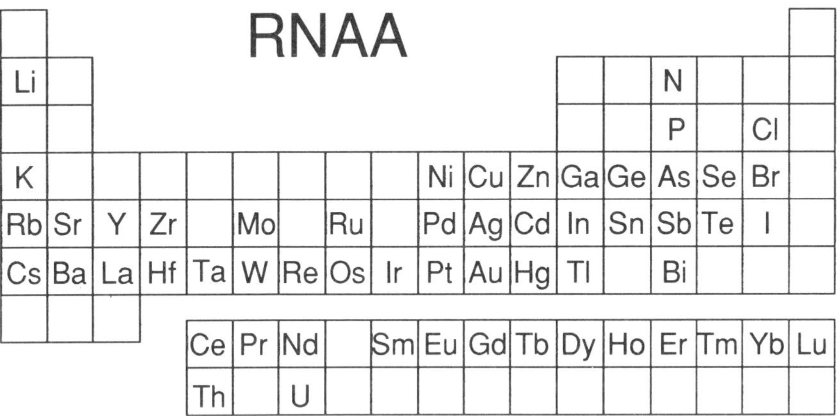

Figure 9. Elements whose concentrations are typically determined by RNAA.

Samples and standards for RNAA experiments at UCLA are irradiated for 25 to 50 hours at flux of 8×10^{13} n sec^{-1} cm^{-2}. Upon return from the reactor, a known amount (~10 mg) of each element of interest is added to the samples as carriers. The samples are then fused with Na_2O_2 to break down the sample and equilibrate it with the added carriers.

The samples then go through a series of chemical separations, depending on what elements are to be determined. A radiochemical scheme for determining up to 10 elements (Ni, Ge, Zn, Ga, Ru, Cd, In, Ir, Au, and U) in geological and cosmochemical samples is given by Sundberg and Boynton (1977). When the separations are complete, the elements are usually in solution.

Counting of RNAA samples is traditionally done on well-type (NaI(T1) detectors. The counting solution is placed into a vial, which is then placed in the detector well for counting. With the increase of efficiency of currently available HPGe detectors, these are sometimes replacing the NaI(T1) detectors. The much higher resolution of the HPGe detector means that it is unnecessary to produce a perfectly clean separation of the element to be determined. In fact, in some cases as with REE's, elements are separated as groups, making counting on a HPGe detector imperative.

The basic principles of data analysis are the same as in the INAA procedure. But the spectra are much less complex as usually only a single peak area is determined, except in the case of group separations.

V. APPLICATIONS

Neutron activation analysis is well-suited for studies of extraterrestrial materials such as lunar rocks and meteorites. Samples of these available for study are usually small, ≤300 mg. In some cases INAA is the only practical method for determining the maximum trace-element information when the same sample needs to be analyzed later for isotopic information due to lack of material for separate studies.

Table 1 shows INAA data determined by the author in the first-discovered lunar meteorite, Allan Hills A81005. The total mass of this lunar meteorite from the highlands region of the Moon was only 31 g and only 113 mg was made available for INAA study. Despite the small size, 40 element concentrations were determined in the sample, including 13 REE's. Another recently discovered lunar meteorite, the 30 g Elephant Moraine 87521, was originally classified as a eucritic meteorite from initial cursory petrographic studies. Analysis of a 270-mg sample by INAA (Warren and Kallemeyn, 1989) yielded some 38 elemental concentrations. The data clearly showed it not to be a eucrite, but instead to be the first-discovered lunar meteorite from a mare region of the Moon.

One of the most interesting facets of the study of chondritic meteorites are the ubiquitous chondrules found in many of them. The study of chondrules is very important in discussions on the origin of chondrites. But their small size makes comprehensive study of them difficult. Rubin and Pernicka (1989) measured 17 element concentrations in 20 chondrules from the Sharps chondrite. The chondrules ranged in size from 0.7 to 3.5 mg. One aspect of this research is that without detailed petrographic data on each of the chondrules, the usefulness of the compositional data alone for a serious discussion of chondrule origin was limited. Because the INAA technique employed was nondestructive, each chondrule could be thin-sectioned for the necessary petrographic descriptions.

Sorensen and Grossman (1989) used the INAA technique to study the trace-element partitioning behavior of minerals in the Catalina Schist Complex, an example of subduction-zone metasomatism. This information is very important when forming models for mass transfer from subduction slab to mantle wedge in the deep regions of subduction zones. To provide this data they determined 28 elements in mineral separates of garnet, amphibole, clinopyroxene, plagioclase, rutile, sphene, apatite, zoisite, and clinozoisite.

In the field of oceanography Frank Kyte at UCLA is studying 325 samples of a North Pacific pelagic clay piston core (private communication). As part of this ambitious effort, over 30 element concentrations are being determined in the samples by combined INAA and RNAA studies. The data should help reveal and refine the understanding of what controls element distributions in pelagic clay sediments.

It is clear that neutron activation analysis is an important analytical tool in the study of geologic and extraterrestrial samples. Concentrations of more than 30 elements can routinely be determined in typical samples. The technique is especially well-suited to the compositional study of small samples; that is, mineral separates, chondrules, and rare extraterrestrial samples. In the case of the INAA method, it also provides a nondestructive means of compositional analysis of precious samples, or samples that need to be used for subsequent analysis by other methods.

REFERENCES

Chou C.-L (1980). Radiochemical neutron activation analysis. In: *Neutron Activation Analysis in the Geosciences: Mineralogical Society of Canada Short Course Handbook* (Muecke, G.K., Ed.), Vol. 5, pp. 133–166.

Kallemeyn G. W. and Warren P. H. (1983). Compositional implications regarding the lunar origin of the ALHA81005 meteorite. *Geophys. Res. Lett.*, **10**, 833–836.

Rubin A. E. and Pernicka E. (1989). Chondrules in the Sharps H3 chondrite: evidence for intergroup compositional differences among ordinary chondrite chondrules. *Geochim. Cosmochim. Acta,* **53**, 187–195.

Sorensen S. S. and Grossman J. N. (1989). Enrichment of trace elements in garnet amphibolites from a paleo-subduction zone: Catalina Schist, southern California. *Geochim. Cosmochim. Acta,* **53**, 3155–3177.

Sundberg L. L. and Boynton W. V. (1977). Determination of ten trace elements in meteorites and lunar materials by radiochemical neutron activation analysis. *Anal. Chim. Acta,* **89**, 127–140.

Warren P. H. and Kallemeyn G. W. (1989). Elephant Moraine 87521: the first lunar meteorite composed of predominantly mare material. *Geochim. Cosmochim. Acta,* **53**, 3323–3330.

THERMAL ANALYSIS

Patrick K. Gallagher

I. Introduction 211
II. Thermogravimetry and Thermomagnetometry 215
III. Evolved Gas Detection and Evolved Gas Analysis 223
IV. Differential Thermal Analysis and Differential Scanning Calorimetry 227
V. Miscellaneous Methods of Thermal Analysis 239
VI. Applications to Geochemistry 242
VII. Concluding Remarks 252
Note Added in Proof 252
References 254
Additional Readings 256
Appendix 257

I. INTRODUCTION

Thermal analysis is a term applied to an open-ended collection of powerful and versatile methods meeting the following definition set forth by The International Confederation for Thermal Analysis and Calorimetry (ICTAC) and adopted by IUPAC and ASTM: "Thermal analysis is a group of techniques in which a physical property of a substance, and/or its reaction products, is measured as a function of temperature whilst the subject is subjected to a controlled temperature program. It is the specific property being measured that identifies and distin-

Advances in Analytical Geochemistry
Volume 1, pages 211–257

ISBN: 1-55938-332-1

guishes the specific technique.'' The definition is very general and even isothermal measurements may be defined as a ''controlled temperature program''. Under these circumstances, virtually all experimental scientists and engineers have been thermoanalysts at some point in their careers.

The most common methods are summarized in Table 1. This listing is only a slight modification of similar tables occurring in general texts on thermal methods (Wendlandt, 1986; Brown, 1988). Because this chapter is concerned with the application of the methods to geochemistry, the discussion will concentrate on the most relevant techniques for that purpose as indicated in the table of contents and the first nine lines of Table 1. Although the majority of applications are presently in the area of polymers and plastics there are, nevertheless, substantial use of these powerful and versatile methods in most areas of science and technology. The early usage of several of the techniques, particularly differential thermal analysis (DTA), was concerned with minerals and clays and there is a strong tradition of thermal methods remaining in geochemistry (Mackenzie 1970, 1972).

Following a brief discussion of some common aspects, the instrumentation, methodology, and precautions associated with each of the selected methods will

Table 1. Principal Thermoanalytical Methods

Property	*Technique*		*Acronym*
Mass	Thermogravimetry		TG
Apparent Mass[a]	Thermomagnetometry		TM
Volatiles	Evolved Gas Detection		EGD
	Evolved Gas Analysis		EGA
	Thermal Desorption		
Radioactive Decay	Emanation Thermal Analysis		ETA
Temperature	Differential Thermal Analysis		DTA
Heat[b] or Heat Flux[c]	Differential Scanning Calorimetry		DSC
Dimensions	Thermodilatometry		TD
Mechanical Properties	Thermomechanical analysis		TMA
	Dynamic Mechanical Analysis		DMA
Acoustical Properties	Thermosonimetry	(emission)	TS
	Thermoacoustimetry	(velocity)	
Electrical Properties	Thermoelectometry	(resistance)	
		(voltage)	
		(current)	
		(dielectric)	
Optical Properties	Thermooptometry	(spectroscopy)[d]	
	Thermoluminesence	(emission)	
	Thermomicroscopy	(structure)	
	Thermoparticulate Analysis		TPA

[a] Changed induced by an imposed magnetic field gradient.
[b] Power compensated DSC.
[c] Heat flux DSC.
[d] Absorption, fluorescence, Raman, etc. Non-optical forms of spectroscopy, e.g., NMR, ESR, Mössbauer, etc. are also applicable.

be considered first. After these sections, several applications in the field of geochemistry will be described. The techniques are conceptually simple and easily understood. The cleverness lies in the recognition of when these methods can be successfully applied to help understand the material or process of interest. Therefore, it is particularly important to have a significant discussion of applications to completely appreciate the subject. The Appendix contains a current listing of the major manufacturers of thermoanalytical apparatus. There is also a supplementary reading list on general thermal analysis following the list of references.

It is seldom that a single thermoanalytical technique will adequately describe the process or solve the problem. Generally several thermal methods are used to attack the problem and even then, depending upon its complexity, it may be necessary or advisable to utilize supplementary techniques such as microscopy or X-ray diffraction. For the purposes of saving time and assuring that the methods can be closely correlated, it has become common to utilize several techniques concurrently on the same sample subjected to the same temperature program. Such measurements are referred to as "simultaneous" and usually expressed in a hyphenated form such as TG-DTA or TG-DTA-EGA. An older and less common approach is to have multiple samples sharing a common thermal environment; each sample being used for a separate measurement. This approach is referred to as "combined" or "concurrent" measurements.

Most temperature programs are based upon a uniform heating rate or some predetermined relationship of temperature as a function of time. There is a growing tendency, however, to utilize feedback from the process to control the heating process. As an example, a prescribed rate of weight loss, pressure change, or shrinkage rate can be predetermined by the investigator and the furnace controller driven by the output from the parameter sensor to maintain that preset rate of change (Rouquerol, 1989). Figure 1, indicates a schematic representation of this approach to thermal analysis (Gallagher, 1991). Examples of its use, effects, and advantages will be described occasionally later in the chapter.

The development of computers has had an enormous impact on scientific instrumentation and thermal analysis has been no exception. Some brief general discussion of this influence is in order before considering the individual techniques. Senior investigators in the field will have been introduced to thermal analysis when the chart recorder was the predominant form of data collection. The tediousness of reading numerous points off of chart paper combined with the significant errors associated with paper slippage, dimensional changes due to humidity and temperature, and line width problems or skips due to ink flow problems all conspired to limit the scope and accuracy of thermoanalytical studies. This was particularly true in the area of kinetics where many data points are very desirable.

The initial application of digital data acquisition progressed through several stages starting with punched paper tapes (Gallagher and Schrey, 1970), to magnetic tapes (Gallagher and Schrey, 1972), to the current stage of dedicated

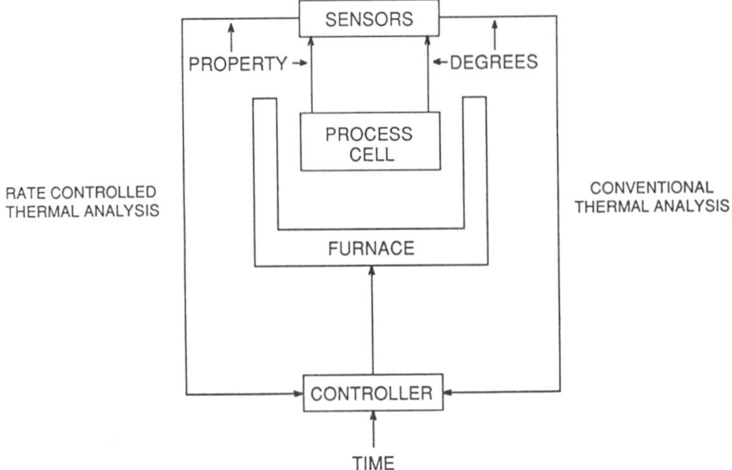

Figure 1. Comparison of conventional and controlled rate thermal analysis (Gallagher, in press).

PC usage. The manufacturers of thermal analysis instrumentation did not become involved until the latter stages of development, but they have now completely embraced the approach. There are many benefits that have resulted from this incorporation of the computer that the new generation of investigators has come to assume as natural. Some of these major advances are greatly improved accuracy, speed, and convenience; markedly improved programming and control of the temperature; large arrays of data that facilitate, smoothing, differentiation, curve fitting, etc.; and the ability to store, recall, display, and plot data or calculations. These benefits have greatly increased the usage of these thermal methods and quality of the results.

The drawbacks of completely relying upon computer processing are a loss of flexibility in the instrumentation dictated by limitations in the manufacturer's software, and/or possible difficulty of accessing the necessary electrical signals for imputing into one's own data acquisition systems. The user is, of course, responsible for verifying the adequacy and accuracy of any software used for the particular application. The precise nature of the information stored in the data arrays must be known before any further use by the investigator. As an example, these arrays may contain programmed temperature instead of the measured temperature, or the weight data may have some offset or background correction subtracted away. Hopefully, the data will be in an appropriate format and the computer operating system compatible so data may be portable to other computers and software depending upon the user's inclinations.

The operator also must remember that elegant data acquisition and manipulation systems will not compensate for inadequate instrumentation. The fundamental quality of the thermoanalytical hardware, experimental parameters, sample, and sample handling procedures will ultimately be the major factors in determining the accuracy of the data and the conclusions based thereon.

A more arguable disadvantage is the loss of intimate contact with the data collection and analysis process. From a learning standpoint, one is more detached from the instrumental parameters and underlying principles when there is a computer interface involved. These disadvantages, however, are overwhelmed by the many advantages associated with the involvement of computers, and particularly dedicated PCs, in the tasks of thermal analysis.

II. THERMOGRAVIMETRY AND THERMOMAGNETOMETRY

Thermogravity (TG) is concerned with following changes of mass as a function of time and/or temperature, while thermomagnetometry (TM) is equivalent except that the sample is subjected to a magnetic field gradient. The latter technique is only of value if the reactants, intermediates, and/or products are magnetic. It will be discussed at the end of this section.

There are a wide variety of sensors available for the detection of changes in mass; however, the simple balance remains the overwhelming choice. A general schematic is presented in Figure 2 for several general types of TG instruments. The sample is placed directly on the end of the beam or suspended from either above or below a balance beam. The sample resides in a controlled temperature device and is subjected to the desired thermal program. All three versions of sample placement occur in the current commercially available instrumentation.

Most of the mass due to the sample holder and suspension is tared, and even a portion of the sample weight may be tared when using heavy samples. The actual change in mass that takes place, however, is generally measured by some form of restoring force necessary to maintain the beam position essentially constant. The current applied to a carefully constructed torque motor is the typical method used.

The Cahn balance mechanism, shown in Figure 3, has been the forerunner of several similar electronic balances. The beam position is sensed photoelectricly using a light source, beam flag, and photodetector. The current applied to the motor, required to hold the photodetector output constant, is proportional to the weight change. Balances of this type are rugged and sensitive to weight changes on the order of 0.1 μg. The range of linear response in the torque motor is limited and it will be necessary to tare out a portion of very large changes in weight depending upon the particular device being used. Balances in most of the modern thermogravimetric equipment function similarly.

The ultimate sample size is determined by the strengths of the balance beam and torque motor. Some manufacturers make several models having different capacities and sensitivities. An alternative way of increasing the sample size at the expense of a proportional reduction in sensitivity is to have several positions

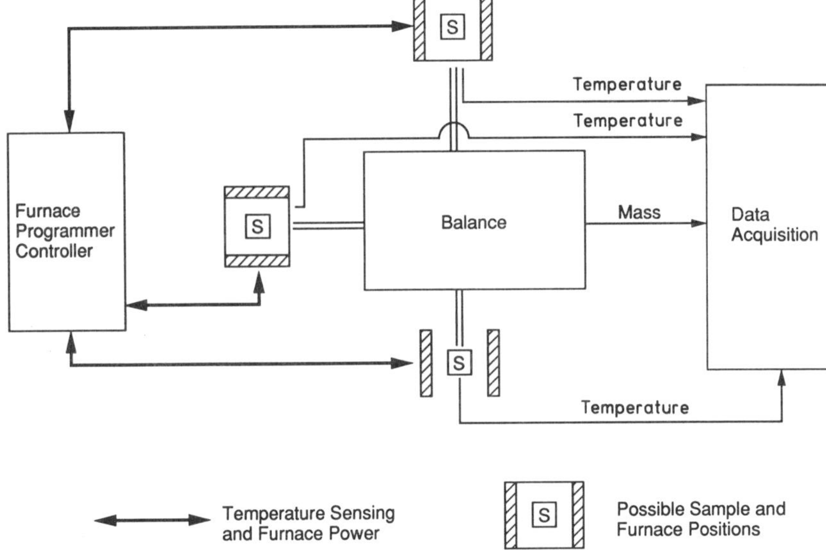

Figure 2. General schematic for a thermogravimetric apparatus.

along the beam from which to suspend the sample. The closer to the fulcrum position of the beam, the lower the distorting force generated and the larger the sample that can be accommodated.

Control of the sample temperature is dictated by the sample size and more importantly the range of temperature involved. The furnace or oven is generally

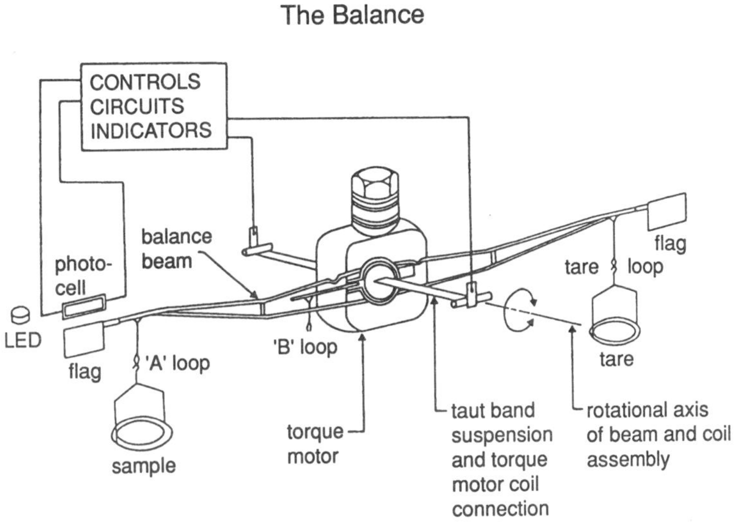

Figure 3. Schematic diagram of a Cahn Electrobalance (Brown, 1988).

heated by using electrical current passing through resistive elements. The resistive elements may be Nichrome or Kanthal alloys up to about 1200 °C, or platinum-rhodium alloys, molybdenum disilicide, stabilized zirconia, silicon carbide, graphite, molybdenum, or tungsten at higher temperatures. The latter three materials must be used in inert atmospheres to prevent their oxidation. Radiant heaters have also been used and are capable of very high heating rates (up to 500 °C min^{-1}) in the region from room temperature to 1200 °C. Figure 4 illustrates one of the possible configurations using focused infrared radiation to heat the sample (Ulvac Sinko Riko Co.).

Gaseous heat exchangers have been employed, particularly at subambient temperatures. The boil-off rate from a liquid nitrogen revivor is a convenient means of temperature control in the region from about 90 K to room temperature.

Obviously the temperature sensors must be matched to the temperature range of the particular equipment in use. Most manufacturers offer several models or modules designed to operate efficiently over a specific range of temperature. Because studies on purely organic materials represent most of the market, the greatest variety and availability of equipment is in the temperature range below 1000 °C. There is, nevertheless, an ample choice of equipment capable of higher temperatures. The next increment of temperature is up to 1500 to 1700 °C. Above this temperature range the selection narrows, cost increases, convenience decreases, and flexibility or versatility is less. Materials for construction and temperature sensors are more limited. The major suppliers for the very high temperature market are European; for example, Netzsch and SETARAM.

The general trend is toward smaller sample sizes with miniature furnaces. This conserves sample, provides better resolution of events, reduces power requirements, minimizes thermal and mass transport problems, and allows for higher heating and cooling rates, thereby reducing the overall time required for the measurement process. There are, however, situations that need larger sample sizes; for example, inhomogeneous samples, specific sample geometries, thin film samples on relatively massive substrates, corrosion studies, etc. A dual heating system is a means of reducing the errors in TG, particularly those asso-

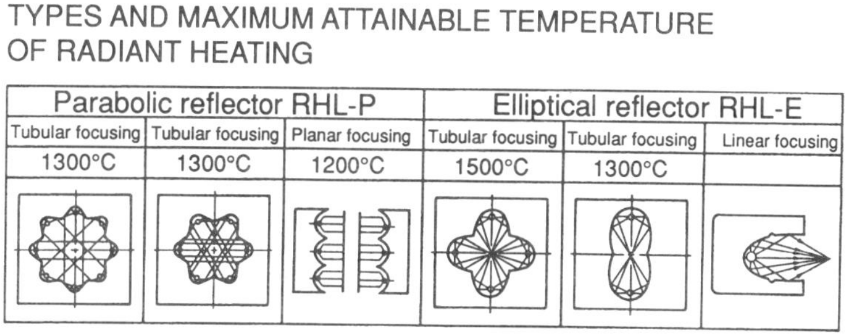

TYPES AND MAXIMUM ATTAINABLE TEMPERATURE OF RADIANT HEATING

Parabolic reflector RHL-P			Elliptical reflector RHL-E		
Tubular focusing	Tubular focusing	Planar focusing	Tubular focusing	Tubular focusing	Linear focusing
1300°C	1300°C	1200°C	1500°C	1300°C	

Figure 4. Various arrangements used for IR radiant heating (Ulvac Sinko Riko Co.).

ciated with buoyancy. This modification utilizes a symmetrical balance arrangement with an identical heating system on the tare side as well as on the sample side. The tare side, naturally, must duplicate as closely as practical the sample side, except for an inert substitute sample having a similar size and density. The flow rate and nature of the gases must also be as identical as possible. The two similar furnace systems are then heated using the same temperature program to minimize successfully the necessary corrections.

Most modern thermobalances provide two separate entrance ports for the flow-through atmosphere. One port is utilized for purging the balance chamber to prevent the back diffusion of any of the active atmosphere or product gases into the balance chamber where the corrosion of sensitive parts might occur. This is usually achieved at a relatively low flow rate. The active atmosphere is introduced well downstream of the balance before flowing around the sample. Since the two flows are merged, the total flow rate and composition are the sum of the two flows. The actual space velocity in the region of the sample also must consider any change in sample temperature that alters the gaseous volume and, hence, the flow rate.

The atmosphere to which the sample is actually exposed is highly dependent upon the flow pattern and the sample container. The effectiveness of the sweep gases for introducing a reactant gas, or for carrying away any product gas, is determined by its ability to contact the sample and permeate it if necessary. The reaction interface generally moves from the sample surface inward; however, it is uncertain in powders whether the effective surface is that of the powder pile or that of the individual particles. It is primarily determined by the thermal and mass transport associated with the process and the potential reversibility of the reaction. These factors will be discussed in somewhat greater detail later in this chapter.

There is a wide range of sample holders available. Figure 5 displays some possibilities and indicates how the choice of sample holder can influence the resulting TG curve (Paulik and Paulik 1986). The decomposition of $CaCO_3$ is readily reversible and, consequently, the equilibrium decomposition temperature is established by the partial pressure of CO_2 and P_{CO_2} in the environment. The sample holder, in Figure 5a, consists of several stacked trays containing the sample in a thin layer of powder having ready access to the flowing gas. Since the product, CO_2, is swept rapidly away, the P_{CO_2} never reaches a high level and the calcite decomposes at a lower temperature, approximately 500 °C. The other extreme is represented by the labyrinth type container depicted in Figure 5d. Although the atmosphere in the chamber initially corresponds to the sweep gas, it very quickly is replaced by the product gases after only a small portion of the overall decomposition. If the decomposition is reversible, then for calcite, the temperature of the decomposition is driven several hundred degrees higher as indicated by the comparison between Figures 5a and 5d. This latter situation is often called decomposition in a ‘‘self generated atmosphere’’. Other sample

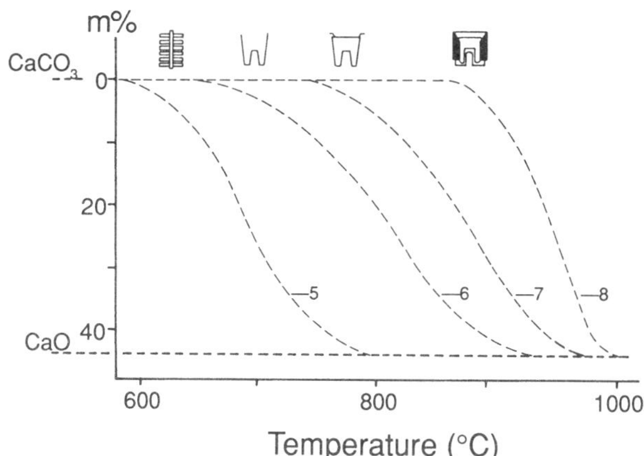

Figure 5. Decomposition of $CaCO_3$ using different sample holders (Paulik and Paulik, 1986).

holders have been designed to achieve the same effect; for example, a simple close fitting cylinder and piston arrangement (Garn and Kessler, 1960). As seen in Figure 5, there is a complete range of intermediate behavior possible between the two extremes.

Experiments can be performed over a wide range of pressures, from high vacuum to 100 atm at 1000 °C using commercially available equipment. Apparatus has been custom-constructed which is capable of much higher pressures; for example, 3000 atm (Sabrowsky and Deckert, 1979). Several highly sensitive devices have been developed to follow weight changes in vacuum systems for determining deposition rates of thin films. Those based upon changes in the resonance frequency of piezoelectric crystals have a sensitivity of nanograms; however, they have a very limited operating temperature. Several of the commercial electronic microbalances described earlier have enclosures routinely capable of 10^{-6} Torr.

The major sources of error and experimental variables for TG are listed in Table 2. If the balance has been properly calibrated, the primary factors affecting the accuracy of the mass are listed in the left portion of Table 2. Buoyancy depends upon several factors such as the temperature, composition and pressure of the atmosphere, and the density of the sample including its hangdown and holder. If other than electronic tare is used, then the density of the tare arrangement is also a factor.

Two general approaches are used to minimize the effects due to buoyancy. The first involves subtracting an identical run made without the actual sample. An inert specimen of similar size and density may be used to simulate the sample for the greatest accuracy. Most manufacturers' software will allow for the storage and subsequent subtraction of one or several blank runs. An alternative

Table 2. Major Factors Affecting Thermogravimetry

Mass	*Temperature*
Buoyancy	Heating Rate
Atmospheric Turbulence	Thermal Conductivity
Condensation and Reaction	Enthalpy of the Process
Electrostatic and Magnetic Forces	Sample, Furnace, and Sensor Arrangement
Electronic Drift	Electronic Drift

method employs a dual furnace, flow, and sample hangdown system so that the tare side undergoes virtually the same effects. A simple approximate calculation can be performed in advance to determine whether buoyancy corrections are required for the degree of accuracy desired.

Atmospheric turbulence can arise from several sources. Convection currents set up by temperature inhomogeneities is the most prevalent form of disturbance. Baffles and special geometrical scaling are the principal means of minimizing these effects. The nature of the sample suspension in addition to the flow rate and pattern, the thermal conductivity, and the pressure of the atmosphere also have major influences on these effects. Wendlandt (1986) describes these factors in some detail.

Any condensation on or reaction with the sample suspension or holder will obviously change the mass being measured. Apparent increases in mass may be due to volatile components condensing on cooler parts of the sample suspension system, or parts of the suspension system may react; for example, oxidize with the atmosphere. Apparent mass losses can be observed if portions of the suspension system vaporize. The latter is particularly important for high temperature studies demanding maximum sensitivity, such as studying small changes in oxygen stoichiometry (Bracconi and Gallagher, 1979).

A frustrating experience is to have a suspended sample swing over to the adjacent wall and cling there due to electrostatic forces. This tendency is strongest in dry atmospheres. Conductive coatings electrically grounded, conductive sprays, and harmless amounts of ionizing radiation are some of the methods employed to minimize this problem.

A tubular furnace with electrical resistive elements is the most common form of heating used. Unless such a furnace is deliberately made in a special manner (winding in a bifilar fashion), the flow of current will set up a magnetic field gradient near the sample. Such a field is proportional to the current flow and will be greatest at start up or high temperatures. If magnetic samples, intermediates, or products might be involved, one must check to assure that the specific furnace/sample arrangement employed will not perturb the measurements.

The short-term stability of the electronic components and system is seldom a limitation in modern apparatus; however, the long-term stability is sometimes a limitation to the meaningful duration of an experiment. Because the current generation of electrobalances exhibits remarkable stability, this only becomes a factor for the most demanding experiments. Electrically compensating devices are available for some balances to help reduce the portion of the drift induced by changes in the balance temperature.

The temperature of the sample is generally more difficult to establish unequivocally than the mass. Factors or variables listed in the second column of Table 2 are some of the basic causes of this problem. The electronic drift influences are similar to those described in the previous paragraph for the amplifier circuits but there are also different circuits involved, such as cold junction compensation for the thermocouple or slow physical or chemical changes in the sensor itself. Unless the sensor has reacted with the gas phase, however, these changes are usually small compared to those arising from the other factors.

The four remaining factors described in Table 2 are highly interactive. Errors in sample temperature arising from a poor sample, furnace, and sensor arrangement will be markedly changed by variations in the heating rate, thermal conductivity or flow rate of the atmosphere, and enthalpy of the events transpiring. Increasing the heating rate will accentuate the temperature lag between the sample and nearby temperature sensor. Raising the thermal conductivity of the gas phase will improve the thermal transport between the various components of the system, but increasing the flow rate may cause the situation to deteriorate depending upon the flow pattern and the reactivity of the gas. The influence due to mass transport has been discussed earlier with reference to the sample holder and other hardware parameters. The perturbations due to the enthalpy of the process are deferred until the section on DTA, where the effects and their correlations are more obvious.

Thermogravimetric data is presented in several forms. The most common form is as percent weight loss as a function of temperature. Normalization in the form of weight percent allows several experiments or curves to be compared directly, regardless of the specific sample weight. Data can be plotted as a function of time for isothermal experiments or for those utilizing isothermal periods during the program. Older schemes plot both weight percent and temperature versus time. This has the advantage of allowing verification of the temperature program, but is more cumbersome for the readers during normal discussion of the results.

The derivative is frequently plotted in the form of rate of weight loss as a function of temperature or time. Such treatment is refered to as differental thermogravimetry (DTG). The major advantages are increased resolution of events, simpler comparison with other differential techniques, and more direct kinetic significance; that is, rate phenomena. The disadvantages are increased noise (a differentiation process is a major noise amplifier unless smoothing is employed) and the added difficulty imposed in calculating the stoichiometry based on values of the integrated weight percent.

The derivative was traditionally generated by an electrical RC circuit which served as an analog computer. The advent of digital data acquisition has eliminated the need for these circuits and provides the information via software routines with much greater convenience and flexibility.

A set of TG and DTG curves are presented in Figure 6 for the thermal decomposition of $CaC_2O_4{\cdot}H_2O$ as a general illustration. This selection is made because of the three well-resolved steps in the overall process corresponding to the loss of H_2O, CO, and CO_2 respectively. The plateau regions associated with the range of thermal stability, therefore, correspond to the starting material, anhydrous calcium oxalate, calcium carbonate, and calcium oxide, respectively.

Thermomagnetometry (TM) is only a slight modification of thermogravimetry in which a magnetic field gradient is imposed in the region of the sample. This will alter the apparent mass of a magnetic sample depending upon the strength of the field gradient, the magnetic properties of the specimen, and the temperature. A strong magnetic field is generally not necessary and may even be undesirable (Gallagher and Gyorgy, 1986). Not all thermobalances are easily adapted to allow a permanent or electromagnet in suitable proximity to the sample. Interactions between the magnet and other parts of the thermobalance must also be considered. If TM is to be performed, these considerations should be invoked at the time of equipment selection. Several examples will be presented latter in this chapter.

TM can be used to calibrate the temperature scale for a thermobalance. Several metals and alloys having well-defined transition temperatures are used for this

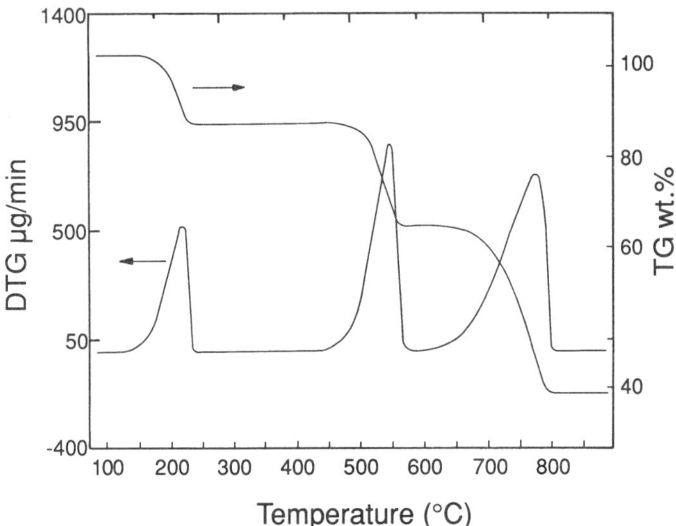

Figure 6. TG and DTG curves for the thermal decomposition of $CaC_2O_4{\cdot}H_2O$ at 20 °C min^{-1}.

purpose (Gallagher and Schery, 1970; Noren et al., 1970; Zhong and Gallagher, 1991). The temperature at which the apparent magnetic influence vanishes corresponds to the transition temperature. Advantages of this method are that several samples may be run simultaneously so multiple calibration points can be achieved in a single experiment. The effect being measured occurs directly at the sample position and thus is independent of the sensor location or relationship. It, therefore, provides an excellent indication of the direct temperature at the sample. There is negligible enthalpy associated with the transition, thus there are no distortions due to that factor.

An alternative method of temperature calibration for TG involves the use of a "fusible link" formed from a thin wire having a well-defined melting temperature (McGhie et al., 1983). A small weight can be suspended from the link and the weight change associated with its falling occurs at the melting temperature. The magnetic and fusible link methods are in good agreement with each other (Gallagher and Gyorgy, 1986). Simultaneous TG/DTA allows standards developed for DTA to also be used to calibrate TG (Zhong and Gallagher, 1991; Hongtu and Laye, 1989). The numerous difficulties associated with temperature measurement during TG necessitate that careful calibration be performed under nearly identical conditions to those of the final experiments for accurate results.

Before leaving the section on instrumentation for TG, there are several other points deserving of mention. Simultaneous techniques will be discussed when the second technique is introduced except X-ray diffraction since it is not treated by itself. Wiedemann and Bayer (1973) have developed a unique apparatus capable of accomplishing the difficult task of simultaneous TG/X-ray diffraction, a combination that should be of considerable interest for geochemical studies. Thermobalances have been modified for operation with several samples weighed sequentially to provide multiple TG curves during the same temperature program (Ferguson et al., 1972). A commercial device is available (LEECO Co.) for multiple, high throughput, quality control applications.

Another useful modification is the separation of balance and sample chambers to allow the use of highly corrosive atmospheres. This has been accomplished by a clever magnetic coupling (Gast, 1971). Finally, the use of a Knudsen cell with a thermal balance is particularly important for determining the vapor pressure of materials; see for example the discussion by Wiedemann and Bayer (1987) on this topic.

III. EVOLVED GAS ANALYSIS

The change in mass measured by TG is a very precise method of evaluating the amount of volatile species evolved, but unfortunately does not provide any information regarding the chemical identity of the species involved. Frequently, knowledge of the decomposition chemistry is sufficiently obvious to dictate the nature of the gaseous products, such as for the decomposition of a simple carbonate or hydrate. In many other instances, however, the situation is unclear or, at least, less certain. Under these circumstances it is particularly useful to be able to analyze the volatile products.

Evolved gas detection (EGD) represents several methods that demonstrate the evolution of volatiles without specifically identifying their composition. Examples of techniques used for EGD are to follow gas density or thermal conductivity. Changes in these properties indicate something has contaminated the normal flow gas, but only defines that the average value of the property is less or greater than that for the normal flowing atmosphere. Another example would be the use of a flame ionization dector to indicate the presence of organic vapors in the exhaust stream from the reactor without revealing their exact nature. Simple unresolved absorption of light within a broad spectral range also might be used to indicate the release of gases into the normal flow.

The usefulness of EGD is primarily confined to those situations in which the specific nature of the gas has previously been determined and the present interest is confined to the temperature, time, or amount of the release. Knowing the nature of the gas allows these methods to become quantitative tools for following the evolution. Under these conditions the technique has become evolved gas analysis (EGA).

EGA includes those techniques that provide either qualitative or quantitative information concerning changes of the gaseous species in the atmosphere surrounding or flowing past the sample. This information could be either additive or subtractive in nature; that is, gases evolved or consumed. The sensors used for such studies may provide information regarding many gases; for example, FTIR spectroscopy, gas chromatography, or mass spectroscopy. The sensors may also be very specific in nature applicable to a single gaseous species or class of species; for example, dew point detectors for moisture, specific radioactive decay, or a spectroscopic technique tuned for a single band, mass number, etc.

An example of the use of a specific sensor is shown in Figure 7 for the dehydration and decomposition of $Ba(OH)_2 \cdot {}_2H_2O$ (Gallagher and Gyorgy, 1980). The sensor is a capacitance device whose output is directly proportional to the dew point of the gas stream. A close approximation of the water lost can be made if the flow rate of the gas is known. The dew point can be converted to the partial pressure of water and, after compensation is made for the intrinsic water background of the carrier gas, integrated over the time of the evolution. The same device can be used to follow the reduction of oxides by hydrogen via the formation of water vapor (Gallagher et al., 1981; 1982).

Some techniques, such as gas chromatography, are unable to continuously analyze the gas stream and therefore are inherently discontinuous. Other times it may be simpler, or for other reasons desirable, to sample intermittently. One such reason would be to trap or store the evolved products over a range of temperature to enhance the effective sensitivity by analyzing more concentrated

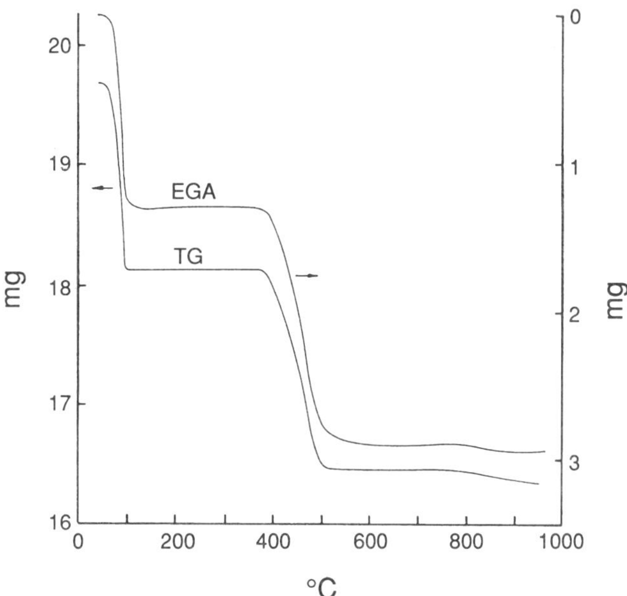

Figure 7. Comparison of the corrected EGA with the DTG curve for the thermal decomposition of $Ba(OH)_2 \cdot H_2O$ (Gallagher and Gyorgy, 1980).

samples. Another reason might be to free an expensive gas analysis instrument to be utilized for other measurements during the overall process, or to allow it to be physically separated from the thermal analysis module.

A cell may be designed solely for EGD or EGA but, more frequently, the technique is run simultaneously with another thermoanalytical technique such as TG or DTA. In this mode the exhaust gas flowing from the other thermal device is taken as an input into the EGD or EGA system. The interface is relatively simple if both systems are at atmospheric pressure. The connecting line must be sufficiently (1) heated to minimize chemisorption and condensation of the evolved gases, (2) inert to prevent any reaction or catalytic conversion of relevant gases, and (3) short to reduce the delay time for correlation between the "simultaneous" measurements. Significant pulses of backpressure may arise during switching gas chromatographic valves, thus it is advisable to use some protection against these surges disturbing the other simultaneous thermal measurements (Gallagher et al., 1982).

A typical FTIR spectrometer cell with heated windows is shown in Figure 8 (BioRad Co.). This is readily linked to another thermal analysis module via a heated connection. Experimental results for $CaC_2O_4 \cdot H_2O$ are presented in Figure 9 (BioRad Co.). These can be compared directly with the DTG curve in Figure 6. The integrated Gram–Schmidt EGA curve shows peaks at the same temperatures

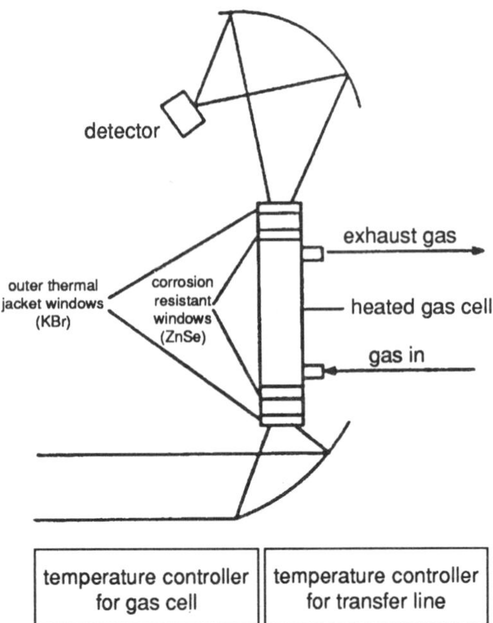

Figure 8. Schematic of a simultaneous TG-EGA (FTIR) apparatus (BioRad Co.).

as for the DTG; however, the relative areas under the curve in the FTIR analysis (Figure 9) are not directly comparable because of differences in the optical absorptivity for each species. Within each peak, the relative fraction decomposed can be determined based upon the fractional area; even though a comparison cannot be made from one peak to that for a different gas.

If the thermal analysis modules are at different pressures, a more complex interface is required to prevent any interferences between techniques. The difference in pressures usually arises when mass spectrometry is used. Intermittent sampling of stored gases can be used to avoid the problem, but a variety of special interfaces have been designed from simple heated capillaries to more elaborate differentially pumped systems. In addition to the requirements described in the previous paragraph, it is necessary to accommodate the pressure drop from near atmospheric pressure in the simultaneous device to less than about 10^{-5} torr in the mass spectrometer.

A solution involving the formation of a molecular beam through the use of differential pumping and very carefully aligned small orifices is given in Figure 10 (Kaisersberger, 1980). This system has the advantage that the beam of gases from the simultaneous TG/DTA have a minimum contact with the hot walls. This minimizes reactions with the walls and condensation of readily condensible species before they reach the mass spectrometer. The sensitivity suffers substan-

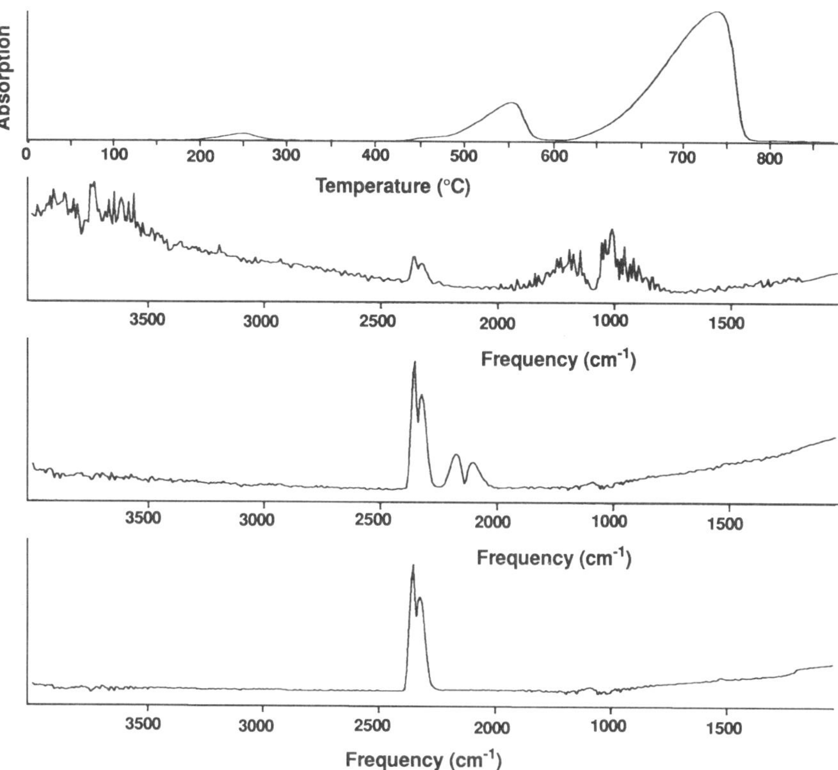

Figure 9. EGA (FTIR) curves for the thermal decomposition of $CaC_2O_4{\cdot}H_2O$ at 20 °C min^{-1}.

tially, however, in comparison with a dedicated mass spectroscopic EGA in which the furnace is directly in the vacuum system in close proximity to the ionizing elements of a quadrupole mass spectrometer (Gallagher, 1984). The disadvantages of the latter approach are that simultaneous measurements are not obtained and the reaction is performed in vacuum. The vacuum accentuates problems associated with the reduced thermal conductivity and complicates comparisons of the observed behavior with complimentary measurements made at atmospheric pressure.

IV. DIFFERENTIAL THERMAL ANALYSIS AND DIFFERENTIAL SCANNING CALORIMETRY

Le Châtlier (1887a) developed thermal analysis during the later part of the last century when he investigated the temperature of clays as they were heated at a uniform rate. It remained for Sir Roberts-Austen to apply the technique to metals

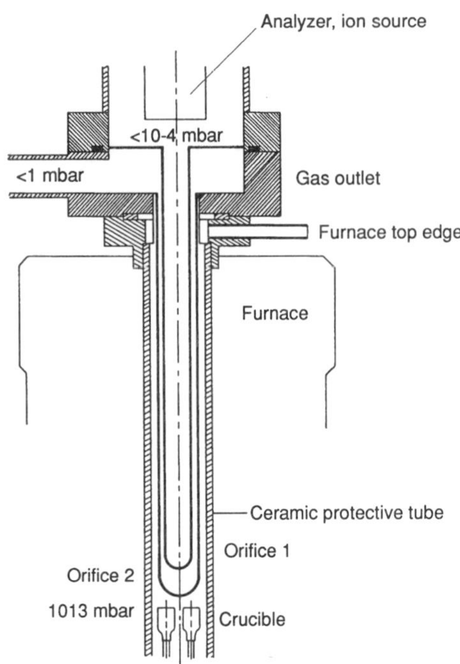

Figure 10. A simultaneous TG-EGA (MS) interface (Kaiserberger, 1980).

using a differential thermocouple and thereby create DTA (LeChâtelier, 1887b; Roberts-Austen, 1899). The technique remains essentially the same today. A sample and inert reference material are heated at a controlled rate (usually linear) and the difference in temperature between the sample and reference is measured as a function of temperature or time.

Figure 11 shows a schematic of the apparatus and a comparison of thermal analysis versus DTA. If the sample undergoes a transformation or reaction that evolves (exothermic) or absorbs (endothermic) heat, the temperature sensor, T, associated with the sample will detect an increase or decrease in temperature relative to the normal rate of temperature rise (thermal analysis), or relative to the sensor following the temperature of the reference material (DTA) during the time of the event. Similarly, if there is a change in heat capacity or thermal conductivity of the sample, this will be detected as a change in slope for thermal analysis, or as an offset in the baseline of the plot of ΔT versus T or time (DTA).

In principle, the equivalent to DTA curves can be mathematically generated by differentiating the thermal analysis curve directly. This, however, does not provide the quality and reliability that DTA does. Differences arising from fluctuations in the flow of gas or irregularities in the rate of heating are much more likely to cause misleading results. The true differential mode of operation tends to aleviate these experimental difficulties.

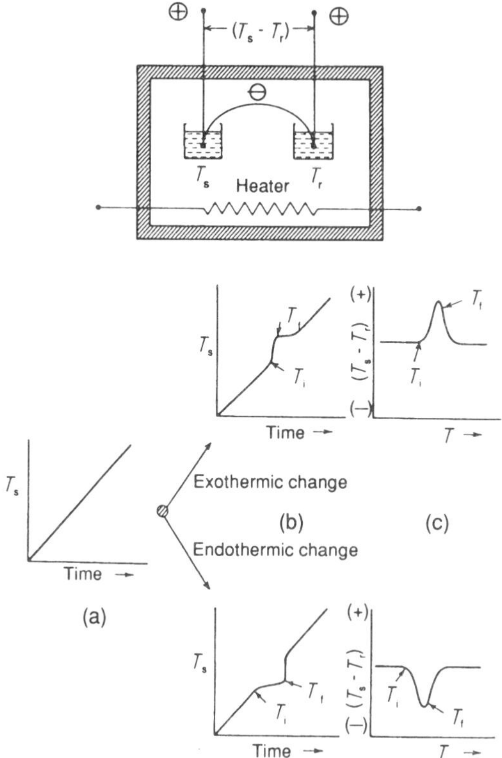

Figure 11. General schematic for a DTA apparatus (Wendlandt, 1986).

The basis of DTA is the departure in temperature of the sample from that which would "normally" have been obtained in the absence of any change in cnthalpy, heat capacity, and thermal conductivity. Consequently, its existance demonstrates the influence that these thermal effects can have upon the actual sample temperature compared to that of a temperature sensor nearby or in the windings of the furnace. That is why any extensive discussion of the effects of enthalpy upon the sample temperature was deferred to this point.

In TG experiments, where one is relying upon a nearby sensor for the temperature, there is obviously some error during those times when the events of interest are occuring. If the DTA experiments were flawless, then the ΔT value obtained could be simply added to the observed value of T at each time as a correction. While this is not complete compensation, it is clearly a step in the proper direction and offers another reason for simultaneous TG/DTA.

A typical set of DTA curves is presented in Figure 12 for the same material whose TG and DTG curves are depicted in Figure 6 and EGA curves in Figure 9. The correspondence of events between the three figures indicates their relation-

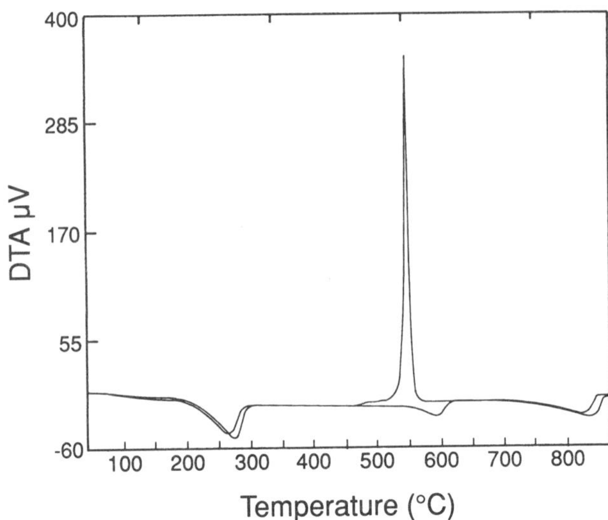

Figure 12. DTA curves for the thermal decomposition of $CaC_2O_4{\cdot}H_2O$ at 20 °C min^{-1}.

ships. The loss of water in the initial step of the decomposition is an endothermic process indicated by the negative value of ΔT. The amplitude and width of the peak is determined by the period of time needed for the dehydration and the time it takes to return to the baseline. Both of these responces are determined by the sample size, heating rate, nature and rate of gas flow, the thermal characteristics of the particular apparatus, and enthalpy of change in heat capacity of the process. These factors will be considered shortly after completion of the discussion of Figure 12.

The second stage of the overall decomposition corresponds to the evolution of CO. This step demonstrates the profound influence that extrinsic processes can have upon the measurements. The decomposition of the anhydrous oxalate is inherantly an endothermic process as shown by the curve in nitrogen. The curve in oxygen, however, shows a very strong exotherm at that point. This is due to the oxidation of the CO formed during the decomposition by the oxygen present in the atmosphere.

The oxidation takes place at the surface and in the intersticies of the powder where the heat is conveyed to the temperature sensor in a poorly defined manner. The ΔT signal is normally unable to keep up with the actual temperature rise. When the sample can be observed at this stage, it is generall at a red glow indicating an actual temperature far in excess of the measured temperature.

The metastability of CO itself adds to the uncertainty in the process. Below about 800 °C, CO is unstable with respect to disproportionation into CO_2 and elemental C. The extent to which this takes place depends upon the catalytic nature of the environment, the temperature, and the competing oxidation process. Any disproportionation will provide some heat and carbon to the sample. These will be assimilated in the overall thermoanalytical responses. The latter will also impart a grey, brown, or black color to the sample.

The final stage of the decomposition process represented in Figures 6, 9, and 12 corresponds to conversion of the $CaCO_3$ to CaO and CO_2. Like the loss of water earlier, this is reversible and the observed temperature will be affected by the mass transport considerations discussed previously in Section II. Although the reaction has a typical enthalpy (about 40 Kcal mol^{-1}) for endothermic reactions in this temperature range, it also serves to indicate self-cooling which can result from endothermic processes. Consideration of this energy in relation to the heat capacities of the product and reactants makes it clear that thermal transport must also be considered (Caldwell et al., 1973; Gallagher and Johnson, 1973).

In principle, the area under the DTA peaks should be proportional to the amount of energy absorbed or released. There are many experimental difficulties, however, basing such measurements on ΔT and making direct comparisons from one experiment to another. The thermal characteristics of the instrument are a strong function of temperature due to variations in thermal transport and sensor sensitivity with temperature. The concept of differential scanning calorimetry (DSC) was developed to minimize these quantitative limitations. There are two fundamental categories of DSC instruments. The first to use the term DSC operates on the principle of *power compensation* (O'Neill, 1964; Watson et al., 1964). A schematic of the sample compartment and general operation is given in Figure 13 (Watson et al., 1964). In contrast to DTA, the value of ΔT is maintained at 0 by the proper application of power to the individual heaters associated with the sample and reference compartments. The power required to offset the thermal events replaces ΔT as the parameter measured as a function of temperature or time. The temperature sensors are Pt resistance units rather than thermocouples.

If the power is applied to the sample compartment, it has a positive value comparable to the endothermic value of ΔH or ΔC_p for the process. The integral of power with respect to time is a direct evaluation of the energy absorbed or released by the event. If the event is exothermic, then the power is applied to the reference side and given the opposite sign. It must be remembered that the sign conventions for DTA and DSC are opposed, since an exothermic process gives a positive value of ΔT but a negative value of ΔH.

The second type of DSC has been given the name of heat flux DSC. It operates in a mode similar to traditional DTA except that the critical values of thermal resistance and temperature sensitivity as a function of temperature have been carefully calibrated and built into the software and hardware of the instrument. Both types of DSC can give excellent results. Some instruments will act in both the DTA and DSC modes depending on a simple switch setting. More extensive discussions of the differences between the traditional DTA and both modes of

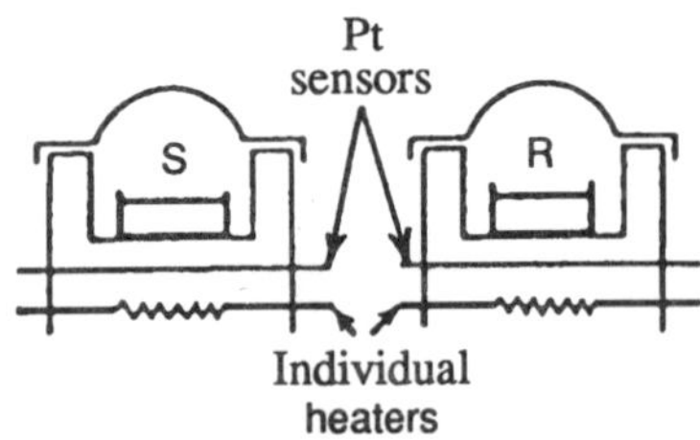

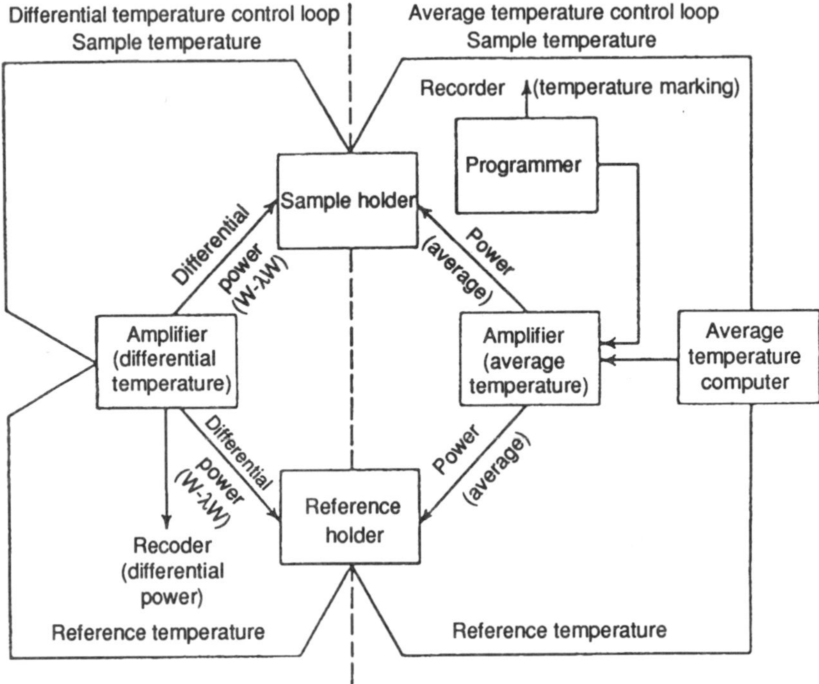

Figure 13. Schematic of a power compensated DSC instrument and its operation (Watson et al., 1964).

DSC are readily found in standard texts on thermal analysis (Mackenzie, 1970, 1972; Wendlandt, 1986; Brown, 1988; Wunderlich, 1990).

DTA and DSC have many of the experimental goals, variables, and problems in common. Table 3 is a listing of the classes of events for which DTA and DSC are suitable experimental tools. Those reactions and transformations which are reversible, of course, will show the opposite effect on cooling. In addition to these events involving a relatively abrupt change in ΔC_p, there is the opportunity to measure C_p

Table 3. Events Detected by DTA and DSC and Their Effects During Heating

Transformation	Observation	Reaction	Observation
1st Order	Endothermic	Decomposition	Either
Higher Order	Step in Baseline	Liquid-Solid	Either
Vaporization	Endothermic	Solid-Solid*	Either
Fusion	Endothermic	Polymerization	Exothermic
Metastable to Stable	Exothermic	Chemisorption and Catalysis	Exothermic

* Also liquid-liquid or gas-gas reactions.

itself or thermal conductivity as functions of temperature (Wendlandt, 1986; Brown, 1988).

Some of the processes listed in the right hand portion of Table 3 were evident in Figure 12. The three endothermic decompositions are present in N_2, but the exothermic oxidation of CO overwhelms the intermediate decomposition in the presence of O_2. Had the starting oxalate been a complex of several metals, then a higher temperature endothermic solid - solid reaction might have followed, such as in the decomposition of $BaTiO(C_2O_4)$ where the products of the oxalate decomposition react at a higher temperature to form $BaTiO_3$ (Gallagher and Thomsen, 1965).

A DTA curve for a glass that depicts several transitions is shown in Figure 14 (Vogel and Gallagher 1985). The glass transition is indicated by the shift in base

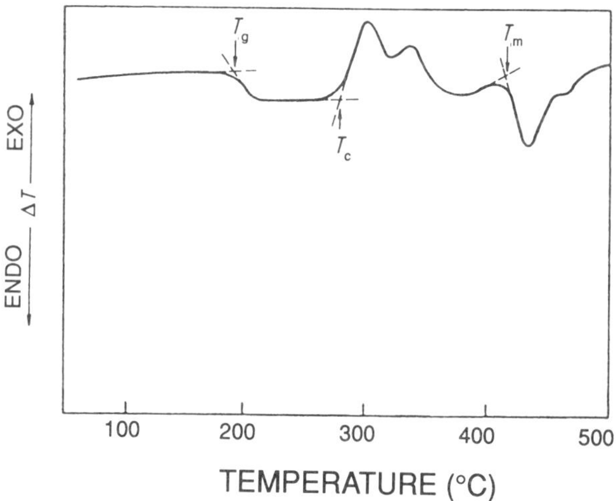

Figure 14. DTA curve for a glass ($0.5BeF_2$ $0.3KF$ $0.1CaF_2$ $0.1A1F_3$), 25mg, at 20 °C min^{-1} in N_2 (Vogel and Gallagher, 1985). Tg = glass transition temperature, T_c = crystallization temperature, and T_m = melting point.

line representing the ΔC_p associated with that higher order transition. The two exothermic amorphous to crystalline transitions represent metastable to stable transformations forming distinct ordered phases that subsequently melt producing the endothermic transitions at higher temperature.

The first three categories of reactions in Table 3 indicate either an exothermic or endothermic response. Virtually all decompositions and reactions that occur during rising temperatures are endothermic unless oxidation or metastability is involved. Examples of these latter categories is the oxidation of the CO depicted in Figure 12, or the oxidation of the potential solid product FeO to a higher oxide during decomposition of $FeCO_3$.

The polymerization, gas-solid, and catalytic reactions are examples of metastable to stable reactions that become kineticly possible as the temperature is increased. They are irreversible under the same conditions. Other types of metastable to stable reactions can occur, such as a supersaturated solution exsolving a second phase when nucleation has finally occurred.

There are many commercial and customized versions of sample holder and sensor configurations, some of which are depicted in Figure 15 (Wendlandt, 1986). These arrangements provide for different gas flow patterns, sample sizes, and sample containers. As shown earlier in Figures 5 and 12, the gas flow is important because of the thermal transport aspects, but primarily because of the frequency of reverse reactions and their dependence on the partial pressure of the gaseous product and the possibility of other gas solid reactions such as oxidation.

The general procedure is to allow the preheated gas to flow by and over the sample as for most of the examples in Figure 15. It is possible, however, to actually flow the atmosphere through a powdered sample as illustrated in Figure 15e. Although this provides for the maximum sweep of product gases, other than working in an continuously evacuated system, it is subject to several disadvantages. The finer starting powder or solid product may also become entrained in the gas, and the flow patterns between the sample and reference are almost certain to change during the experiment, thereby altering the baseline.

Another approach is similar to the "self-generated" atmosphere described in the TG section. In TG, the volatile products must escape in order to have a change in weight. This limitation, however, is not a factor in DTA or DSC so that sealed containers may be used to either protect the sample from reaction with the flowing atmosphere, or to simulate a closed environment. One of many such designs is shown in Figure 15h. Besides the use of sealed glass or welded metal tubes prepared by the individual investigators, many manufactures provide sample pans which can be cold-welded using a press that they also supply. Other steel containers with various seals are manufactured that are capable of holding much higher pressures but which add very significantly to the overall sample heat capacity reducing the sensitivity.

The common temperature sensor is a thermocouple selected to adaquately cover the temperature range of interest with satisfactory sensitivity. The sensor

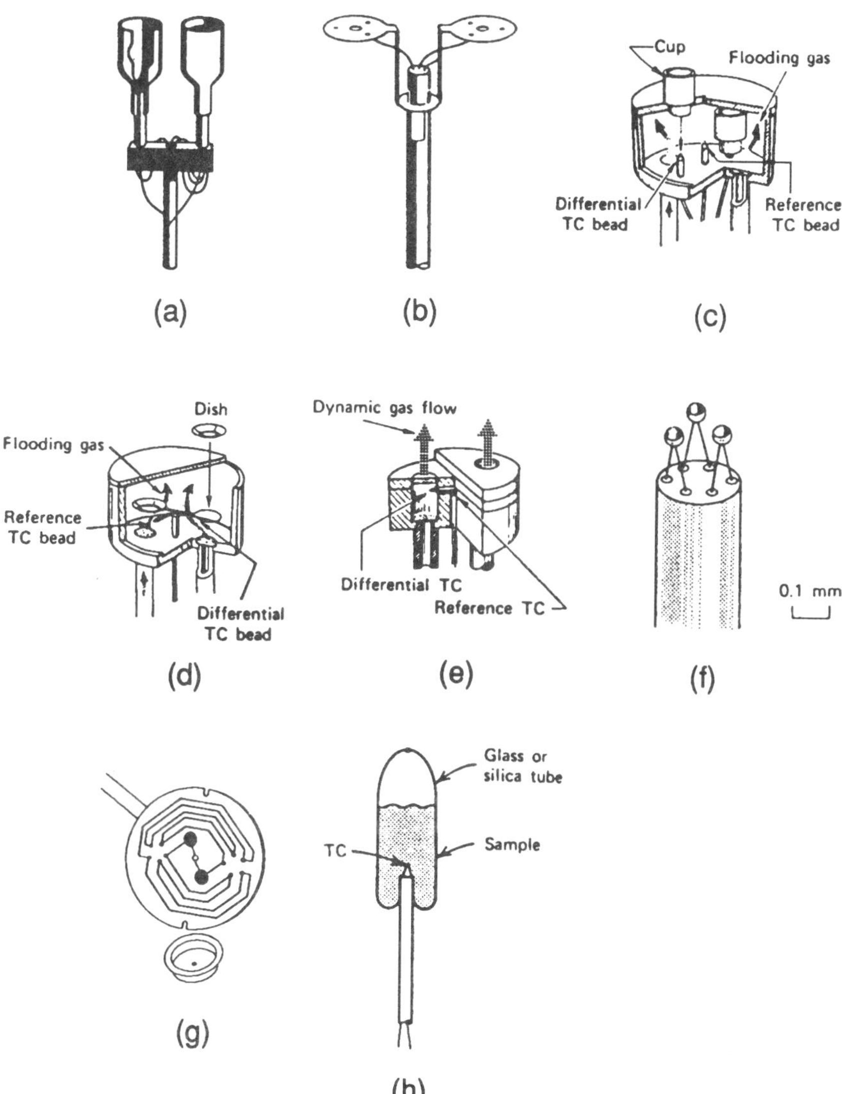

Figure 15. DTA sample holder and sensor arrangements (Wendlandt, 1986).

may be in direct contact with the sample, but is much more often in contact with the sample or reference container. In Figure 15d an interesting ring arrangement for the thermocouple is shown. A micro-DTA arrangement is depicted in Figure 15f and the use of a thin-film thermopile is indicated in Figure 15g.

There are many variables which will affect the results of DTA and DSC. These are classified into four categories, presented in Table 4. The first category is

Table 4. Classes of Variables in DTA and DSC

Construction – Thermal Linkages – Sensor Response
Atmosphere – Reactions – Thermal Conductivity
Sample Size – Packing – Particle Size
Heating Rate – Hysteresis

concerned with the design and construction of the instrument. Such factors as: (1) the choice of the sample, reference, and furnace arrangement; (2) the materials of construction; and (3) the selection, size, and placement of the temperature sensors all play a significant role in the thermal transport between the furnace, sample, and reference materials. These factors, along with the sensitivity of the sensor, strongly influence the value of ΔT accompanying any change in the relative thermal properties of the sample and reference. The greater the thermal linkage between the sample and reference the smaller the value of ΔT, but the faster the return to the base line.

Influences of the flow patterns and reactions of the atmosphere have already been discussed. It also should be emphasized that the thermal conductivity of the gaseous media significantly effects both the sample temperature (Caldwell et al., 1977) and the value of ΔT (Berg, 1975). Working in vacuum or in very low molecular weight gases, such as hydrogen and helium, can have dramatic effects. The humidity of the atmosphere can alter the decomposition temperatures of hydrates and hydroxides as well as oxidize more sensitive materials. The complete exclusion of oxygen and water vapor is difficult and extensive purification and careful plumbing may be required.

The effects of sample size and heating rate on the sample temperature and rate of reaction have been discussed. The sample packing influences the ability of gaseous products to escape or reactants to reach the reacting interface. The packing of the sample and reference powders, however, also alters the thermal conductivity of system. Improved packing has the same effect as any other means of reducing the thermal resistance between the sample and reference. Diluting the sample with inert material, such as that used as the reference, has been used to reduce the value of ΔT when desirable. It also has the advantage of improving the base line.

Faster heating rates will increase the apparent temperature lag between the process and the sensor and decrease the resolution similar to that discussed earlier in connection with TG. There is an added factor when considering DTA and DSC, however. The value of ΔT, and hence the sensitivity, is dependent on the rate of the reaction and, consequently, increasing the heating rate increases the sensitivity. This is similar to the behavior of EGA but unlike that of TG where the weight loss is not a function of the heating rate unless other processes occur. When DTG is used, however, the heating rate has the same effect as in DTA.

The choice of a specific instrumental design and the experimental parameters selected represent a compromise between the quest for maximum resolution, sensitivity, and accuracy. Some of these compromises are indicated in Table 5 (Wendlandt, 1986). The safest approach is to initially run under conditions which maximize the sensitivity in order to insure the likelyhood of observing all events, and then to alter subsequent experimental parameters to optimize the results for the particular goals of the study. This is a potential drawback of simultaneous measurements; that is, the inability to simultaneously optimize the experimental conditions for each technique.

DTA and DSC curves come in many varieties and the ability to have well-defined points of reference and accepted methods for establishing baselines are important. Some of these aspects are depicted in Figure 16 (Brown, 1988). The departure from the baseline represents the beginning of the thermal event, and is generally accepted as the relevant point for comparison with the ''true'' temperature of the event in question. Selecting the departure point is very subjective and depends on the degree of amplification. Consequently, the extrapolated onset is more commonly used. This represented by point A in Figure 16a.

Construction of the baseline is frequently another subjective process. Guidelines have been presented; see, for instance, texts in the field (Mackenzie, 1970, 1972; Pope and Judd, 1977; Wendlandt, 1986; Brown, 1988; Wunderlich, 1990). Several instrumental parameters can give rise to a sloping baseline; for example, uneven heating of the sample and reference or poor matching of the thermal properties of the sample and reference materials. Most modern instruments have hardware and/or software features which can minimize these instrumental factors.

Changes in the thermal properties of the sample during the event, however, must be reflected in the measurements in order to obtain the true picture of the process. These changes in ΔC_p appear as a change in baseline after the event. A variety of approaches to account for this when drawing the baseline are summarized in Figure 16. In Figure 16f, the base line is adjusted proportional to the rate of change of the curve. The specific method used is a judgement decision and the operator must be aware of how the particular software employed operates in this respect.

Table 5. Operational Compromises for DTA and DSC

Parameter	*Maximum Resolution*	*Maximum Sensitivity*
Sample Size	Small	Large
Heating Rate	Slow	Fast
Sample – Reference	Linked	Isolated
Particle Size	Small	Large
Atmosphere	High Conductivity	Low Conductivity

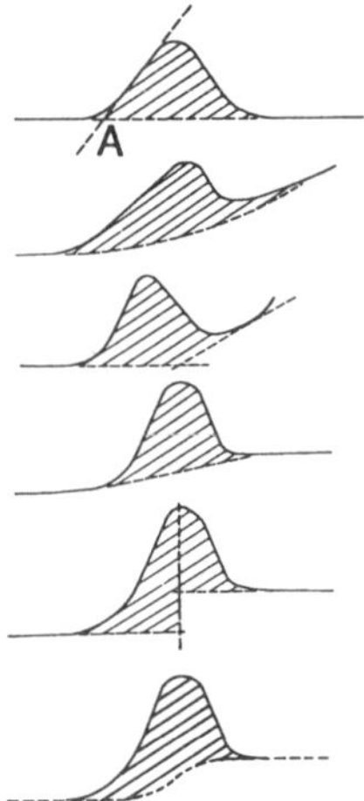

Figure 16. Typical DTA or DSC curves illustrating several methods of defining the baseline (Brown, 1988).

From the previous discussion it is apparent that calibration of the temperature axis for the particular set of experimental parameters used is essential. Table 6 lists a number of the materials commonly used for this purpose. There are the melting points used to define the current international temperature scale and a series solid-state crystallographic transformations that have been recommended and extensively studied (Garn, 1975) by the International Confederation for Thermal Analysis and Calorimetry (ICTAC). Figure 17 illustrates the use of two of these solid state transitions and one magnetic standard for the calibration of a

Table 6. Some Standard Materials for DTA and DSC

Material	*Melting Points*[a] *Temperature °C*	*Enthalpy J/g*	*Material*	*Solid$_1$ – Solid$_2$*[b] *Temperature °C*	*Enthalpy J/g*
In	156.5985	28.42	KNO_3	127.7	50.48
Sn	231.928	59.2	$KClO_4$	299.5	99.32
Pb	327.502	23.16	Ag_2SO_4	412	59.85
Zn	419.527	112.0	SiO_2	573	12.1
Al	660.323	400.1	K_2SO_4	583	48.49
Ag	961.78	104.7	K_2CrO_4	665	51.71
Au	1064.18	63.7	$BaCO_3$	810	89.00
Cu	1084.62	205.4	$SrCO_3$	925	133.23

[a] International Temperature Scale 1990 for temperatures and R. Hultgren, et al., Selected Values of the Thermodynamic Properties of the Elements 1973 for the enthalpies.

[b] NBS – ICTA Certificates GM-758, GM-759, GM-760 for the temperatures and I. Barin, Thermochemical Data of Pure Substances, Verlagsgesellschaft, Weinheim, 1989 for the enthalpies.

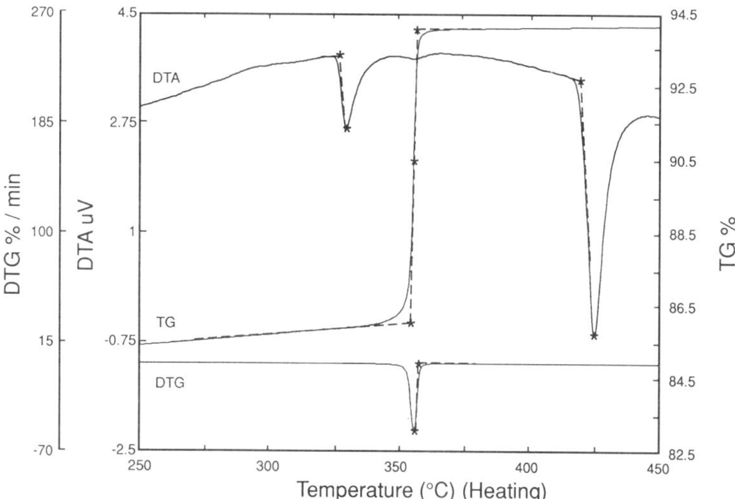

Figure 17. An example of the use of magnetic TG standards and DTA standards to calibrate the temperature for a simultaneous TG-DTA instrument (Zhong and Gallagher, 1991).

simultaneous TG-DTA instrument (Zhong and Gallagher, 1991). The ability to use DTA standards to calibrate TG is advantageous since these materials are better defined and more universally applicable than magnetic standards.

V. MISCELLANEOUS METHODS OF THERMAL ANALYSIS

The preceding discussions concentrated on the three major classes of thermoanalytical techniques: TG/TM, EGA/EGD, and DTA/DSC. Table 1 alluded to many more methods. There are several of those that are occasionally used in geochemical research and therefore deserve further mention.

Emanation thermal analysis (ETA) can be considered a rather special subclass of EGA in that the radioactivity of the evolved gas is measured. This is invariably due to the deliberate introduction of a specific radioactive parent or gas into the sample and, consequently, the radioactivity in the gas phase can be directly related to the rate of release of a specific known species similar to using the dew point as an analytical method for water. Introduction of ^{228}Th or ^{224}Ra by coprecipitation will provide a source of a emitting ^{220}Rn. Alternatively, radioactive gases can be directly injected into the samples via ion bombardment.

A general discussion of this method and some applications can be found in the proceedings of a recent workshop on the topic (Balek, 1987). The primary applications to date have been studies of surface morphology and porosity, defect chemistry and diffusion, crystallization and sintering, and solid-state reactions and catalysis. The apparatus is simple and sensitive, but the evidence is indirect and the interpretation or correlation is frequently complex.

Thermodilatometry and thermomechanical analyses (TMA) are widely used techniques; however, they have been sparingly applied to geological problems. Changes in volume or, more commonly, length are detected by a variety of devices from relatively crude dial micrometers to sophisticated laser generated interference fringe techniques. The usual method utilizes displacement measured by a linear voltage differential transformer (LVDT).

The displacement of the probe in contact with the sample can be used to measure thermal expansion, softening, or sintering. Changes in thermal expansion occur abruptly with discontinuity at a first-order phase transition. Higher order transformations or reactions lead to changes in slope. Sintering is indicated by the sample shrinkage over a range of temperature, while a more rapid shrinkage occurs at the softening point due to penetration of the probe into the specimen. For this reason a variety of probes are used for different purposes. As an example, Earnest (1983) has used TMA to study the residues from coal liquifaction processes. Again, the reader is referred to the basic texts for a more detailed presentation of the topic (Wendlandt, 1986; Brown, 1988; Wunderlich, 1990).

In contrast, thermosonimetry is a relatively seldom used technique, but many of its applications have been related to minerals, rocks, etc. Decompositions and transformations give rise to several mechanical processes which produce sound. The method involves listening through a stethoscopic device for the emission of sounds from a sample as it is being heated or cooled. A sound pipe such as a polished silica rod is placed in good contact with the subject. At the other end, outside of the furnace the rod is carefully placed in contact with a piezoelectric sensor. A diagram of such an apparatus is shown in Figure 18 (Lonvick, 1987).

The apparatus can function in two modes. The simple integrated sonic intensity can be collected to detect events, and the actual sonic spectrum can be obtained to assist in its interpretation. Observations have indicated substantial differences using the latter mode for decompositions of nominally the same material, but there has been little success to date in unravelling the complexities of the sonic spectrum. The sounds can arise from decrepitation phenomena associated with decomposition and the escape of volatile components or crystallization. More subtle emissions can arise from the release of stresses or the movement of dislocations accompanying phase transformations or simple mismatches of thermal expansion in highly anisotropic and polyphasic materials.

Of the electrically based techniques mentioned in Table 1, dielectric analysis (DA) appears to be the most generally promising approach. Rejeshwar et al. (1978) have utilized the method to make simultaneous DTA-DA measurements on oil shales. The dielectric measurements probe the motions at a molecular level and offer insights difficult to obtain by other techniques. The correlation between the thermal properties and the electrical properties represents a useful combination. The frequency range from 1 to 10^7 Hz was studied up to 800 °C in a care-

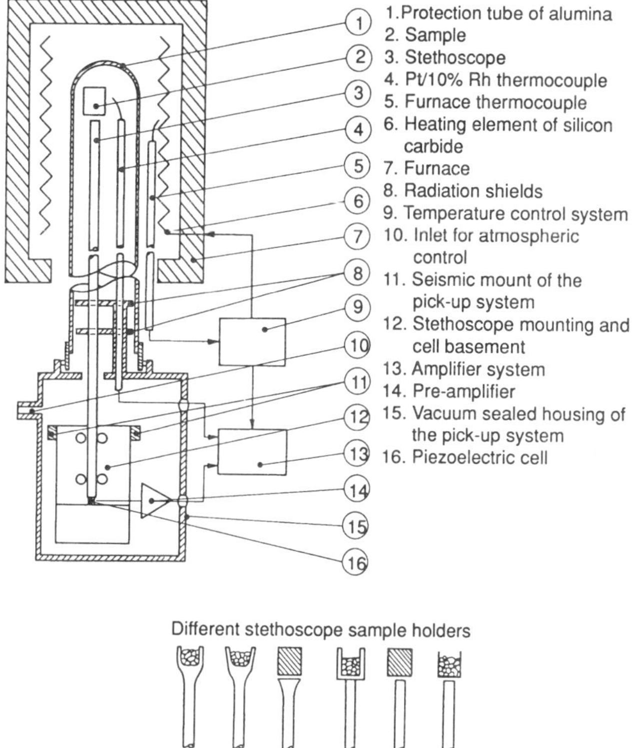

Figure 18. A typical apparatus for thermosonimetry (Lonvick, 1987).

fully controlled flowing atmosphere. A better understanding of the transformations and thermal reactions of the shale was obtained through comparisons of the changes in dielectric constant and loss with the DTA curves.

The microscope is one of the geologist's most useful tools. Coupling this with a programmable hot stage provides an even more powerful means of studying geological processes and phenomena. The value and techniques associated with thermomicroscopy are straight forward and need no further elaboration here.

Thermoluminescence is another of the thermo-optical techniques which finds wide application in geology and geochemistry. It measures the optical emission from specimens as they are heated, generally in the range from room temperature to about 500 °C. The process is irreversible and arises from the defects trapped in the material during irradiation. Natural irradiation by naturally occurring radioactive sources allow the method to be used for geological dating. The recombination of the trapped holes and electrons emits light that is integrated by a photomultiplier tube and displayed as a function of temperature or time (Wendlandt, 1986; MacDougall, 1968).

VI. APPLICATIONS TO GEOCHEMISTRY

It is necessary to describe, at least briefly, some applications of thermal analysis in order to have a full appreciation of its power and versatility. Warne (1989) listed the primary areas of Earth sciences in which thermal analysis has made significant contributions. These are presented in Table 7. A small number of examples are selected to illustrate each of the methods discussed earlier.

Dolomite is a naturally occurring mineral composed of a solid solution of magnesium and calcium carbonates, $Mg_{1-X}Ca_XCO_3$. Recent work (Wiedemann and Bayer, 1985) indicates that the separation in the decomposition of the magnesium and calcium portions are poorly resolved in air. The strong tendency of the calcium carbonate decomposition to be reversible, however, provides a clear resolution, as seen in Figure 19a, when the decomposition occurs in flowing carbon dioxide. The reversibility is particularly evident in Figure 19b where the total decomposition was conducted in nitrogen and the atmosphere switched to carbon dioxide. At that time the calcium oxide reverts to the carbonate but the magnesium oxide does not. Control of the atmosphere, therefore, allows TG to analyze the individual magnesium and calcium contents of naturally occurring and synthetic dolomites.

Fossil fuels are of ever increasing importance. Thermoanalytical methods are valuable in assessing both the chemical composition and the calorific value of the coals, shales, etc. The use of TG and TM for the proximate analysis of coal is an excellent example of how these simple, relatively rapid techniques have markedly enhanced technology of the field. Aylmer and Rowe (1984) and Hyman and Rowe (1982), have demonstrated a scheme for determining the quantities of moisture, volatiles, fixed carbon, ash, and iron minerals using TG and TM in controlled atmospheres.

The scheme is summarized in Figure 20. Samples are ground to <120 mesh. Weights from 50 mg to several grams are used depending upon the particular

Table 7. Application Areas of Thermal Analysis: Examples in the Earth Sciences (Warne, 1989).

Minerals
Mineral Mixtures
Isomorphous Substitution Series
Soils
Ceramics – Cements
Concrete Aggregates
Building Stones
Industrial Raw and End Product Materials
Fossil Fuels
Wastes – Mine, Coal Washery, Ash, Gases
Environmental Aspects

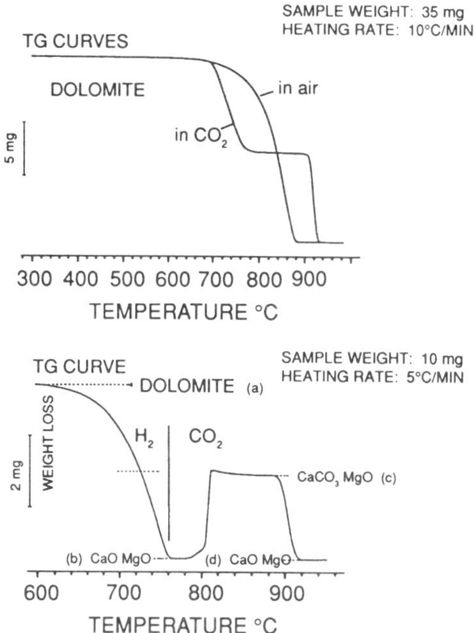

Figure 19. TG curves for the decomposition of dolomite in several atmospheres (Wiedemann and Bayer, 1985).

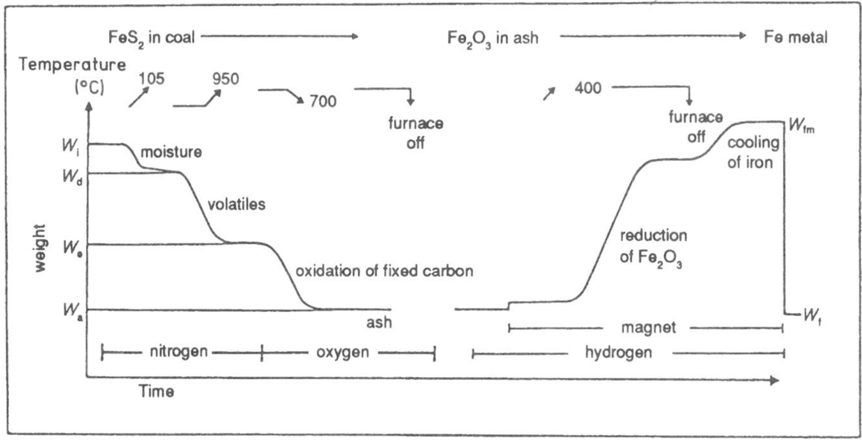

Figure 20. Proximate analysis of coals and lignites using TG and TM in controlled atmospheres (Aylmer and Rowe, 1984).

balance. The sample is heated to 105 °C in an inert atmosphere and the loss of moisture is recorded. The temperature is raised to 750 to 950 °C and held while the volatiles are lost. When essentially constant weight has been achieved, the atmosphere is switched to oxygen and the fixed carbon burned. The remaining material represents the inorganic ash.

The iron present in this ash is as Fe_2O_3 and the exact amount can be determined by TM. A magnet is positioned and there is a small apparent weight gain attributed to the parasitic antiferromagnetism of the Fe_2O_3. The temperature is raised to 400 °C in flowing hydrogen or forming gas and the apparent weight gain recorded. Although the sample is being reduced to iron, there is an apparent weight gain due to the greatly enhanced magnetic attraction. The sample is cooled in hydrogen to prevent reoxidation. Finally, the magnet is removed and the actual iron content calculated using the magnetic susceptibility and the changes of apparent weight.

The results based on this straight forward and rapid analysis are in good agreement with the more complex and time-consuming ASTM-Fisher method of coal analysis. A comparison of results for 25 to 30 samples is presented in Figure 21. The 45° angle of the lines indicate the close correlation of the methods.

Other examples of the use of TG and TM in studies related to geochemistry are the analysis of meteorites (Hyman and Rowe, 1986) and studies of mechanisms, such as that for the decomposition of siderite (Gallagher and Warne, 1981). The purpose of the latter study was to determine if magnetite [iron(II,III) oxide] formed as an intermediate during the oxidative decomposition of siderite [iron(II) carbonate] to form hematite [iron(III) oxide]. Figure 22 shows the TG and TM curves for a highly pure siderite sample in flowing oxygen and nitrogen.

The TG curves in Figure 22a indicate a final weight loss corresponding to hematite, and show a brief period of weight loss in nitrogen suggesting the temporary presence of a more chemically reduced species than hematite. The TM curves in Figure 22b clearly show that magnetite is present as an intermediate during the decomposition in nitrogen but not in oxygen. The large apparent weight gain is associated with the formation of a magnetic material having a transition temperature of 585 °C. This corresponds to the ferrimagnetic spinel structure of magnetite. This was subsequently confirmed through the use of Mössbauer spectroscopy (Gallagher et al., 1981).

Impure specimens of siderite did not show the apparent weight gain during their decomposition because the impurities were directly incorporated into the spinel during its formation. These impure spinels have magnetic transition temperatures below the decomposition temperature of the carbonate. Consequently, the spinel phase was nonmagnetic at the time of its creation. Upon cooling, however, each material showed a magnetic transformation characteristic of its impurity content. This clear proof of the direct incorporation of the impurities during the initial formation of the spinel phase was an added benefit from the TM study.

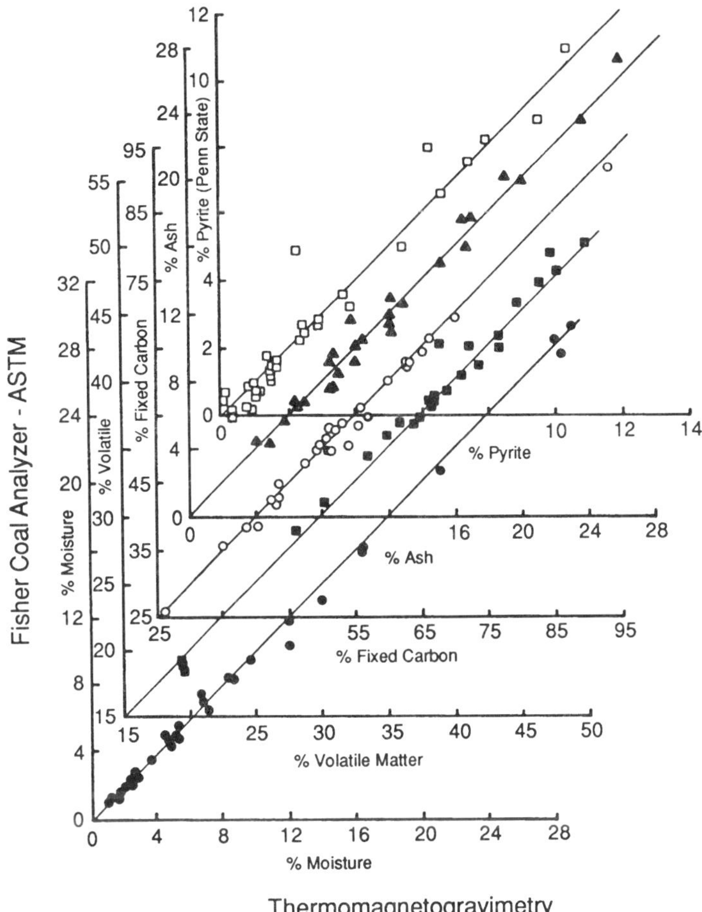

Figure 21. Comparison of the results for the proximate analysis of coals and lignites based on TG and TM with that obtained by the standard ASTM Fisher coal analyzer (Aylmer and Rowe, 1984).

Some applications of EGA to Earth science have recently been reviewed by Morgan et al. (1988). These studies are concerned with simultaneous FTIR and mass spectroscopic techniques applied to establish the decomposition products of various rocks, clays, and minerals. An interesting application of EGA to follow and compare the formation of nickel ferrite ($NiFe_2O_4$) from several starting materials was performed by MacKenzie and Cardile (1990). The intensity of the mass spectroscopic peak for oxygen is shown in Figure 23 as a function of temperature in different atmospheres for several different sources of iron oxide, with and without NiO.

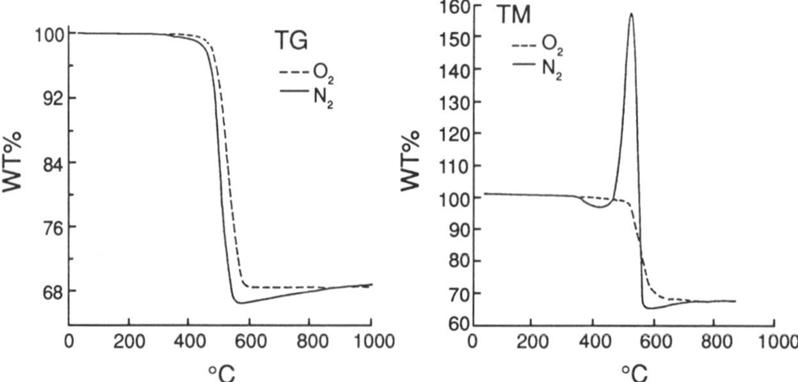

Figure 22. TG and TM curves for relatively pure siderite in oxygen and nitrogen (Gallagher and Warne, 1981).

In the presence of an oxidizing atmosphere, hematite does not lose oxygen up to 1200 °C and no oxygen evolution is necessary to form the spinel phase, $NiFe_2O_4$ (curves C and D). In the inert atmosphere (curves A and B) there is oxygen loss to form magnetite around 1000 °C and somewhat earlier in the presence of NiO. When magnetite is used in place of hematite as a starting material, its oxidation is evident (curves G and H) by the decrease in the oxygen content of the oxidizing atmosphere around 300 °C. At higher temperatures the behavior is similar to that described earlier for hematite. In the inert atmosphere (curves E and F) there is no early oxidation and some loss of oxygen at the higher temperatures due to traces of hematite. The ironsands, which contain titanium as well, behave (curves I–L) qualitatively similar to the compositions containing synthetic magnetite. These conclusions were also consistent with the X-ray diffraction and Mössbauer spectroscopic results.

The usefulness of DTA is nicely demonstrated by its ability to measure the iron contents in high-iron ferroan dolomites and ankerites at levels as low as 1 wt% (Warne et al., 1981). In addition, the presence of uncombined iron as siderite is also detected at low levels. The key to the method's success is the increased resolution and peak intensity achieved through the use of a flowing carbon dioxide atmosphere. Figure 24 shows a series of DTA curves for various concentrations of iron in a series of dolomite/ankerite. The presence of the central peak is associated with the iron content and becomes evident as early as 1 wt%.

Siderite exists as a separate iron carbonate mineral and can be clearly distinguished by its distinctive DTA pattern in Figure 25. It would appear that the sensitivity for the presence of siderite or ankerite in each other is also around 1 wt%. The ability to shift and sharpen the peaks arising from reversible reactions is a powerful technique for many mineral applications.

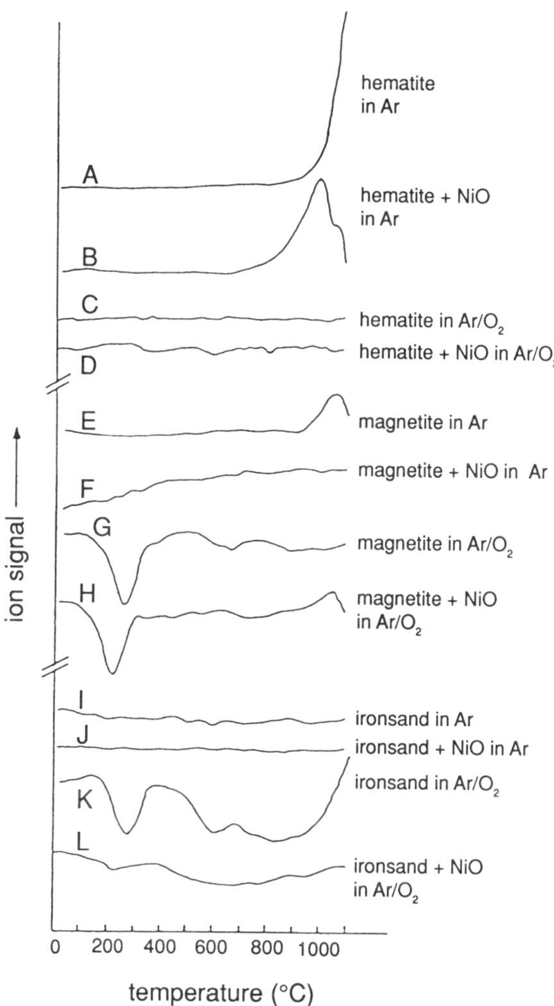

Figure 23. Mass spectroscopic EGA curves for the evolution of oxygen from selected mixtures of iron oxides and NiO (MacKenzie and Cardile, 1990).

One of the most fascinating and challenging applications of thermal analysis in the fields of modern geology and geochemistry is to nonEarth sciences; that is, extraterrestrial studies. Miniaturization, reduction of power requirements, sample collection, and efficient data storage or transmission are the instrumental hurdles to overcome.

Gooding et al. (1989) and Gooding (1989) have recently described proposed studies to investigate Mars and comets. Their initial approach is to utilize state-of-the-art DSC and EGA to fingerprint relevant water bearing geological mate-

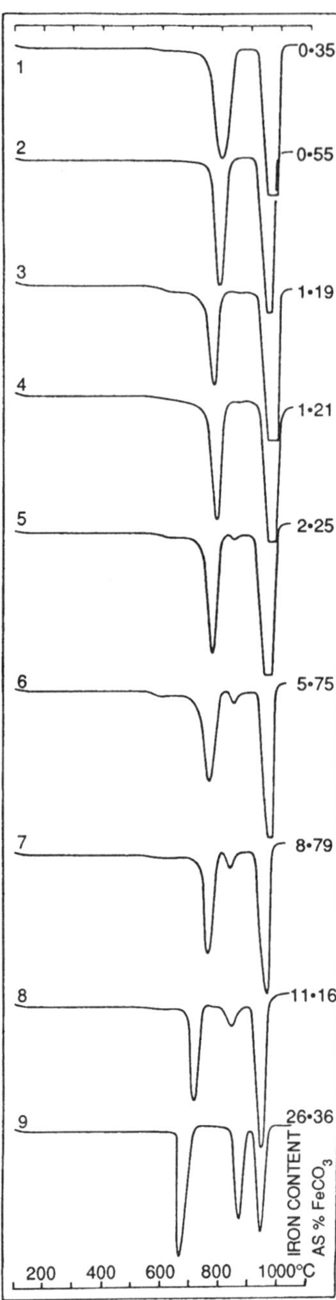

Figure 24. DTA curves in flowing CO_2 for naturally occurring samples of ferroan dolomite and ankerite (Warne et al. 1981).

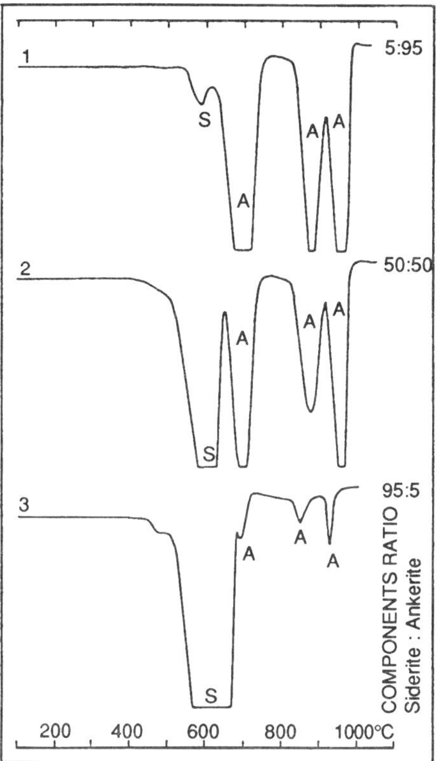

Figure 25. DTA curves in flowing CO_2 for synthetic mixtures of siderite and ankerite (Warne et al., 1981).

rials. Figure 26 summarizes the potential use of DSC and EGA to study a variety of water bearing materials that might be encountered. Some of the initial goals (1) and conclusions (2, 3) are listed below:

1. DSC fingerprinting of geological materials important to testing hypotheses in planetary science.
   a. Nontronite vs. "palagonite" on Mars.
   b. Distinguish types of meteorites on comets and asteroids.
2. Define the sensitivity of DSC for hydrated materials.
   a. 1 wt% or better for well-crystallized samples.
   b. 10 wt% or worse for poorly crystallized hydrates and zeolitic materials.
3. Detectability can be significantly improved through deconvolution analysis of the DSC data.

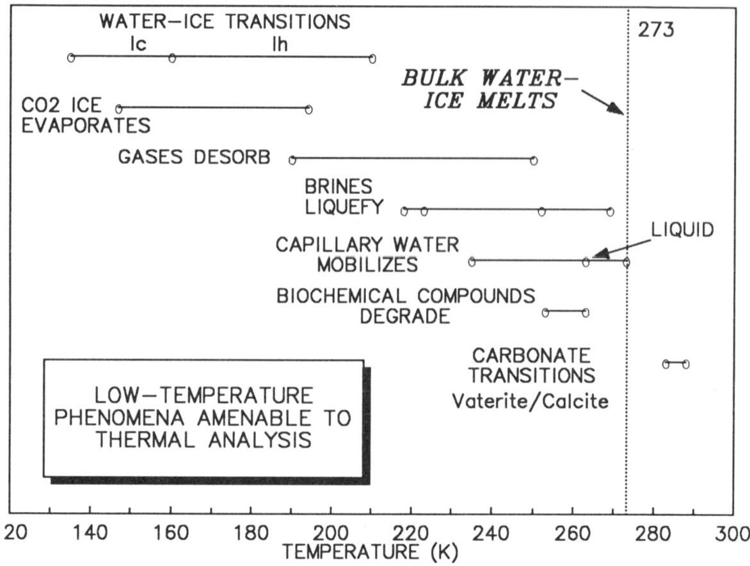

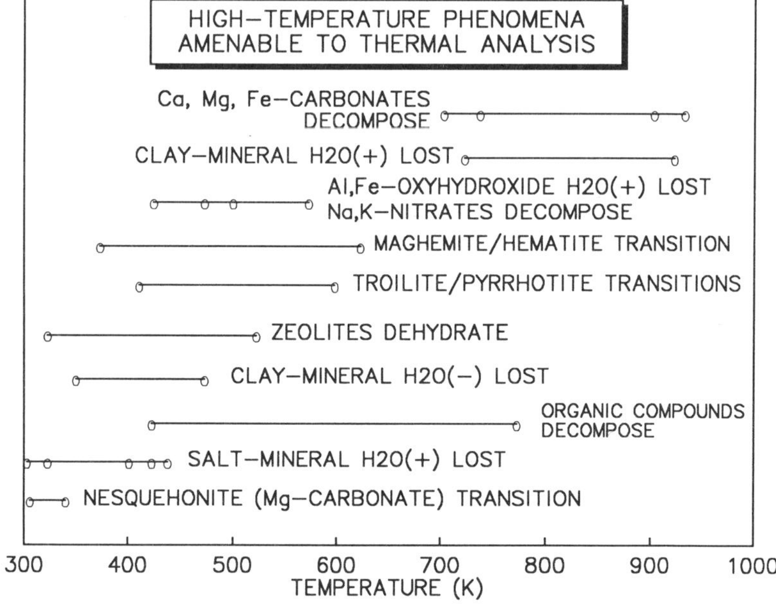

Figure 26. Transition temperatures applicable to thermal analysis studies in planetary science. "VL Oxidants" were discovered on Mars by the Viking Lander biological experiments (Gooding, 1989).

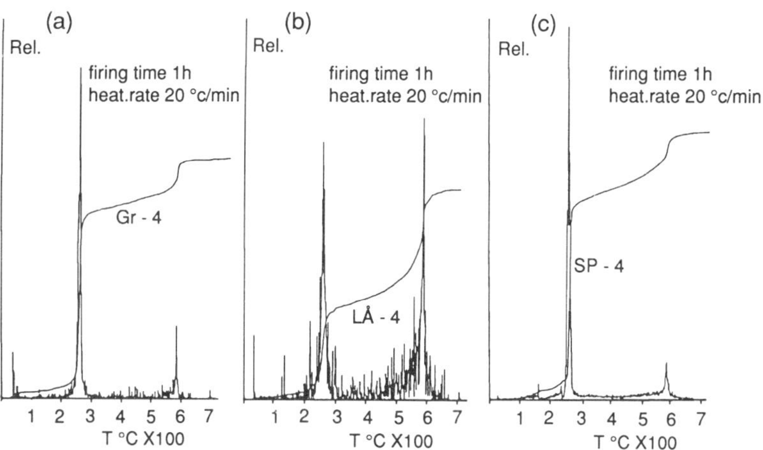

Figure 27. Differential and integral TS curves, at 20 °C min^{-1}, of different quartz specimans. The samples had been preheated for an hour at 1470 °C (Lonvick, 1987).

There have been a number of interesting studies of SiO_2 made utilizing thermosonometry. Quartz transforms slowly around 1470 °C to form crysobilite. The time at elevated temperature determines the amount of the high-temperature form. Figure 27 presents both differential and integral TS curves for different sources of quartz that had been given identical high-temperature treatments (Lonvick 1987). The transition around 280 °C is associated with the amount of crystobilite present in the quartz samples. This method has been used to detect as little as 1 wt% of crystobilite in synthetic micro silicas (Heggestad et al. 1984). Differences in TS curves have also been used to distinguish between quartz taken from different veins (Robertson, 1973).

There are many forms of spectroscopy which can be combined with traditional thermoanalytical methods to provide a more complete understanding of the mechanisms and structural changes taking place during the thermal decomposition and reactions of geological materials. MacKenzie and his group have recently published an extensive series of studies using solid-state NMR (Al and Si), Mössbauer (Fe), and IR spectroscopies in conjunction with X-ray diffraction and thermal analysis to investigate a wide range of clays and minerals (MacKenzie and Berezonwski, 1984; MacKenzie et al., 1986; Brown et al., 1987; MacKenzie et al., 1987; MacKenzie and Cardile, 1988; MacKenzie et al., 1988; MacKenzie et al., 1989). The spectroscopies provide information on the local environments, bonding, and oxidation states of key atoms during the course of the processes under study. Integrating the broader results leads to a much superior description of the events.

VII. CONCLUDING REMARKS

It is hoped that this brief selection of applications has served to indicate the utility and power of thermal analysis. Because of their great versatility and simplicity, these methods find application in virtually every field of chemistry. Not only do they provide analytical information, as their name implies, but they can also measure some physical properties and describe many physiochemical processes and events.

It is a rapidly growing, and hence, evolving technology. New techniques and modifications develop daily as a result of the extensive use of the methods in diverse areas. Innovative adaptation to investigate new phenomena and materials will continue. More sensitive and convenient hardware and software are imminent. Perhaps the most important aspect is the ever increasing use of simultaneous techniques and studies using combinations of techniques. Rarely does a single method provide the complete and unequivocal answer.

The methods deserve a place in every geochemist's collection of tools for the characterization of materials and processes. Because of their simplicity, the ingenuity or cleverness is associated with your recognition of how these methods can help you to solve your problems.

NOTE ADDED IN PROOF

Since this manuscript was prepared in the Spring of 1991, some new instrumentation has been introduced. Rate controlled TG is now comercially available through TA Instruments, Inc. They refer to this modification as HIREZ™ and it provides the opportunity to enhance resolution and possibly speed up data collection by rapidly passing through ranges with none or very little weight change. These techniques involving feedback control have been around for many years, particularly with regards to thermal expansion. Their commercial introduction in TG, however, will no doubt stimulate further development and use.

Modulated differential scanning calorimetry, MDSC™, is an entirely new approach to DSC that has recently been introduced. The concept originated within the laboratories of ICI and is currently being developed by TA Instruments. Shortly thereafter, Seiko Instruments also began marketing a similar method, oscillating DSC or ODSC. A sine wave is imposed on the normally linear heating ramp so that portions of each cycle are at different heating and cooling rates, although the general overall trend is a linear change in average temperature (see Figure 28a).

The amplitude and period of the modulation along with the average heating rate are set by the operator. A wide range of instantaneous heating and cooling rates is established within each experiment by such a process. It was described earlier how the sensitivity of DTA and DSC techniques decreased with decreasing heating or cooling rates. It is possible, however, in MDSC™ to use slow

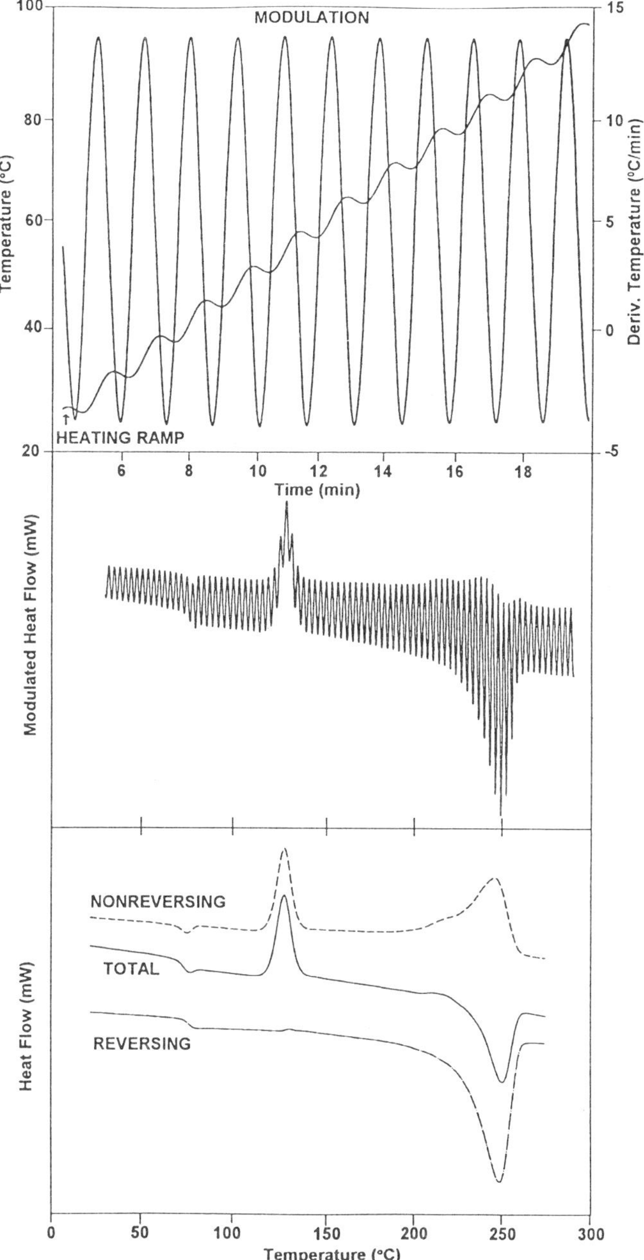

Figure 28. Example of MDSC™ (TA Instruments, Inc.). (**a**) Typical temperature-time profile. (**b**) Raw data for a MDSC™ scan of quenched PET. (**c**) Deconvolution and analysis of the curve in (**b**).

average rates with good sensitivity because the superimposed oscillation assures reasonably rapid instantaneous rates. Initial applications have been directed at polymers, although obvious possibilities exist in the field of geosciences as well.

One of the major contributions of this technique and analysis is that the total heat flow can be separated into two additional signals. A typical ''raw'' heat flow curve in MDSC™ is shown in Figure 28b. The material is quenched PET. Subsequent deconvolution of this ''raw'' data using a discrete Fourier transform yields several pieces of information besides a curve equivalent to the conventional DSC curve (see Figure 28c). One of these is a curve which represents the component of total heat flow that is heating rate-dependent; that is, that which is in phase with the modulated heating. The second curve corresponds to the heat flow that is dependent on only the absolute temperature; therefore, that which is out of phase with the modulated heating. These two components of the heat flow are designated as ''reversing'' and ''nonreversing'' respectively.

In PET, the change in C_p at the glass transition and the endotherm due to melting are reversible. It must be kept in mind, however, that the kinetics of these reversible processes may alter their appearance as the frequency of modulation increase to a point that the process can not follow the temperature cycle. As the number of nuclei for crystallization are consumed, the process becomes less reversible and a non-reversible component grows. The endothermic relaxation at T_g and the exothermic (metastable to stable) crystallization are nonreversing and appear as a separate curve.

The second major advantage of MDSC™ is that heat capacity data can be determined in a single run and with somewhat greater precision and accuracy. The precision is excellent and comparison with literature values indicates good accuracy as well (TA Instruments, Inc). The ultimate usefulness and acceptance of MDSC™ remains to be determined. The technique is so new that an inadequate data base exists in the literature at this time for a proper evaluation. Certainly, preliminary results suggest a bright future.

The final point to be made is to call the readers attention to a recent publication that should be of considerable interest. It is entitled *Thermal Analysis in the Geosciences* and is co-edited by W. Smyykatz-Kloss and S. St. J. Warne.

REFERENCES

Aylmer, D. and Rowe, M. W. (1984). *Thermochim. Acta*, **78**, 81.

Balek, V. (1987). *Thermochim. Acta*, **110**, 222 and following papers.

Berg, L. G., Equnov, V. P. and Kiyaev, A. D. (1975). *J. Therm. Anal.*, **8**, 11.

BioRad Co.

Bracconi, P. and Gallagher, P. K. (1979). *J. Am. Ceram. Soc.*, **62**, 171.

Brown, I. W. M., MacKenzie, K. J. D. and Meinhold, R. H. (1987). *J. Mater. Sci.*, **22**, 3265.

Brown, M. E. (1988). *Introduction to Thermal Analysis*. Chapman and Hall, London.

Caldwell, K. M., Gallagher, P. K. and Johnson, D. W. (1977). *Thermochim. Acta*, **18** 15.

Earnest, C. M. (1983). In *Analytical Calorimetry* (Gill P. S. and Johnson J. F., Eds.). Plenum Press, NY, Vol. 5.
Ferguson, J. M., Livesey, P. M. and Mortimer, D. (1972). In: *Thermal Analysis*. (Wiedemann H. G., Ed.). Birkhauser Verlag, Basel, Vol. 1, p. 197.
Gallagher and F. Schrey, P. K. (1970). *Thermochim. Acta*, **1**, 465.
Gallagher, P. K. (1984). *Thermochim. Acta*, **82**, 325.
Gallagher, P. K. (1991) *Mikrochim. Acta.*, **1991II**, 391.
Gallagher, P. K. and Gyorgy, E. M. (1980). In: *Thermal Analysis* (Wiedemann H. G., Ed.). Birkhauser Verlag, Basel, Vol. 1, p. 113.
Gallagher, P. K. and Johnson, D. W. (1973). *Thermochim. Acta*, **6**, 67.
Gallagher, P. K. and Schrey, F. (1972). In: *Thermal Analysis* (Wiedemann H. G., Ed.). Birkhauser Verlag, Basel, Vol. 2, p. 623.
Gallagher, P. K. and Thomsen, J. (1965). *J. Am. Ceram. Soc.*, **48**, 644.
Gallagher, P. K., Gyorgy, E. M. and Jones, W. R. (1981). *J. Chem. Phys.*, **52**, 3847.
Gallagher, P. K., Gyorgy, E. M. and Jones, W. R. (1982). *J. Ther. Anal.*, **23**, 185.
Gallagher, P. K., Johnson, D. W. and Vogel, E. M. (1976). In: *Catalysis in Organic Syntheses-1976* (Rylander P. A., and Greenstreet, H., Eds.). Academic Press, NY, p. 113.
Gallagher, P. K., West, K. W. and Warne, S. St. J. (1981). *Thermochim. Acta*, **50**, 41.
Gallagher, P. K. and Gyorgy, E. M. (1986). *Thermochim. Acta*, **109**, 193.
Gallagher, P. K. and Warne, S. St. J. (1981). *Thermochim. Acta*, **43**, 253.
Garn, P. D. (1975). *J. Therm. Anal.*, **7**, 593.
Garn, P. D. and Kessler, J. (1960). *Anal. Chem.*, **32**, 1563.
Gast, T. (1971) In: *Vacuum Microbalance Techniques* (Czanderna A. W., Ed.). Plenum Press, NY, Vol. 6, p. 59.
Gooding, J. L. (Sept. 1989). *Proceedings 18th NATAS Meeting* Vol. 1, San Diego, CA, p. 223.
Gooding, J. L., Kettle, A. J. and Lauer, H. V. (Sept. 1990). *Proceedings 19th NATAS Meeting* Vol. 1, Boston, MA, p. 10.
Heggestad, H., Holm, I. L., Lonvick, K. and Sandberg, B. (1984). *Thermochim. Acta*, **72**, 205.
Hongtu, F. and Laye, P. G. (1989). *Thermochim. Acta*, **153**, 311.
Hyman, M. and Rowe, M. W. (1982). *J. Chem. Ed.*, **59**, 424.
Hyman, M. and Rowe, M. W. (1986). *Meteoritics*, **21**, 1.
Kaisersberger, E. (1980). In: *Thermal Analysis* (Wiedemann H. G., Ed.). Birkhauser Verlag, Basel, Vol. 1, p. 245.
LeChâtelier, H. (1887a). Compt. Rend. hebd. Seanc. Acad. Sci. Paris, **104**, 1443 and 1517.
LeChâtelier, H. (1887b). *Bull. Soc. Fr. Miner.*, **10**, 204.
Lonvick, K. (1987). *Thermochim. Acta*, **110**, 253.
MacDougall, D. J. (1968). *Thermoluminescence of Geological Materials*. Academic Press, NY.
MacKenzie, K. J. D. and Berezowski, R. M. (1984). *Thermochim. Acta*, **74**, 291.
MacKenzie, K. J. D. and Cardile, C. M. (1988). *Thermochim. Acta*, **130**, 259.
MacKenzie, K. J. D. and Cardile, C. M. (1990). *Thermochim. Acta*, **165**, 207.
MacKenzie, K. J. D., Berezowski, R. M. and Bowden, M. E. (1986). *Thermochim. Acta*, **99**, 273.
MacKenzie, K. J. D., Bowden, M. E., Brown, I. W. M. and Meinhold, R. H. (1989). *Clays & Clay Minerals*, **37**, 317.
MacKenzie, K. J. D., Brown, I. W. M., Cardile, C. M. and Meinhold, R. H. (1987). *J. Mater. Sci.*, **22**, 2645.
MacKenzie, K. J. D., Cardile, C. M. and Brown, I. W. M. (1988) *Thermochim. Acta*, **136**, 247.
Mackenzie, R. C. (1970 & 1972). *Differential Thermal Analysis*. Academic Press, NY, Vols. 1 & 2.
McGhie, A. R., Chiu, J., Fair, P. G. and Blaine, R. L. (1983). *Thermochim. Acta*, **67**, 241.
Morgan, D. J., Warrington, S. B. and Warne, S. St. J. (1988). *Thermochim Acta*, **135**, 207.
Noren, S. D., O'Neill, M. J. and Gray, A. P. (1970). *Thermochim. Acta*, **1**, 29.
O'Neill, M. J. (1964). *Anal. Chem.*, **36**, 1238.
Paulik, F. and Paulik, J. (1986). *Thermochim. Acta*, **100**, 23.

Pope, M. I. and Judd, M. D. (1977). *Differential Thermal Analysis*. Heyden & Son, London.
Rajeshwar, K., Nottenburg, R. N. and Dubow, J. D. (1978). *Thermochim. Acta*, **26**, 1.
Roberts-Austen, W. C. (1899). *Metallographist*, **2**, 186.
Roberts-Austin, W. C. (1899). *Proc. Inst. Mech. Eng.*, **1**, 35.
Robertson, R. (1973). *Scot. J. Sci.*, **1**, 175.
Rouquerol, J. (1989). *Thermochim. Acta*, **144**, 209.
Sabrowsky, H. and Deckert, H. G. (1978). *Chem. Eng. Tech.*, **50**, 217.
Ulvac Sinko Riko Co., Yokahama, Japan and Kennebunk Port, ME.
Vogel, E. M. and Gallagher, P. K. (1985). *Mater. Lett.*, **4**, 5.
Warne, J. St. J. (Sept. 1989). Mettler Award Plenary Address, NATAS Meeting, San Diego, CA.
Warne, S. St. J., Morgan, D. J. and Milodowski, A. E. (1981). *Thermochim Acta*, **51**, 105.
Watson, E. S., O'Neill, M. J., Justin, J. and Brenner, N. (1964). *Anal. Chem.*, **36**, 1238.
Wendlandt, W. W. (1986). *Thermal Analysis*, 3rd ed. Wiley Interscience, NY.
Wiedemann, H. G. and Bayer, G. (1985). *Thermochim. Acta*, **121**, 479.
Wiedemann, H. G. and Bayer, G. (1978). *Topics in Current Chem.*, **77**, 256.
Wiedemann, H. G. and Bayer, G. (1973). *Z. Anal. Chem.*, **266**, 97.
Wunderlich, B. (1990). *Thermal Analysis*. Academic Press, NY.
Zhong, Z. and Gallagher, P. K. (1991). *Thermochim. Acta.*, **186**, 199.

ADDITIONAL READINGS

Books

Brown, M. E. (1988). *Introduction to Thermal Analysis: Techniques and Applications*. Chapman and Hall, London.
Brown, M. E., Galway, A. K. and Dollimore, D. (1980). *Chemical Kinetics* Vol. 22 *Reactions in the Solid State*. Elsevier Scientific, Amsterdam.
Duval, C. (1976). *Thermal Methods of Analytical Chemistry*. Elsevier Scientific, Amsterdam.
Earnest, C. M. (1984). *Thermal Analysis of Clays, Minerals, and Coal*. Perkin-Elmer, Norwalk, CT.
Earnest, C. M. (1988). *Compositional Analysis by Thermogravimetry*. ASTM, Philadelphia.
Elving, P. J., Murphy, C. B., and Kolthoff, I. M., Eds. (1983). *Treatise on Analytical Chemistry* Part I Vol. 12 *Thermal Methods*. John Wiley & Sons, New York, 1983.
Garn, P. D. (1965). *Thermoanalytical Methods of Investigation*. Academic Press, New York.
Hemminger, W. and Hohne, G. (1984). *Calorimetry: Fundamentals and Practice*. Verlag Chemie, Basel.
Keattch, C. J. and Dollimore, D. (1975). *An Introduction to Thermogravimetry*. Heyden & Son, London.
Koch, E. (1977). *Non-Isothermal Reaction Analysis*. Academic Press, New York.
Pope M.I. and Judd, M. D. (1977). *Differential Thermal Analysis*. Heyden & Son, London.
Ramachandran, V. S. (1969). *Applications of Thermal Analysis in the Cement Industry*. Chemical Publishing, New York.
Shull, R. D. and Joshi, A., Eds. (1992). *Thermal Analysis in Metallurgy*. TMS, Warrendale, PA.
Smykatz-Kloss, W. and Warne, S. St. J., Eds. (1991). *Thermal Analysis in the Geosciences*. Springer-Verlag, Berlin.
Svehla, G., Ed. (1981). *Wilson & Wilson's Comprehensive Analytical Chemistry Vol. 12 Thermal Analysis*. Elsevier Science, Amsterdam.
Turi, E., Ed. (1981). *Thermal Characterization of Polymeric Materials*. Academic Press, New York.
Wendlandt, W. W. (1986). Thermal Analysis. John Wiley & Sons, New York.
Wunderlich, B. (1990). *Thermal Analysis*. Academic Press, New York.

Proceedings Issues

International Congress on Thermal Analysis (ICTA), 1st through 9th (various publishers).
European Symposium for Thermal Analysis and Calorimetry (ESTAC), 1st through 5th (various publishers).
North American Thermal Analysis Meetings (NATAS), 10th through 20th. NATAS.
Proceedings in Vacuum Microbalance Techniques (several). Heyden & Sons and Plenum Press, London.

Journals and Abstract Services

Journal of Thermal Analysis, J. Simon (Ed.). John Wiley & Sons, Chitchester.
Thermochimica Acta, W. W. Wendlandt (Ed.). Elsevier Scientific, Amsterdam.
Chemical Abstracts Selects: *Thermal Analysis*. ACS, Columbus.

APPENDIX

Some Manufacturers of Thermal Analysis Equipment (All are international in operations)

American

Cahn Co., Cerritos, CA
Deltatherm Co., Denver, CO
Harrop Co., Columbus, OH
Holometrix Co., Cambridge, MA
LECO Co., St. Joseph, MI
Orton Foundation, Columbus, OH
Perkin-Elmer Co., Norwalk, CN
Ruska Co., Houston, TX
TA Instruments, Inc., New Castle, DE
Theta, Instruments Inc., Port Washington, NY

European

Hungarian Optical Works, Budapest, Hungary
Linseis Co., Selb, Germany
Mettler Co., Greifensee, Switzerland
Netzch-Getätebau Gmbh., Selb, Germany
Polymer Laboratories Ltd., Loughborough, U.K.
SETARAM, Caluire, France

Asian

Seiko Instruments, Tokyo, Japan
Shimadzu Instruments, Columbia, MD
Ulvak Sinku-Riko, Yokohama, Japan

INDEX

Abundance ratios, of radioisotopes, 3
Accelerator mass spectrometry
  accuracy, 33-35
  applications, 2-3, 99-100
  background, 3, 29, 37
  Bragg curve spectrometer, 20
  contamination in, 30
  counting efficiency, 36-37
  and decay counting, comparison of, 36-38, 74-75
  detector system, 16-22
    ΔE, 16
    ΔE-E gas counter, 16-17
    isobar suppression, 17-18
    passive absorber, 20-21
    semiconductor, 20
    time-of-flight, 16, 21-22
  efficiency, 31-33
  electrostatic deflector, 5, 15
  energy drifts in, 11
  future of, 100
  gas-filled magnet for, 25, 26
  gas stripper, 12
  gas stripping
    complete, 25-26, 27
    efficiency, 12-13, 29, 32
    yield, 12-15
  injector, 8-9
  instrumental background
    isobar separation, 31
    mass separation, 30-31
  instrumentation, 3-28
    multiple filtering system, 4-5
    sample loading and changing system, automated, 6-7
    sensitivity, 3-4
  ion source, 5-7
  isobar separation, 16-20, 31
  laser techniques with, 26-27
  mass analysis, high energy, 15-16
  mass separation, 30-31
  measuring procedure, 35-36
  normalization procedure, 22-25
    fast-switching, 23-24
  performance, 28-36
  performance limits, 29, 32
  precision, 33-36
  principles, 3-5
  for radiocarbon dating, standard material for, 36
  results, uncertainty in, 33-35
  RF fields, for acceleration and background suppression, 27-28

secondary ion, 98-99
sensitivity, 29
source efficiency, 29, 32
sputtering tail, 8-9
stable isotope detection, 28
systematic error in, 33-35, 37-38
tandem accelerators, 8-15
advantages and disadvantages of, 8
current source, 10
principle, 4-5
technique, 3-38
throughput, 29, 33
transport losses, 32
Wien filters, 15-16
Alpha counting, and ion counting, comparison, 139
Aluminum-26
accelerator mass spectrometry
applications, 2, 81-84, 99
performance limits of, 29
sample preparation, 81
sample purification, 81
sensitivity, 81
technique, 21, 37
in atmosphere, production and distribution, 80
in dating, 57
in extraterrestrial samples, 82-84
in geophysical studies, applications, 43
half-life, 80
in hydrosphere, production and distribution, 80
in lithosphere, production and distribution, 81
in marine archives, 82-83
in meteorites
production and distribution, 81
studies, 84
in polar ice, studies, 84
production and distribution, 51, 81-82
radioisotope to stable isotope ratio in geophysical reservoirs, 58
in rocks, 82
in uranium and thorium ions, studies, 84
Ankerites, thermal analysis, 246, 248
Archaeology, carbon-14 in, 79
Argon-39, in geophysical studies, applications, 43
Atmosphere
aluminum-26 in, production and distribution, 80
beryllium-7 in, production and distribution, 59
beryllium-10 in, production and distribution, 59
carbon-14 in
applications, 75-76, 77
production and distribution, 70-71, 74-75
chlorine-36 in
production and distribution, 85-86
studies, 88-89
depth, cosmic ray fluxes as function of, 41-43
iodine-129 in, production and distribution, 94
methane in, 76
Atmospheric $\Delta^{14}C$, 70-71, 74-77
Atmospheric chemistry, carbon-14 in, 80
Atmospheric circulation, studies, beryllium-10 in, 70
Atmospheric turbulence effects, thermogravimetry of, 220
Atom buncher, for RIMS of krypton, 148
Atomic selection rules, and RIMS isotope ratio measurement, 137

Basalts, ocean floor, dating, 138-139
Beryllium-7
  accelerator mass spectrometry, 61
    technique, 26
  in atmosphere, production and distribution, 59
  half-life, 59, 61
Beryllium-10
  accelerator mass spectrometry
    applications, 2, 99
    performance limits of, 29
    sample preparation, 60
    sample purification, 60
    sensitivity, 60-61
    special techniques, 60-61
    technique, 14, 20-21, 25-26, 30, 33, 37
  applications, in geophysics, 52-53, 61-70
  in atmosphere, production and distribution, 59
  in atmospheric circulation, studies, 70
  in biogenic reservoirs, studies, 70
  in dating, 55-57
  in diamonds, studies, 70
  in extraterrestrial samples, applications, 69
  geochemical behavior, in seawater, 64-65
  in geophysical studies, applications, 43, 53
  half-life, 59, 61
  in hydrology, 70
  in hydrosphere, production and distribution, 59
  in lithosphere, production and distribution, 59-60
  in loess, studies, 70
  in marine archives, 82-83
    applications, 63-68
  in meteorites
    production and distribution, 60
    studies, 69
  in particle settling velocities, studies, 70
  in polar ice, applications, 61-63
  production and distribution, 51-52, 59-60, 81-82
  radioisotope to stable isotope ratio in geophysical reservoirs, 58
  in rocks, 82
    applications, 68
  in sediments, applications, 65-69
  in soils, applications, 68
  in tracing geochemical pathway, 57-58
  in uranium and thorium ions, studies, 84
Beryllium-10/beryllium-7 ratio
  applications, in geophysics, 53
  determination, 61
Bone
  calcium-41 in, 92-93
  dating, calcium-41 in, 93
Bragg curve spectrometer, 20

Cahn Electrobalance, 215-216
Calcium
  atomic energy levels, 159
  photon energies, 159
  X-ray fluorescence emission, wavelength, 159
Calcium-41
  accelerator mass spectrometry
    applications, 2, 99
    performance limits of, 29
    sample preparation, 92
    sample purification, 92
    sensitivity, 92
    technique, 20-21, 28, 37
  anthropogenic, production and distribution, 92
  applications, in geophysical studies, 92-93
  in atmosphere, production and distribution, 91

in bone, 92-93
in extraterrestrial samples, studies, 93-94
in geophysical studies, applications, 43
half-life, 91
in lithosphere, production and distribution, 91-92
in meteorites
production and distribution, 92
studies, 93-94
production and distribution, 91-92
radioisotope to stable isotope ratio in geophysical reservoirs, 58
in rocks, 92-93
Carbon-14
accelerator mass spectrometry
applications, 74-75, 99
performance limits of, 29
sample preparation, 73
sample purification, 73
sensitivity, 73
special techniques, 73
technique, 6, 13, 19-20, 30, 32-33, 36-37, 53
anthropogenic, production and distribution, 72
applications
in archaeology, 79
in geophysics, 52-53
in atmosphere
applications, 75-77
production and distribution, 70-71, 74-75
in atmospheric chemistry, 80
concentration, calculation, carbon cycle model used for, 70-72
dating, 4, 54-55, 57, 74
decay counting techniques, and accelerator mass spectrometry, comparison, 74-75
in geophysical studies, 74-75
in groundwater, studies, 79
half-life, 70
in hydrosphere, production and distribution, 71
in ice core dating, 80
in lithosphere, production and distribution, 71-72
in marine archives, applications, 76-78
in meteorites
production and distribution, 72-73
studies, 79
production and distribution, 43, 51-52, 70-72, 74-75
radioisotope to stable isotope ratio in geophysical reservoirs, 51, 58
in rock varnish dating, 80
in situ production, studies, 79
in soil organic carbon, studies, 80
in varved lake sediments, studies, 80
Carbonate deposits, dating, 138
Carbon cycle model, for carbon-14 concentration calculation, 70-72
Carbon monoxide
disproportionation, 230-231
oxidation, 230, 234
Catalina Schist Complex, trace-element partitioning behavior of minerals in, 208
Cesium-sputter ion source, for accelerator mass spectrometry
frit-type, 5-6
high intensity, 6-7
Chlorine-36
accelerator mass spectrometry
applications, 2, 99
performance limits of, 29
sample preparation, 86-87
sample purification, 87
sensitivity, 87

technique, 20, 25, 27-28, 30-31, 33, 37, 53
anthropogenic, production and distribution, 85-86
applications
in geophysical studies, 87-91
in geophysics, 53
in atmosphere
production and distribution, 85-86
studies, 88-89
in dating, 55-57
in evaporite deposits and brines, studies, 91
in evapotranspiration rate studies, 88-89
in extraterrestrial samples, studies, 90-91
geochemical behavior, 87
in geophysical studies, applications, 43, 46-47
in groundwater, 87-88
half-life, 85
in hydrology, 87-88
in hydrosphere, production and distribution, 85
in lithosphere, production and distribution, 85
in meteorites
production and distribution, 88
studies, 90-91
from nuclear weapons fallout, as tracer
applications, 48
for long-term liquid and vapor movement in desert soils, 88, 90
production and distribution, 51
radioisotope to stable isotope ratio, 88, 89
in geophysical reservoirs, 58
as tracer for long-term liquid and vapor movement in desert soils, 88, 90
in rocks, studies, 89-90
stratospheric, 89
subsurface production, 88
as tracer, for oil field structure studies, 91
Coals, thermal analysis, 242-245
Compton scatter, as compensation for absorption effects, in X-ray fluorescence spectrometry, 170, 173-175
Compton scattering, 202, 204
Corals, dating, 138
Cosmic rays
absorption mean free path, 40-41
characteristic energies, 38
fluxes, 38-39
depth dependence, 38-41
as function of atmospheric depth and latitude, 41, 43
as function of depth below Earth's surface, 41, 43
geomagnetic field and, 43-45, 52-53
solar modulation of, 45-46
temporal variations of, 43-46
variations, 43
galactic, 38
mouns, 41
nuclear reactions induced by
radionuclide formation in, 38-46
secondary particle production, 38-40
solar, 38
Cosmochemistry, neutron activation analysis in, 193
Critical energies, of X-ray emission, 159-160
Critical wavelengths, of X-ray emission, 159
Crystobilite, thermosonometry, 251
Cyclotron, for accelerator mass spectrometry, 27

Dating
  aluminum-26 in, 57
  beryllium-10, 55-57
  bone, calcium-41 in, 93
  carbon-14, 4, 54-55, 57, 74
  carbonate deposits, 138
  chlorine-36, 55-57
  conventional, 53-55
  corals, 138
  double, 55
  groundwater, krypton-81 in, 147-148
  ice core, carbon-14 in, 80
  iodine-129, 55
  long-lived radioisotopes in, 53-57
  meteorites, 91, 143
  ocean floor basalts, 138-139
  ores, 143
  profiling-based, 55-56
  rocks, 58, 143
  rock varnish, carbon-14 in, 80
  by in situ production, 56-57
  volcanic rocks, 138-139
Decay counting, and accelerator mass spectrometry, comparison of, 36-38, 74-75
Dew point detectors, 224
Diamonds, beryllium-10 in, studies, 70
Dielectric analysis, 240-241
Differential scanning calorimetry, 212, 227-239
  applications, 232-233
    in extraterrestrial studies, 247-250
  baseline, establishing, 237-238
  factors affecting, 235-236
  heat flux mode, 231-232
  modulated, 252-254
  operational compromises for, 237
  oscillating, 252
  power compensation mode, 231-232
  principle of, 231
  standard materials for, 238
Differential thermal analysis, 212, 227-239
  apparatus, 228-229
  applications, 232-233
    in geochemistry, 246, 248-249
  atmosphere, 234
  baseline, establishing, 237-238
  curves, 228-230
    for glass transitions, 233-234
  factors affecting, 235-236
  operational compromises for, 237
  principle of, 228
  sample holder, 234-235
  sensor arrangements, 234-235
  standard materials for, 238
  thermocouple for, 234-235
Differential thermogravimetry, 221-222
Dolomites
  high-iron ferroan, thermal analysis, 246, 248
  thermal analysis, 242-243
Doppler linewidth, and RIMS, 140-141
DSC (*see* "Differential scanning calorimetry")
DTA (*see* "Differential thermal analysis")
Dynamitrons, 10

EGA (*see* "Evolved gas analysis")
Electronic drift, in thermogravimetry, 220-221
Emanation thermal analysis, 212, 239-240
Energy dispersive spectrometer, 157-158, 163
  solid-state detector, 158
Energy drifts, in accelerator mass spectrometry, 11

Enhancement, in X-ray fluorescence spectrometry, 161-162
Evapotranspiration rates, studies, chlorine-36 in, 88-89
Evolved gas analysis, 212, 223-227
  applications, 224-225
    in extraterrestrial studies, 247-250
    in geochemistry, 245, 247
  discontinuous techniques, 224-225
  interface, 225
  mass spectroscopic, 224, 226-227
  sensors, 224
Evolved gas detection, 224
Extraterrestrial radionuclides, distribution, 49
Extraterrestrial samples (*see also* "Meteorites"; "Rocks, lunar")
  aluminum-26 in, 82-84
  beryllium-10 in, applications, 69
  calcium-41 in, studies, 93-94
  chlorine-36 in, studies, 90-91
  iodine-129 in, studies, 97
  thermal analysis, 247-249, 250
Extraterrestrial studies
  differential scanning calorimetry in, 247-250
  evolved gas analysis in, 247-250

Faraday cup detector, 128, 130-131
Ferrimagnetic spinel, 244
Foraminifera, carbon-14 in, studies, 78
Fossil fuels, thermal analysis, 242-245
Fourier transform infrared spectroscopy
  for evolved gas analysis, 224-227, 245
  instrumentation, 225-226
Frisch grid, 17
FTIR (*see* "Fourier transform infrared spectroscopy")

Gas chromatography, for evolved gas analysis, 224
Geochemical exploration samples, analysis, 156-157
  quality control, 157
Geochemistry
  analytical, requirements, 156-157
  neutron activation analysis in, 193
Geochronology, 138-139
Glow discharge, as atomization source for RIMS, 134, 136
Greenhouse warming, 76
Groundwater
  carbon-14 in, studies, 79
  chlorine-36 in, 87-88
  dating, krypton-81 in, 147-148
  iodine-129 in, 96

Hydrology
  beryllium-10 in, 70
  iodine-129 in, 96
Hydrosphere (*see also* "Groundwater")
  aluminum-26 in, production and distribution, 80
  beryllium-10 in, production and distribution, 59
  carbon-14 in, production and distribution, 71
  chlorine-36 in, production and distribution, 85
  iodine-129 in, production and distribution, 94
Hyperfine structure, and RIMS isotope ratio measurement, 137, 140-141

Ice (*see* "Polar ice")
Ice core dating, carbon-14 in, 80
Infrared spectroscopy (*see also* "Fourier transform infrared spectroscopy")
  and thermal analysis, combination of, 251

Instrumental neutron activation analysis, 194, 199-205
applications, 199
Compton suppression, 205-206
data analysis, 205
experimental procedure, 203-205
γ-γ coincidence spectroscopy system, 205, 207
γ-ray detector
efficiency, 200-202
resolution, 200, 202
γ-ray spectroscopy system for, 199-203
setup, 202-203
Ge(Li) detector, 199
high-purity Ge(HPGe) detector, 199
HPGe detector, 200-201
efficiency, 200-201
peak-to-Compton ratio, 200, 202
HPGe planar detector, 2^2, 204
instrumentation, 199-203
NaI(Tl) γ-ray detectors, 199
sample preparation, 203
specialized procedures, 205
SPECTRA γ-ray analysis program, 205
standards, 204
UCLA neutron activation analysis laboratory procedure, 204-205
Internal isochron technique, 139, 143
Iodine-129
accelerator mass spectrometry
applications, 2, 99
performance limits of, 29
sample preparation, 95
sample purification, 95
sensitivity, 95
technique, 30-31
anthropogenic
production and distribution, 95
studies, 96-97
applications, in geophysical studies, 96-97
in atmosphere, production and distribution, 94
in dating, 55
in extraterrestrial samples, studies, 97
in groundwater, 96
half-life, 94
in hydrology, 96
in hydrosphere, production and distribution, 94
in lithosphere, production and distribution, 94-95
in meteorites
production and distribution, 95
studies, 97
production and distribution, 51, 94-96
radioisotope to stable isotope ratio in geophysical reservoirs, 58
sources, noncosmogenic, 96
Iron
atomic energy levels, 159
photon energies, 159
X-ray fluorescence emission, wavelength, 159
Iron-60
accelerator mass spectrometry
performance limits of, 29
technique, 21, 28
in geophysical studies, applications, 98
half-life, 98
Isotope shift
and mass effects, 140
and RIMS, 140
and volume effects, 140
Isotopic biases, and RIMS isotope ratio measurement, 141

Krypton
analysis, photon burst mass spectrometry for, 148
half-life, 147
resonance ionization mass spectrometry, 147-149
atom buncher for, 148
pre-enrichment, 148
Krypton-81
in groundwater dating, 147-148
production and distribution, 147
Krypton-85, production and distribution, 147

Lakes, particle settling velocity studies, beryllium-10 in, 70
Langmuir evaporation, 134
Laser (*see also* "Resonance ionization mass spectrometry, laser")
linewidth, and RIMS isotope ratio measurement, 137
Laser ablation, as atomization source for RIMS, 134-136
thermal plume generated by, 135-136
Laser desorption, as atomization source for RIMS, 134, 147
Latitude, cosmic ray fluxes as function of, 41, 43
Lead-205, accelerator mass spectrometry, 98
technique, 21, 28
Libyan desert glass, origin, studies, aluminum-26 in, 84
Lignites, thermal analysis, 242-245
LINAC, for accelerator mass spectrometry, 27-28
Lithosphere (*see also* "Rocks")
aluminum-26 in, production and distribution, 81
beryllium-10 in, production and distribution, 59-60
calcium-41 in, production and distribution, 91-92
carbon-14 in, production and distribution, 71-72
chlorine-36 in, production and distribution, 85
iodine-129 in, production and distribution, 94-95
Loess, beryllium-10 in, studies, 70
Loss on ignition, in X-ray fluorescence spectrometry, 179-180
Lunar gardening, studies, aluminum-26 in, 84

Manganese-53
accelerator mass spectrometry
performance limits of, 29
technique, 21
in geophysical studies, applications, 98
half-life, 98
production and distribution, 98
radioisotope to stable isotope ratio in geophysical reservoirs, 58
Manganese nodules, growth rates, determination, 63, 66-68
Marine archives
aluminum-26 in, 82-83
beryllium-10 in, 82-83
applications, 63-68
carbon-14 in, applications, 76-78
Mass spectrometers, 128-130
field-free region, 129
ion cyclotron resonance, 129-130
magnetic sector, 128
multiple slit/detector form, 128-129
single slit/detector form, 128
quadrupole, 129
time-of-flight, 129
reflectron geometry, 129

Mass spectrometry, 116-117 (*see also* "Resonance ionization mass spectrometry"; "Secondary ion mass spectrometry")
advantages and disadvantages of, 117
analytical
applications, 116-117
of isotope concentration, 117
of isotopic distribution, 117
background, sources, 117
detectors, 130-131
Daly, 130
electron multiplier, 130
Faraday cup, 128, 130-131
for evolved gas analysis, 224, 226-227
Faraday cup detection, 128, 130-131
fringe fields, 128
instrumentation, 117, 128-130
ion scattering and, 117
isobaric interferences, 117
photon burst, for krypton analysis, 148
pulse counting, 128
Matrix effect correction factor, for X-ray fluorescence spectrometry, 164
evaluation
compensation (comparative) methods, 165, 170-175
numerical (mathematical) methods, 165, 175-183
Matrix effects, 161-162
total, theoretical expressions for, 169-170
Metastable to stable transformation, detection, by DTA and DSC, 233-234
Meteorites
aluminum-26 in
production and distribution, 81
studies, 84
beryllium-10 in
production and distribution, 60
studies, 69
calcium-41 in
production and distribution, 92
studies, 93-94
carbon-14 in
production and distribution, 72-73
studies, 79
chlorine-36 in
production and distribution, 88
studies, 90-91
chondritic, 208
dating, 91, 143
iodine-129 in
production and distribution, 95
studies, 97
lunar (Allan Hills A81005), neutron activation analysis of, 196, 208
neutron activation analysis of, 194, 208-209
thermal analysis, 244
Methane, atmospheric, 76
Mössbauer spectroscopy, 244, 246, 251
Muons, 41

Neutrinos, solar, measurement of Earth's exposure to, 146-147
Neutron activation analysis, 193-209 (*see also* "Instrumental neutron activation analysis"; "Radiochemical neutron activation analysis")
applications, 193, 208-209
detectability of element by, 198-199
element concentration from induced radioactivity, 197-198
elements studied by, 194

neutron irradiation of samples, 195
sensitivity, 193, 198-199
theory, 195-199
and X-ray fluorescence analysis, comparison, 194
Neutrons
epithermal, 195
fast, 195
thermal, 195
Nickel-59, accelerator mass spectrometry
performance limits of, 29
technique, 21, 28
Nickel ferrite, thermal analysis, 245, 247
Nuclear clocks, 138
Nuclear magnetic resonance, solid-state, and thermal analysis, combination of, 251
Nucleon-nucleon reactions, 41-43
Nucleon-nucleus reactions, 41-43

Ocean circulation studies, carbon-14 in, 76-78
Ores, dating, 143

Paleoenvironmental studies, 53
Particle accelerators
tandem, 8-15
principle, 4-5
van de Graaff, 10
single-ended, 28
Particle sputtering, as atomization source for RIMS, 134, 135
Pelagic clay sediment, element distribution in, neutron activation analysis studies, 208-209
Pelletrons, 10
Photon burst mass spectrometry, for krypton analysis, 148
Polar ice
aluminum-26 in, studies, 84
beryllium-10 in, applications, 61-63
Potassium-40, 46

Quartz, thermosonimetry, 251

Radioactive decay law, 197
Radiocarbon dating (*see also* "Carbon-14, dating")
accelerator mass spectrometry for, standard material for, 36
Radiochemical neutron activation analysis, 194, 199, 205-208
applications, 199
elements determined by, 205, 207
HPGe detector, 207-208
NaI(Tl) detector, 207
Radionuclides
anthropogenic, 48
atmospheric origin, distribution, 49-51
cosmogenic, 38-46
in dating, 53-57
deposition, 53
distribution in nature, 49-51
extraterrestrial origin, distribution, 49
long-lived, application to geophysics, 52-57
primordial, 46-47
production, variations in, 52-53
radiogenic, 46-47
in situ origin, distribution, 51
sources, in nature, 38-48
transport, 53
Resonance ionization mass spectrometry, 116-154 (*see also* "Mass spectrometry")
abundance sensitivity, 119
apparatus, 118-119
atomization sources, 134

continuous wave, 136-138, 141
in geochemical studies, applications, 116
ionization efficiency, 123-127, 141-143
ionization processes, 119
ion scattering and, 119
isobars and, 119
for isotope geochemistry, 137
accuracy, 137
precision, 137
reproducibility, 137
isotope ratio measurement, 128, 137
krypton, 147-149
laser, 120
continuous wave, 125-128, 133, 136-138
diode, 138
pulsed, 122-125, 133
laser linewidth for, 140-141
magnetic sector analysis, 121
modeling, 120-128
density matrix approach, 120-121
rate equations method, 120-122
multistep photoionization in, 117-119
principles, 117-119
pulsed, 132-136
pulse pile-up problems, 141
rhenium/osmium, 143-146
sample overcoat, 131
sample preparation, 131-132
selectivity, 119
spectral brightness, 120
technetium, 146-147
thermal ionization, minimization, 131-132
thorium, 138-143
time-of-flight analysis, 121
Rhenium/osmium
isobaric interference in, 144
resonance ionization mass spectrometry, 143-146
effective duty cycle, 144
energy levels and transitions used for, 144-146
as tracer system, in geochemical and geophysical studies, 143-144
Rocks (*see also* "Lithosphere")
aluminum-26 in, 82
beryllium-10 in, 82
applications, 68
calcium-41 in, 92-93
chlorine-36 in, studies, 89-90
crustal versus mantle, Rh-Os system as tracer in, 143-144
dating, 58, 143
by in situ radioisotope production, 56-57
erosion rates, determination, 68
lunar, neutron activation analysis, 194, 208-209
nuclides in, 43
volcanic, dating, 138-139
Rock varnish dating, carbon-14 in, 80
Rubidium-87, 46

Secondary ion accelerator mass spectrometry, 98-99
Secondary ion mass spectrometry, 99
Sediments
lake, varved, carbon-14 studies, 80
marine
beryllium-10 in, applications, 65-69
carbon-14 studies, 80
Siderites, thermal analysis, 244, 246, 248-249
Siderophile, definition, 143
Si-detectors, for accelerator mass spectrometry, 20

Silicon-32
accelerator mass spectrometry
performance limits of, 29
technique, 21, 25
applications, in geophysical studies, 97
half-life, 97
radioisotope to stable isotope ratio in geophysical reservoirs, 58
Sintering, 240
Sodium-22, accelerator mass spectrometry, 98
Soils
beryllium-10 in, applications, 68
organic carbon, carbon-14 studies, 80
Spallation reactions, 41-43
Spectroscopy, and thermal analysis, combination of, 251
Sunspot numbers
Maunder minimum, 46, 52
from 1610 to 1988, 46-47
Sunspots, Schwabe cycle, 52

Tandetron, 10-11
Technetium
production and distribution, 146
resonance ionization mass spectrometry, 146-147
Tektites, origin, studies, aluminum-26 in, 84
Thermal analysis, 211-257
applications, 212
to geochemistry, 242-251
computers in, 213-214
controlled rate, 213-214
definition, 211-212
and differential thermal analysis, comparison, 228
digital data acquisition, 213-214
enthalpy effects, 231
feedback used in, 213, 252
instrumentation, 214-215
manufacturers, 257
methods, 212
simultaneous measurements, 213
Thermal filament, as atomization source for RIMS, 134
Thermal plume, generated by laser ablation, 135-136
Thermobalances, 215-216
ports for flow-through atmosphere, 218
temperature scale, calibration, 222-223
Thermodilatometry, 212, 240
Thermogravimetry, 212, 215-223
applications, to geochemistry, 242-246
atmospheric turbulence effects, 220
balance for, 215-216
buoyancy effects, 219-220
condensation and reaction effects, 220
corrosive atmosphere for, 223
data analysis and presentation, 221-222
and differential thermal analysis, simultaneous, instrument calibration for, 238-239
and differential thermogravimetry, simultaneous, 223, 226, 228
electronic drift in, 220-221
electrostatic and magnetic forces affecting, 220
factors affecting, 219-220
heating system, 217-218
instrumentation, 215-217
temperature range, 217
pressure range, 219
rate controlled, 252
sample holders, 218-219
sample size, 215-217
sample temperature, control, 216-217

self-generated atmosphere, 218-219
temperature calibration for, 222-223
temperature variables, 221
for vapor pressure determination, 223
and X-ray diffraction, simultaneous, 223
Thermoluminescence, 241
Thermomagnetometry, 212, 215, 222-223
applications, to geochemistry, 242-246
Thermomechanical analyses, 212, 240
Thermomicroscopy, 241
Thermosonimetry, 212, 240-241
applications, in geochemistry, 251
Thorium, resonance ionization mass spectrometry, 138-143
transitions used in, 141-143
Thorium-232, 46
Time-of-flight analysis, resonance ionization mass spectrometry, 121
Time-of-flight detector system, for accelerator mass spectrometry, 16, 21-22
Time-of-flight mass spectrometers, 129
reflectron geometry, 129
Tin-126, accelerator mass spectrometry, 98
technique, 25, 28
Transform limit, of pulsed laser linewidths, 132
Tree rings, isotope concentration in, applications, 52
Tritium
applications, in geophysical studies, 97
half-life, 97
from nuclear weapons fallout, as tracer for long-term liquid and vapor movement in desert soils, 88, 90

Uranium-235, 46
Uranium-238, 46
Uranium-series disequilibrium, 138-139
Uranium-238/thorium-230 disequilibrium, 138

Van de Graaff accelerator, 10
single-ended, 28

Wavelength dispersive spectrometer, 162-163
sequential, 157
simultaneous, 157-158
Wien filters, 15-16

X-radiations
absorption, 160
absorption edges, 160
attenuation, 160
mass attenuation coefficient, 160
X-ray fluorescence, emission, theory of, 158-159
X-ray fluorescence spectrometry, 155-192
absorption effects
Compton scatter as compensation for, 170, 173-175
theoretical expressions for, 165-168
alpha coefficients, 176-180
oxide systems, 180-182
analytical reproducibility, 190-191
analytical strategies, 183-191
background intensity, determination, 186
blind duplicates for, 188
calibration, 187-190

Claisse-Quintin model, 181-182
compensation for loss on ignition, 179-180
compensation (comparative) methods
  Compton scatter, 170, 173-175
  dilution, 170-172
  experimental measurements, based on, 165
  standard addition, 170, 172-173
computerized, 158, 163, 188-189
counting monitor, 187
data reduction process, 184, 188-190
detection limits, 190-191
energy dispersive spectrometer, 157-158, 163
enhancement effects, theoretical expressions for, 168-169
instrumentation, 157, 162-163
intensity measurements, 184-188
internal standard method, 172-173
matrix effect, 161-162
matrix effect correction factor, 164
  evaluation, 165, 170-183
mineralogical effects, 171-172
net intensities, 185
numerical (mathematical) methods, 165, 175-183
  alpha coefficients, 176-180
  concentration model, 182-183
  empirical coefficient methods, 182-183
  fundamental parameters approach, 176
  influence coefficient methods, 176-177
  intensity model, 182-183
  iteration, 176
particle size effects, 171-172
precision, 191
primary radiation source, 157
quantitative, 156
  principles, 164-170
reference standards
  dual role of, 189-190
  geological, 187
secondary fluorescence (enhancement), 161-162
secondary X-radiations, 157
specimens
  fused disc, 177, 184-185
  infinite thickness, 184-185
  infinite width, 185
  preparation, 184-185
  pressed pellet, 180, 184-185
spectral overlap, 185
third element effect, 181
total count time, 185-186
total matrix effects, theoretical expressions for, 169-170
wavelength dispersive spectrometer, 157-158, 162-163
X-ray physics, 158-162
X-ray spectra, 158